Springer Series in Astrostatistics

Springer Series in Astrostatistics,
More information about this series at http://www.springer.com/series/1432

Springer Series in Astrostatistics

Astrostatistical Challenges for the New Astronomy: *ed. Joseph M. Hilbe*
Astrostatistics and Data Mining: *ed. Luis Manuel Sarro, Laurent Eyer, William O'Mullane, Joris De Ridder*

Asis Kumar Chattopadhyay • Tanuka Chattopadhyay

Statistical Methods for Astronomical Data Analysis

Springer

Asis Kumar Chattopadhyay
Department of Statistics
University of Calcutta
Calcutta, India

Tanuka Chattopadhyay
Department of Applied Mathematics
University of Calcutta
Calcutta, India

ISSN 2199-1030 ISSN 2199-1049 (electronic)
ISBN 978-1-4939-1506-4 ISBN 978-1-4939-1507-1 (eBook)
DOI 10.1007/978-1-4939-1507-1
Springer New York Heidelberg Dordrecht London

Library of Congress Control Number: 2014945364

Printed on acid-free paper

Springer is part of Springer Science+Business Media (www.springer.com)

To the Astrostatistics Community

Preface

Universe is deadly adventurous and man's eternal quest to unfold its mysteries will never cease. Astronomy, perhaps the oldest observational science, has got its spectacular emergence with the advent of theoretical astrophysics and it has started to spread in India after some important contributions by Indian scientists in 1920s and later. Astronomy in the recent past has developed a lot with the launch of several missions like GALEX (Galaxy Evolution Explorer), Kepler Space Telescope, Hubble Space Telescope (HST), etc. through which terabytes of data are available for preservation. Hence several virtual archives like SDSS (Sloan Digital Sky Survey), MAST (Multimission Archive at STSCI), Vizier, EDD, LEDA, Chandra, etc. have been developed to preserve the one-time snapshots of various astronomical events.

During the last two decades galaxy formation theory and their related star formation histories have drawn interests among the astrophysicists to a great extent to uncover these mysteries using the reach treasure of virtual archives. While digging the pathway, a new branch ASTROSTATISTICS (or Statistical Astronomy) has emerged since 1980s. It is a blending of statistical analysis of astronomical data along with the development of new statistical techniques useful to analyze astrophysical phenomenon. The target is not only to explore the formation and evolutionary history of galaxies but also to uncover the unknown facts related to star formation, gamma ray bursts, supernova and other intrinsic variable stars. Where much data are not available, model-based approaches may be adopted.

In this book we have tried to introduce "Astrostatistics" as a subject just like biostatistics. Through the various chapters we have discussed the basic concepts of both Astrophysics and Statistics along with the possible sources of Astronomical data. Subsequently we have entered into different types of applications of statistical techniques already developed or specifically introduced for astrophysical problems. We have discussed on techniques like regression, clustering and classification, missing data problems, simulation, data mining and time series analysis. Along with the discussion, specific examples are given so that readers can easily digest the method and can apply it to their own problem. Finally we have included a chapter on the use of R package which is an open access software. There we have included several examples which will be helpful for the readers. In the appendix we have included some astronomical data sets which we have used in our examples.

We are indeed grateful to many persons and organizations for helping us during the preparation of the manuscript. In particular, we are grateful to Professor Ajit Kembhavi, Director, IUCAA, India, for continuous encouragement and providing new ideas related to "fundamental plane" concept. We are also deeply indebted to Professor Sailajananda Mukherjee, Retired Professor and Former Head, Department of Physics, North Bengal University, India, for carefully reading the Astrophysical part of the book and suggesting significant improvements. We will always remember the cooperation and support we have received from Professor Joseph Hilbe, Emeritus Professor, University of Hawaii, USA. We will also remain grateful to Professors Jogesh Babu and Eric Feigelson, Pensylvania State University, USA, whose contributions inspired us to start work in this area.

We offer our heartiest thanks to our collaborators Didier Fraix Burnet, Emmanuel Davoust, Margarita Sharina, Ranjeev Misra and Malay Naskar. We feel proud of our students Saptarshi Mondal, Tuli De, Abisa Sinha, Pradeep Karmakar and Bharat Warule for their sincere efforts and dedication. Being faculty members of Calcutta University, India, we are really grateful to our Vice-Chancellor Professor Suranjan Das and Pro-Vice-Chancellor (Academic) Professor D.J. Chattopadhyay for their continuous support.

Finally we thank Mr. Aparesh Chatterjee for carefully preparing the typescript of this book.

Calcutta, India
August 15, 2014

Asis Kumar Chattopadhyay
Tanuka Chattopadhyay

Contents

Chapter - 1

Introduction to Astrophysics

1.1 Light and Radiation

The most important carrier of information from all kinds of heavenly bodies is light. Light is an electromagnetic wave which is characterized by an electric field **E** and magnetic field **H**, and the direction of its propagation is at right angles to both these fields. Unlike mechanical waves, e.g. sound waves on the ocean, it can propagate through empty space. So, if Z-direction is the direction of its propagation **(Fig. 1.1)**, then Y and X directions are the directions of electric and magnetic fields. The wave motion can be described by a sine curve and the distance between two consecutive peaks measures the wavelength λ, which is an important characterization of the propagating light.

The number of oscillations of electromagnetic wave per unit time at a given point is called the frequency ν of the electromagnetic radiation so that

$$\lambda\nu = c \tag{1.1}$$

where c is the speed of light and its value in vacuum is 3×10^8 m/s. The entire electromagnetic spectrum is shown in **Fig. 1.2**. It consists of Gamma rays, X-rays, ultraviolet, visible light, infrared, microwaves and radio waves of which visible range covers a very small window (400–700 nm, 1 nm $= 10^{-7}$ cm). For detection of spectra in different regimes appropriate detectors are required, e.g. for visible light optical telescope is required whereas X-ray or infrared telescopes are required to detect X-ray and infrared rays from astronomical sources.

Laws of Radiation in Thermodynamic Equilibrium

When a system filled with gas and radiation is kept isolated from its surroundings, then the system will eventually have equal temperature at all points and we say that the system is in thermal equilibrium. This happens because of frequent collisions among the constituents of the system leading to exchange of energy which helps the system to reach rapidly to a state

A.K. Chattopadhyay, T. Chattopadhyay, *Statistical Methods for Astronomical Data Analysis*, Springer Series in Astrostatistics 3,
DOI 10.1007/978-1-4939-1507-1_1

of thermal equilibrium. In other words when mean free path, which is the path between two successive collisions, becomes sufficiently small (in case of electromagnetic radiation it is the mean free path of photons), the system soon attains thermal equilibrium.

A system is said to be in mechanical equilibrium if the velocity distribution follows Maxwellian velocity distribution.

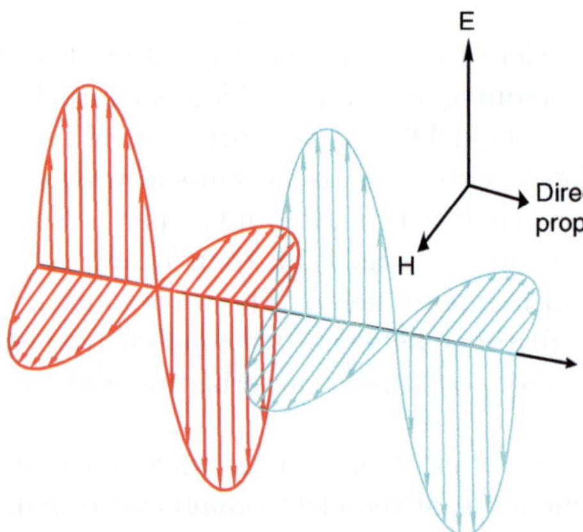

Figure 1.1 Electric field, magnetic field and direction of propagation

A system is said to be in chemical equilibrium if chemical activities have no net change over time.

A system is said to be in thermodynamic equilibrium (TE) when it is simultaneously in thermal, mechanical as well as in chemical equilibrium.

Black Body Radiation

According to Kirchoff's law the ratio of the emissive power to its absorption power at a given temperature T is the same for all bodies. So this ratio is universal function of temperature T of the body and the frequency ν of radiation falling on it. A black body (BB) is something which absorbs all the radiation falling on it. So, absorption power of BB is unity. This shows that the emissivity is universal function of ν and T provided the body is in TE, having a constant temperature T.

Specific Intensity (I_ν)

It is the amount of energy crossing per unit area of the emitting surface, placed perpendicular to the direction of propagation of light in unit time, per unit solid angle, per unit frequency interval.

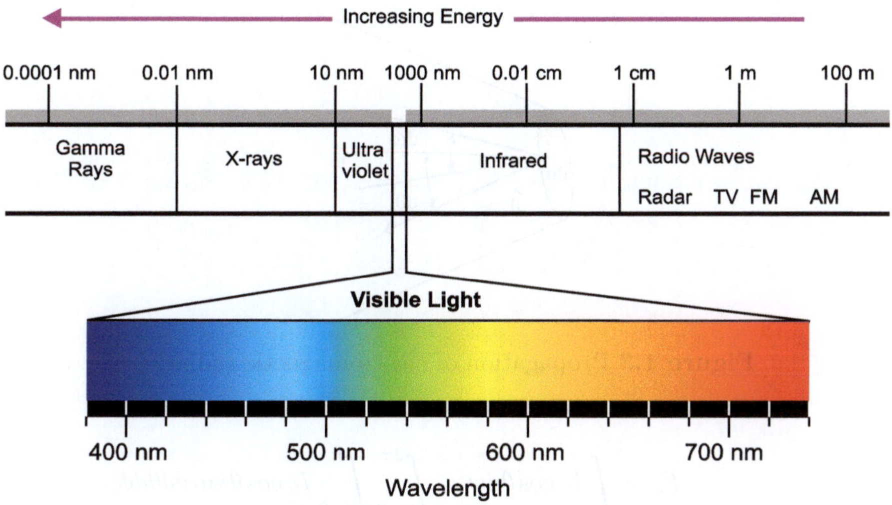

Figure 1.2 Electromagnetic spectrum

If θ be the angle between the direction of propagation $\vec{p}$ and normal $\hat{n}$ to the surface of area $d\sigma$, then (Fig. 1.3)

$$I_\nu = \frac{dE_\nu}{d\sigma \cos\theta dw d\nu} \tag{1.2}$$

Mean Intensity (J_ν)

It is the mean intensity obtained by averaging I_ν over all directions.

$$J_\nu = \frac{\int I_\nu dw}{\int dw} = \frac{1}{4\pi}\int_{\phi=0}^{2\pi}\int_{\theta=0}^{\pi} I_\nu \sin\theta d\theta d\phi \tag{1.3}$$

For isotropic radiation (independent of ϕ and θ)

$$J_\nu = I_\nu = B_\nu \tag{1.4}$$

which is true for BB radiation.

Flux (F_ν)

It is the total amount of energy crossing per unit area in unit time, per unit frequency in all directions.

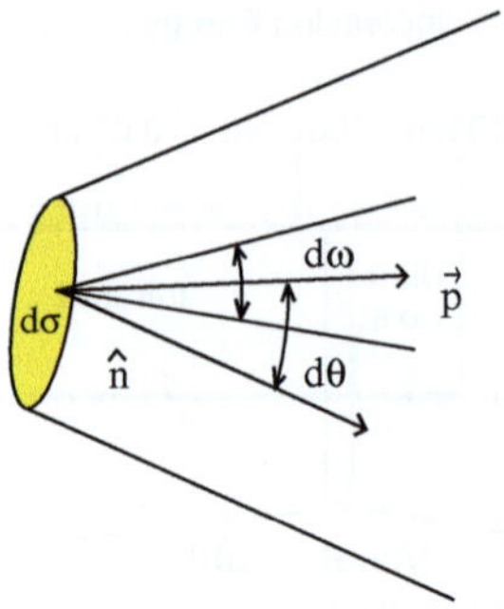

Figure 1.3 Propagation of electromagnetic radiation

$$F_\nu = \int I_\nu \cos\theta dw = \int_{\phi=0}^{2\pi}\int_{\theta=0}^{\pi} I_\nu \cos\theta \sin\theta d\theta d\phi$$

$$= 2\pi \int_0^{\pi} I_\nu(\theta)\cos\theta\sin\theta d\theta \;\; (for \;\; axisymmetric \;\; radiation)$$

$$= 2\pi \int_{-1}^{+1} I_\nu(\mu)\mu d\mu \qquad (1.5)$$

Here, $\mu = cos\,\theta$.

At the surface of BB (which is a hemisphere facing towards the observer)

$$F_\nu = \int_{\phi=0}^{2\pi}\int_{\theta=0}^{\pi/2} I_\nu \cos\theta\sin\theta d\theta d\phi = \pi I_\nu = \pi B_\nu \qquad (1.6)$$

Energy Density (U_ν)

It is the amount of radiation energy received by a cylinder with unit cross section in unit time whose axis is along the direction of propagation per unit frequency per unit volume is

$$dU_\nu = \frac{dE_\nu}{dvd\nu} = \frac{I_\nu d\sigma\cos\theta dw d\nu}{cd\sigma\cos\theta d\nu} = \frac{1}{c}I_\nu dw \qquad (1.7)$$

c being the speed of light.

Considering sources which are isotropically distributed at the same distance, the total energy is

$$U_\nu = \frac{1}{c}\int I_\nu dw = \frac{4\pi}{c}J_\nu = \frac{4\pi}{c}B_\nu = \frac{4}{c}F_\nu \qquad (1.8)$$

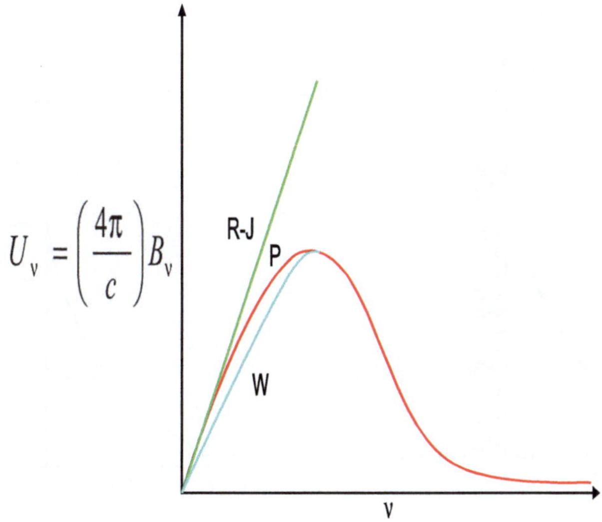

Figure 1.4 Energy density vs frequency plot for black body

Hence,

$$F_\nu = \frac{c}{4} U_\nu \tag{1.9}$$

The various laws of radiation are as follows:

Planck's Law

$$B_\nu = \frac{2h\nu^3}{c^2} \frac{1}{e^{h\nu/kT} - 1} \tag{1.10}$$

It is clear from Fig. 1.4 that Planck's law is consistent with the observed radiation.

Rayleigh–Jeans Law: For $h\nu << kT$

$$B_\nu = \frac{2kT\nu^2}{c^2} \tag{1.11}$$

where k is the Boltzmann constant.

Wien's Law: For $h\nu >> kT$

$$B_\nu = \frac{2h\nu^3}{c^2} e^{-h\nu/kT} \tag{1.12}$$

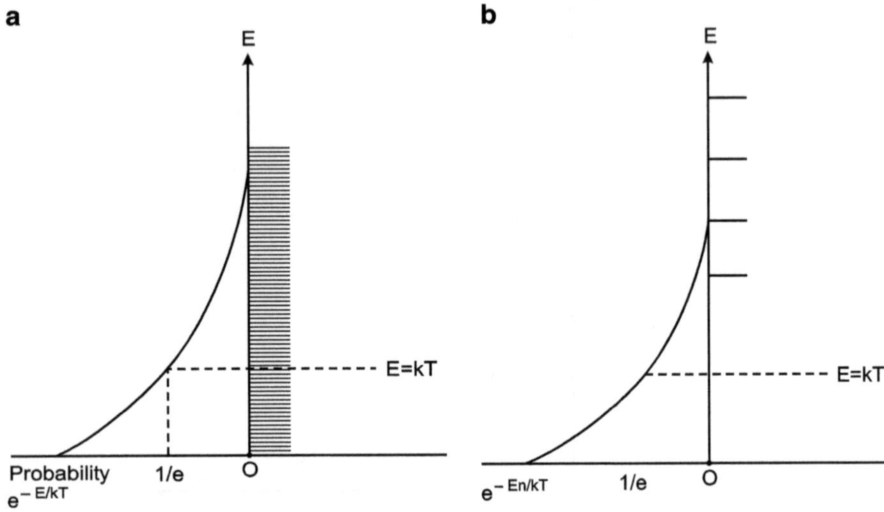

Figure 1.5 Continuous (**a**) and discrete (**b**) energy levels

In classical theory followed by Rayleigh–Jeans law, the energy is continuous and the average energy is

$$\overline{E} = \int_0^{\infty} Ee^{-E/kT} dE / \int_0^{\infty} e^{-E/kT} dE = kT \tag{1.13}$$

where $e^{-E/kT}$ is the probability of having the energy E at temperature T, so for $\overline{E}$ it is $\frac{1}{e}$. In **Fig. 1.5a**, the energy is continuous. In **Fig. 1.5b** the energy is discrete and energy values are so wide (ν large) that no allowed energy values are found near kT. In this case all energy values have negligible probabilities except $E = 0$. So $\overline{E}$ is close to zero rather than kT. According to Planck's law the energy values are discrete and energy at the n th level is

$$E_n = nh\nu$$

$$\text{so}, \overline{E} = \frac{\sum nh\nu e^{-nh\nu/kT}}{\sum e^{-nh\nu/kT}} \tag{1.14}$$

Now,

$$\frac{1}{1-x} = 1 + x + x^2 + \ldots (for \quad x < 1)$$

$$\frac{x}{(1-x)^2} = x + 2x^2 + 3x^3 + \ldots$$

$$\text{So}, \overline{E} = [\frac{h\nu e^{-h\nu/kT}}{(1 - e^{-h\nu/kT})^2}]/[\frac{1}{(1 - e^{-h\nu/kT})}] = \frac{h\nu}{e^{h\nu/kT} - 1} \tag{1.15}$$

Rayleigh–Jeans law diverges at high ν because at high ν, there are still too many degrees of freedom ($\propto\nu^2$) as ν is large as energy spacing is low and each degree of freedom has a finite amount of energy $\frac{1}{2}kT$. But quantum theory predicts that each degree of freedom shares a negligible amount of energy instead of $\frac{1}{2}kT$ so that total energy remains finite. The concept of discreteness thus makes the distribution function a convergent one at high frequency zone which is also clear from Eqs. (1.12) and (1.15), respectively.

Wien's Displacement Law

Planck's law in terms of wavelength λ is

$$B_\lambda = \frac{2hc^2}{\lambda^5}\frac{1}{e^{hc/\lambda kT} - 1} \tag{1.16}$$

Making $\frac{dB_\lambda}{d\lambda} = 0$ gives $\lambda = \lambda_{max}$ for which

$$\lambda_{max}T = 0.2895 \tag{1.17}$$

where λ_{max} is the peak wavelength of the BB radiation at temperature T. This relation is known as Wien's displacement law. It shows that the peak shifts towards shorter wavelength at higher temperature (Fig. 1.6)

Stefan–Boltzmann Law: Energy flux radiated from unit area over all wavelengths is

$$\mathrm{F} = \int F_\nu d\nu = \sigma T^4 \tag{1.18}$$

where σ is the Stefan–Boltzmann constant.

1.2 Sources of Radiation

In the previous section we have discussed so far about the properties of electromagnetic radiation. As we look at the night sky we see several bright spots as sources of energy. These are called stars. With more powerful telescopes we can observe clusters of bright spots or extended objects of various shapes, e.g. elliptical, spiral or irregular. The former class of objects are called star clusters and the latter are called external galaxies. These objects have been discussed in detail in Sects. 1.11 and 1.12, respectively. Star clusters are ensembles of stars and galaxies are vast collection of stars, interstellar gas and dust, cosmic rays and unseen matter pervaded by magnetic field. The Galaxy which we belong to consists of some 10^{11} stars. Sun is just a mediocre member of this group. Nevertheless it is the most important from our point of view because being the nearest star, it is most suitable for detailed physical studies. Unlike other stars seen as a point sources Sun is seen as a disc. This is called Photosphere. Sun is composed of layers of hot gases.

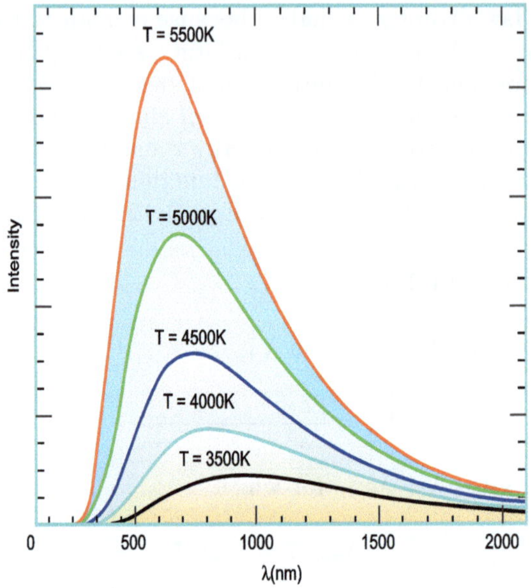

Figure 1.6 Schematic of Wien's displacement law

The other layers starting from Photosphere are Chromosphere and Corona. The visible disc of Sun is actually the base of the Photosphere. Our view is obstructed beyond this layer due to the high opacity of denser layers below it. The temperature at the base of Photosphere is roughly 5,800 K. Above the Photosphere the second major layer is Chromosphere which extends nearly 20,000 km above the Photosphere, with density decreasing and temperature increasing upwards. The Corona is the outermost layer starting from Chromosphere. With the invention of the instrument Coronagraph in 1930 by French physicist B. Lyot it has now become possible to study the spectra of Corona in great detail. The Chromosphere and Corona are dominated by emission spectra whereas Photosphere is dominated by Frounhofer lines which are absorption phenomena. The Coronal temperature is extremely high ($\sim 10^6$). The density of particles (mostly electrons) is of the order of $\sim 10^6$ to $10^8\,\mathrm{cm}^{-3}$ compared to $10^{10} - 10^{12}$ in Chromosphere and $10^{16} - 10^{17}$ in Photosphere.

1.3 Brightness of Stars

As we observe the brightness of stars through telescope we find that those differ from one another. During the second century BC, Hipparchus first observed 1,000 stars and classified them into six groups according to the decreasing order of brightness $B_1, B_2, \ldots, B_6$. This scale of brightness is called apparent magnitude often denoted by "m". In 1830 William Herschel found (through stellar photometry experiment) that a first magnitude star is 100 times brighter than a sixth magnitude star. Later in 1856, N.R. Pogson gave a quantitative scale assuming equal ratios of brightness would give equal differences in magnitude. So if

$$\frac{B_1}{B_6} = 100, \quad \frac{B_1}{B_2}\frac{B_2}{B_3}\frac{B_3}{B_4}\frac{B_4}{B_5}\frac{B_5}{B_6} = 100$$

If

$$\frac{B_i}{B_{i+1}} = x, i = 1, 2, \ldots, 5$$

then

$$x^5 = 100$$

giving

$$x = \sqrt[5]{100} = 2.512.$$

So if B_m and B_n are the brightness of two stars having apparent magnitudes m and n, then

$$\frac{B_m}{B_n} = (2.512)^{(n-m)},$$

yielding

$$\log\frac{B_m}{B_n} = (n-m)\log 2.512 = 0.4(n-m)$$

$$\text{Hence, } \frac{B_m}{B_n} = 10^{0.4(n-m)} \tag{1.19}$$

1.3.1 Absolute Magnitude and Distance

Since the brightness depends strongly on the distance, so two stars having equal brightness but placed at different distances will give different magnitudes, i.e. farther one will be dimmer than the nearer one. This creates ambiguity in considering apparent magnitude. So, for calibration one introduces absolute magnitude which is the apparent magnitude of the star placed at a standard distance of 10 parsecs (1 parsec $= 3 \times 10^{18}$ cm). So for a star having apparent and absolute magnitudes m and M, respectively, from Eq. (1.19),

$$\frac{B_m}{B_M} = 10^{0.4(M-m)}$$

If B_m and B_M are the brightness of two stars at distances d and D, respectively,

$$\frac{B_m}{B_M} = \frac{D^2}{d^2} = 10^{0.4(M-m)}$$

i.e.

$$m - M = 5(\log d - \log D)$$

If we take $D = 10$ parsec, the above relation reduces to

$$m - M = 5\log d - 5 \tag{1.20}$$

This is the fundamental relation involving apparent magnitude, absolute magnitude and distance of a star and if any two are given the third can be computed. $m - M$ is often called the distance modulus. The above relation is true for nearby astronomical objects but for larger distance (redshift $\gg 1$) another term known as "K-correction" is to be added on the RHS of (1.20) to get the absolute magnitude at any particular colour of wavelength. This will be discussed later.

1.3.2 Magnitude–Luminosity Relation

Now from (1.19),

$$n - m = 2.5\log B_m/B_n$$

Let L_m, L_n be the luminosities (energy radiated from a star per unit time) of the two stars at distances d_m and d_n respectively. Then,

$$B_m = \frac{L_m}{4\pi d_{m^2}}, \quad B_n = \frac{L_n}{4\pi d_{n^2}}$$

This yields

$$n - m = 2.5\log(L_m/L_n) + 5\log(d_n/d_m)$$

i.e.

$$(n - 5\log d_n) - (m - 5\log d_m) = 2.5\log\frac{L_m}{L_n}$$

Using (1.20),

$$n - M_n = 5\log d_n - 5$$

and $m - M_m = 5\log d_m - 5$ where M_m and M_n are their absolute magnitudes.

Hence $M_m - M_n = -2.5\log L_m/L_n$

If one of the stars is sun, then $M_n = M_\odot, L_n = L_\odot$,

and

$$M - M_\odot = -2.5\log L/L_\odot \tag{1.21}$$

for any star having absolute magnitude and luminosity M and L, respectively. $M_\odot = +4.83$ and $L_\odot = 3.84 \times 10^{33}\,\mathrm{ergs\,s^{-1}}$.

Thus the above relation gives a conversion law between absolute magnitude (M) and luminosity (L) of a star since $M_\odot$ and $L_\odot$ are known.

1.3.3 Different Photometry Systems

With the development of photo electric photometry it has become possible to measure the magnitudes by modern photo electric methods. Among the various photometric techniques during the last few decades the U, B, V magnitude system, developed by Johnson and Morgan (1953) is most widely used. This system measures the apparent magnitudes of a star in ultraviolet (U), blue or photographic (B) and green or visual (V) regions of the spectrum. The scale is calibrated as, for a class AOV star $U = B = V$. The centres of the bands for U, B, V magnitudes are, respectively, at $\lambda 350$ nm, $\lambda 430$ nm and $\lambda 550$ nm. There are other photometric systems, e.g. Johnson Cousin UBVRI System, Washington CMT_1T_2 system established by Canterna (1976) and Geisler (1990), Sloan Digital Sky Survey (SDSS), ugriz system, etc. The wavelengths and widths of the above broad band system (in Å, 1 Å $= 10^{-8}$ cm) are given in Table 1.1.

UBVRI			Washington			SDSS		
	λ_{eff}	$\Delta\lambda$		λ_{eff}	$\Delta\lambda$		λ_{eff}	$\Delta\lambda$
U	3,663	650	C	3,982	1,070	u'	3,596	570
B	4,361	890	M	5,075	970	g'	4,639	1,280
V	5,448	840	T_1	6,389	770	r'	6,122	1,150
R	6,407	1,580	T_2	8,051	1,420	i'	7,439	1,230
I	7,980	1,540				z'	8,896	1,070

Table 1.1 Different photometric systems

Colour Index of a Star

According to Wien's displacement law, at higher temperature λ_{max} shifts towards shorter wavelength. Therefore stars of higher temperature will emit more in shorter wavelengths (blue–violet) and vice versa. Since colour depends on temperature, stars at different wavelengths will have different colours. The difference between the photographic (magnitude measured on the basis of photographic image using blue filter and blue sensitive emulsions) and photo visual (same but using a yellow filter and yellow sensitive emulsions) magnitudes is known as colour index (**CI**) of a star. In U B V photometry the widely used colours are $B - V$ and $U - B$. A hot star emits more in blue or violet than yellow or red. So B magnitude of a hot star is brighter (numerically smaller number) than its V magnitude (numerically larger number). So for a hot star $B - V$ is negative. Similarly it is positive for a cool star. When light from a star passes through atmosphere blue light is more scattered than red. So the star appears redder. This is known as "reddening". It can be measured numerically comparing a star of same spectral class if the spectral class of the star is known and CI is measured.

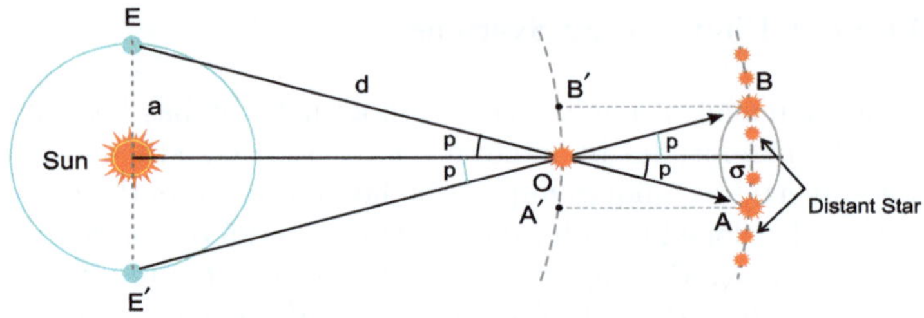

Figure 1.7 Parallax distance relation

1.3.4 Stellar Parallax and Stellar Distances

Parallax is a virtual shift of the original position of an object, due to the view along different lines of sight. Stellar parallax is caused by different orbital positions of earth, i.e. the nearby star under consideration appears to move with respect to the distant stars. Figure 1.7 shows that E and $E^{'}$ are the two positions of earth. When earth is at E, to an observer at earth, the position of a nearby star at O appears to shift to the position A, relative to a distant, hence, apparently fixed star. Similarly at $E^{'}$, the shift occurs at B. Since the observer observes a two-dimensional picture of the sky, the places AB and $A^{'}B^{'}$ appear to coincide. So if the total angular shift is divided by 2 then, parallax p is found for the star, actually at O. From the triangle E(Sun)O we have

$p_{rad} = \frac{a}{d}$, where d is the distance of the star and a is the distance of earth from Sun.
But $1_{rad} = (\frac{180\times60\times60}{\pi})^{''}$ of arcseconds

i.e. $1_{rad} \simeq 206265$ of arcseconds

Therefore, $\frac{p^{''}}{206265} = \frac{a}{d}$ since the parallax is expressed in arcseconds. This finally gives the distance d of the nearby star as

$$d = \frac{206265a}{p^{''}} \tag{1.22}$$

Now, when $p^{''} = 1^{''}, d = 206265a = 3.26\ lightyear = 3\times10^{18}$ cm.

This distance is used as the unit of distance in astronomy and is called 1 "parsec".

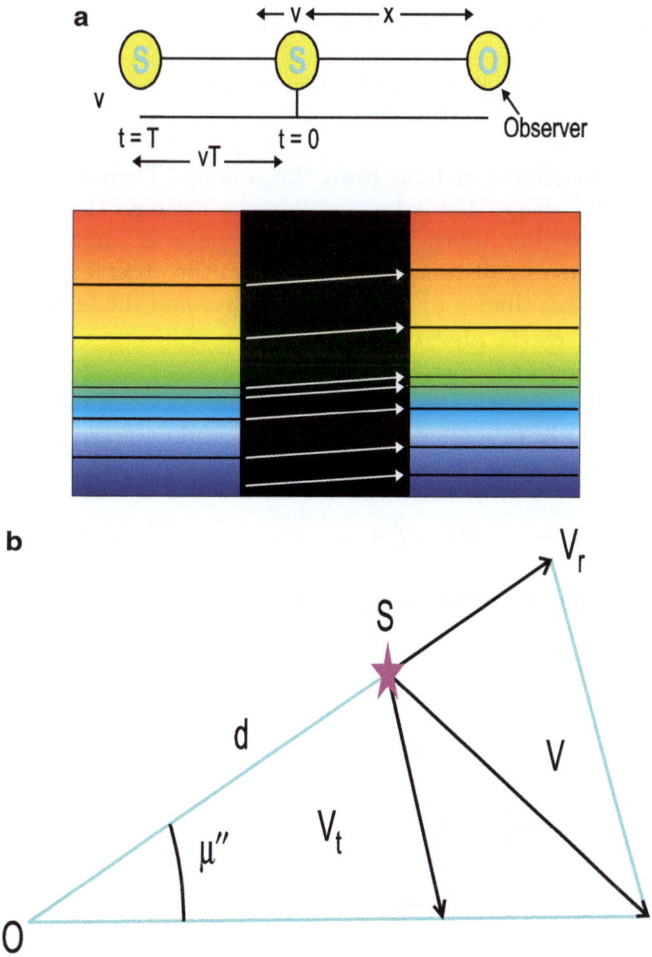

Figure 1.8 (**a**) Doppler Shift and Stellar motions. (**b**) Proper motion of star

The above method of trigonometric parallax is applicable for stars which are astronomically nearer, i.e. approximately within 50 parsec (abbreviated as pc here after). This distance corresponds to $0^{''}.02$. Measurement of still smaller parallaxes introduces various kinds of errors and thus is not appropriate. For larger distances other methods such as "Moving Cluster Method" or various objects, e.g. Cepheid Variables, Supernovae, RR Lyrae Variables, W Virginis stars, brightest red giants in globular star clusters, Mira Variables are used as distance indicators.

1.3.5 Doppler Shift and Stellar Motions

Suppose an observer is at rest and a source emitting light is moving away from the observer at a speed $v_r \ll c$.

Let ν_0 be the frequency of light from the source. Then it sends waves at regular interval $T = \frac{1}{\nu_0}$. Let x be the distance between the source and the observer, at $t = 0$ (Fig. 1.8a). The next wave pulse is emitted after time T and the source moves a distance $v_r T$ away from the observer, within time T. The first pulse takes time $\frac{x}{c}$ to reach the observer and the second pulse takes $\frac{x+v_r T}{c}$ time to reach the observer.

So the first pulse reaches the observer at time, $t_1 = \frac{x}{c}$ and second pulse reaches at time $t_2 = T + \frac{x+v_r T}{c}$. The time interval detected by the observer between two pulses is

$$T' = t_2 - t_1 = T + \frac{x + v_r T}{c} - \frac{x}{c} = \frac{v_r + c}{c} T$$

So, apparent frequency experienced by the observer is

$$\nu' = \frac{1}{T'} = \frac{c}{v_r + c} \frac{1}{T} = \frac{c}{v_r + c} \nu_0$$

$$or, \frac{\nu_0}{\nu'} = \frac{v_r}{c} + 1$$

$$or, \frac{\lambda'}{\lambda_0} = 1 + \frac{v_r}{c} (since \ c = \nu_0 \lambda_0 = \nu' \lambda')$$

$$or, \frac{\lambda' - \lambda_0}{\lambda_0} = \frac{v_r}{c}$$

$$or, \frac{\triangle \lambda_0}{\lambda_0} = \frac{v_r}{c}$$

$$or, v_r = cz \tag{1.23}$$

where $z = \frac{\triangle \lambda_0}{\lambda_0}$ is called the Doppler redshift (Fig. 1.8a) as $\lambda > \lambda_0$. So when a heavenly body is moving away from us then the wavelength emitted from that body gets red shifted. So if z can be measured, the velocity along the line of sight, also called radial velocity of that object can be measured. Redshift of light also takes place due to expansion of the Universe (known as cosmological redshift) or due to deflection of light near a compact object (known as gravitational redshift) but these are beyond our present discussion.

In the above part we have discussed about the radial velocities of heavenly bodies and it is along the line of sight of an observer. There is another kind of motion of celestial objects which is perpendicular to the line of sight direction. This is called transverse component of velocity, v_t. Then the space velocity v of the object is measured as

$$v^2 = {v_r}^2 + {v_t}^2$$

If μ be the corresponding angular shift at the position of the observer O (say) in 1 year (Fig. 1.8b), then μ is called the proper motion of the object.
If the star be at a distance d from the observer,

$$\frac{\mu''}{206265} = \frac{nv_t}{d}$$

where n is the number of seconds in a year.

Then using (1.21) for d,

$$v_t = \frac{\mu''}{206265} . \frac{1}{n} . \frac{206265}{p''} \quad a$$

Using

$$a = 1.49 \times 10^8 \,\text{km} \;\; \text{and} \;\; n = 3.16 \times 10^7 \,\text{s}$$

$$v_t = 4.74 \frac{\mu''}{p''} \;\; \text{km}\,\text{s}^{-1} \tag{1.24}$$

Peculiar Motion

In all velocity measurements, sun is considered as the origin with respect to which the velocities are defined. But sun, as a star also moves. So one should consider the sun's motion also and hence there should be a suitable reference frame with respect to which stellar motions could be defined. The hypothetical origin of this reference system is called "Local Standard of Rest (LSR)".

It is defined as the origin of a reference system such that the motion of all stars within a small neighbourhood of Sun, say, 50–100 parsec, have the mean velocity zero. So, LSR with respect to a group of stars can be considered as the centroid of the system. The motion of an individual star with respect to LSR is called its "Peculiar Velocity".

1.4 Spectral Characteristics of Stars

In eighteenth century (1787–1826), the great German physicist, Joseph Fraunhofer, first observed dark line (often called as absorption lines) superimposed on a continuous background in the spectrum of Sun. The origin of such spectra was an intriguing problem among the astrophysicists unless the structure of atom was explored gradually. In 1885, J.J. Balmer fitted a simple empirical formula to the spectrum of hydrogen atom as

$$\lambda = B\frac{n^2}{n^2 - 4}$$

where B is a constant (viz. 3645.6) and n is an integer taking the values 3, 4, 5, ... etc. The series of different wavelengths for successive values of n are known as Balmer series. In 1890, J.R. Rydberg modified the formula and replaced λ by its corresponding frequency ν by

$$\nu = R_H\left[\frac{1}{2^2} - \frac{1}{n^2}\right], n = 3, 4, \ldots \text{etc.}$$

where R_H (viz. 109,677.58 cm^{-1}) is known as Rydberg constant. Though with the above formulae the wavelengths or frequency of hydrogen atom can be computed for other lines in its spectra still its origin and also the complexities in the spectra of other elements remain a puzzle for a long time. Finally in 1901, Planck introduced his quantum hypothesis and in 1913 Neils Bohr devised the structure of hydrogen atom incorporating Planck's quantum hypothesis and the concept of nucleus in an otherwise empty atom, as a result of Rutherford's famous experiment of scattering of α-particles by a thin gold foil, held in 1911. The various postulates of his theory are as follows:

(1) Electron revolves around the nucleus of hydrogen atom and during the revolution they do not radiate electromagnetic radiation according to Maxwell's theory, even it is accelerated.

(2) Electrons move along certain stationary orbits, defined by their angular momentum,

$$m_e vr = n(h/2\pi), n = 1, 2, 3 \ldots \text{etc.}$$

where m_e is the mass of the electron, v is its velocity at a distance r from the centre of the atom and h is the Planck's constant. n is called the principal quantum number.

(3) When transition of electrons occurs between any two energy levels of the electron then only radiation is emitted/absorbed following the law,

$$E_i - E_f = h\nu(\text{emission})(E_i > E_f)$$

$$E_f - E_i = h\nu(\text{absorption})(E_i < E_f)$$

where E_i, E_f are the initial and final energy values, and ν is the frequency of the corresponding radiation (Fig. 1.9). When we model the stellar bodies we assume it as a BB having a spherical structure. So when light emitted from the surface of a star passes through a cold diffuse gas, under low pressure, then the atoms present in the gas absorb some wavelengths of the incoming radiation of the actual source, resulting in an absorption feature in the spectrum (Fig. 1.10). An observer in front of these two sources, see, what we call "absorption spectra".

But when an observer is at a position which only includes the diffuse cloud but no radiative BB source, then the source now is a glowing gas with some selective wavelengths of radiation consisting of discrete bright lines or emission lines and the observer sees what is called "emission spectra". When the observer is in front of a BB emitter without any intervening diffuse cloud, he/she sees what is called "continuous" spectra.

1.5 Spectral Features and Saha's Ionization Theory

After the discovery of Fraunhofer lines, astrophysicists became interested to study the spectral characteristics of other stars. It is found that though the spectrum of other stars sustains the general characteristics of Fraunhofer lines

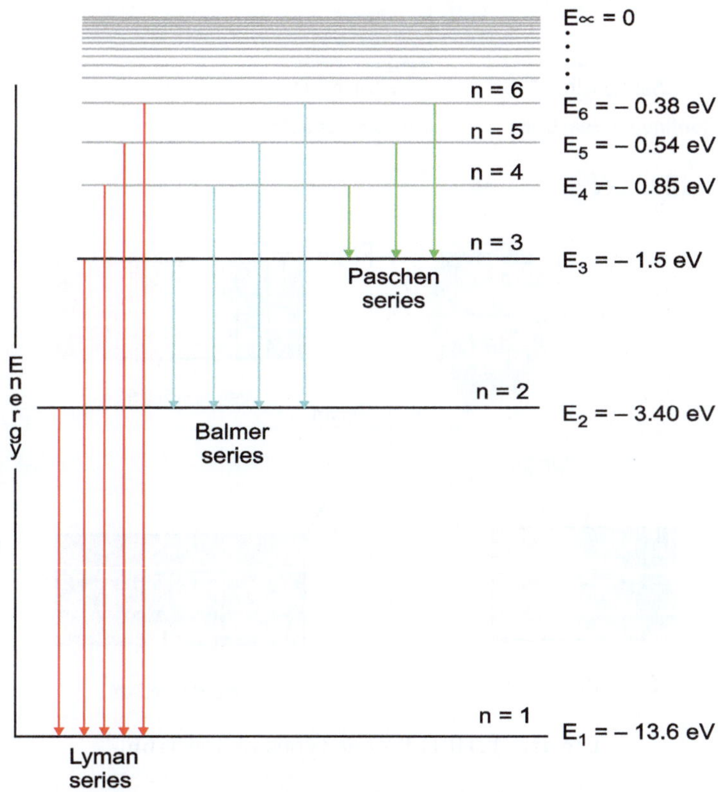

Figure 1.9 Energy levels of hydrogen atom

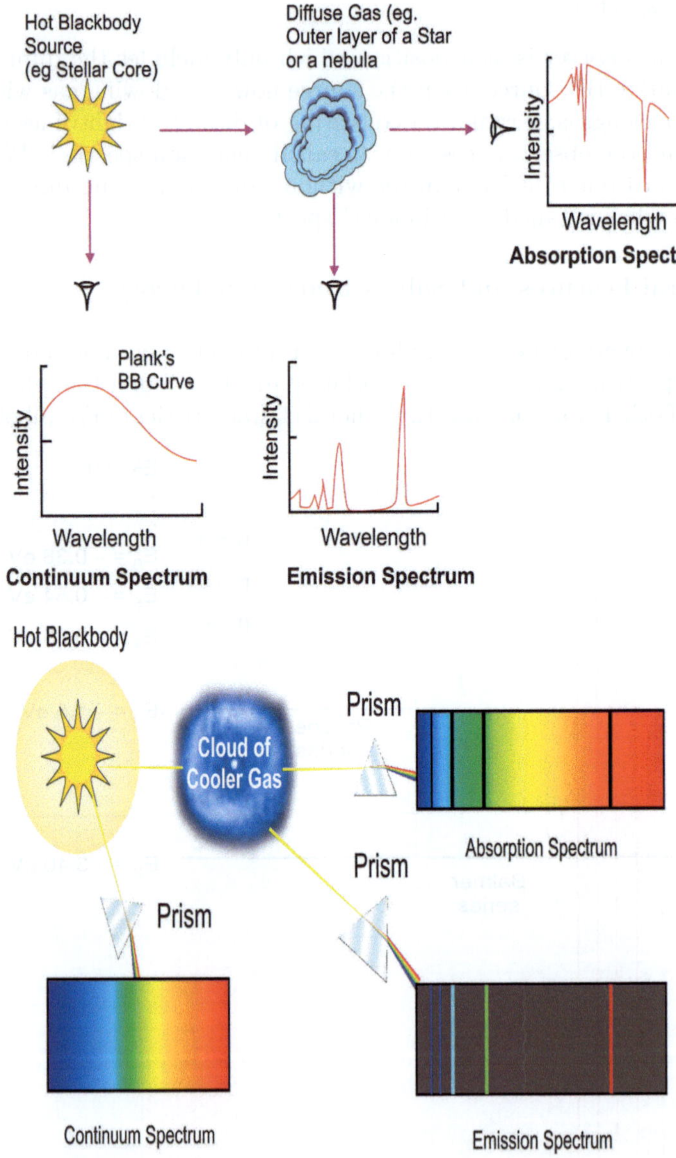

Figure 1.10 Different types of spectrum

but (1) the strengths of the lines vary from star to star on the basis of different elements and (2) strength of any particular element varies continuously in the spectra of different stars, e.g. lines of He I and He II (i.e. He^+) are the principal features for some spectra or these lines are completely absent in others, where lines of neutral or ionized metals become dominant. At that time the atomic structure was not known. So various interpretations were suggested by the scientists. Thus stars showing prominent He lines were suggested to be composed of He. Hence the reason for various spectral features were due to compositional differences in different stars. Another group suggested that the differences were the manifestation of the different stages of stellar evolution. But is was Sir Norman Locker's excellence who for the first time on the basis of laboratory experiment described the spectral features as a temperature sequence. But with the poor knowledge of atomic structure the quantitative interpretation of spectral features was not become very successful. Finally with the advent of atomic theory by Planck (1901), Rutherford (1911) and Bohr (1913), Saha (1920) was able to successfully explain the Fraunhofer lines through his famous ionization theory known as Saha's ionization equation.

Maxwell Distribution

According to Maxwell's hypothesis moving particles of a gas having velocities in the range v_i and $v_i + dv_i$ follow a Gaussian distribution,

$$n(v_i)dv_i = n\left(\frac{m}{2\pi kT}\right)^{1/2} \exp\left(\frac{-m{v_i}^2}{2kT}\right) dv_i$$

where $n(v_i)dv_i$ is the number of particles in the velocity range v_i and $v_i + dv_i$ and n is the total number of particles and m is the mass of a gas particle.

For a spherical shell within radii v, and $v + dv$ the volume is $4\pi v^2 dv$. Then number of particles in the shell having velocity range v and $v + dv$ is

$$n(v)dv = n\left(\frac{m}{2\pi kT}\right)^{3/2} \exp\left(\frac{-mv^2}{2kT}\right) 4\pi v^2 dv \qquad (1.25)$$

Ideal Gas Laws

Consider a cube of gas of unit length consisting of n particles. Then the average number of particles towards any arbitrary face is $\frac{n}{6}$. If v_{rms} is the root mean square velocity of the particles, then the number of particles hitting any face per unit time is $\frac{n}{6}v_{rms}$. After striking a face, a gas particle of mass m is reflected and the reflection takes place in the opposite direction without loss of energy for an ideal gas. So the change in momentum of the gas particle is $2mv_{rms}$. So, pressure on the face of the cube = Force/unit area is nkT.

Problem 1 *Show that the pressure in a cube of perfect gas having* $v_{rms} = \frac{3kT}{m}$ *is* nkT.

$$\begin{aligned} P &= Rate\ of\ change\ of\ momentum \\ &= (2mv_{rms})(\frac{n}{6}v_{rms}) \\ &= \frac{1}{3}mnv_{rms}^2 \\ &= \frac{1}{3}mn(\frac{3kT}{m}) \\ &= nkT \end{aligned}$$

This is the equation of state of a perfect gas.

Boltzmann Law for Excited State for Electron Gas

Let (p_{ex}, p_{ey}, p_{ez}) be the momentum of an electron at a point (x, y, z) following quantum mechanical laws. Then, $dxdydzdp_{ex}dp_{ey}d_{ez}$ = volume element in phase space. The minimum volume of a cell in phase space = h^3, h being the Planck's constant.

Then, number of cells in the volume element $= \frac{d^3rd^3p_e}{h^3}$. Since the electrons have two states of spin up and spin down, the total numbers of electron states in the cells

$$dg(p_e) = \frac{2d^3rd^3p_e}{h^3} = \frac{2dvd^3p_e}{h^3}$$

If the electron density in three-dimensional space is n_e, then the volume per electron is $dv = \frac{1}{n_e}$ and the volume in the momenta space $(p_e, p_e + dp_e)$ is $d^3p_e = 4\pi {p_e}^2 dp_e$, i.e.

$$dg(p_e) = \frac{8\pi {p_e}^2 dp_e}{n_e h^3} \tag{1.26}$$

Then from (1.25)

$$\begin{aligned} n(v)dv &= n\left(\frac{m}{2\pi kT}\right)^{3/2} \exp\left(\frac{-mv^2}{2kT}\right) 4\pi v^2 dv \\ &= n\left(\frac{h^3}{2(2\pi mkT)^{3/2}}\cdot\frac{8\pi v^2}{h^3}.m^3 exp(-mv^2/2kT)dv\right) \\ &= n\left(\frac{h^3}{2(2\pi mkT)^{3/2}}\right) g(v)\exp(-E_v/kT)dv \end{aligned} \tag{1.27}$$

where

$$g(v) = m^3h(v) = \frac{8\pi m^3v^2}{h^3} and\ \ E_v = \frac{1}{2}mv^2$$

So, for any two states of gas with the velocity ranges $(v, v + dv)$ and $(v', v' + dv')$,

$$\frac{n(v)}{n(v')} = \frac{g(v)}{g(v')} \exp[-(E_v - E_{v'})/kT]$$

Let s, o be the excited state and ground state of the gas particles, then

$$\frac{n_s}{n_o} = \frac{g_s}{g_o} \exp[-(E_s - E_o)/kT] = \frac{g_s}{g_o} \exp(-\psi_s/kT) \tag{1.28}$$

Let n be the total number of atoms in all states s, then

$$n = \sum_{s=0}^{\infty} n_s$$

$$\begin{aligned}
\text{so, } g_o \frac{n}{n_o} &= g_o \sum_{s=0}^{\infty} \left(\frac{n_s}{n_o}\right) = g_o \sum_{s=0}^{\infty} \left(\frac{g_s}{g_o}\right) e^{-\psi_s/kT} \\
&= g_o + g_1 e^{-\psi_1/kT} + g_2 e^{\psi_2/kT} + \ldots \\
&= u_p(T)
\end{aligned}$$

Here $u_p(T)$ is called the partition function. So the Boltzmann formula now takes the form following (1.28)

$$\frac{n_s}{n} = \frac{g_s}{u_p} e^{-\psi_s/kT} \tag{1.29}$$

Now an atom is at r th stage of ionization means it has stripped off r electrons. If χ_r be the ionization potential at (r + 1)th stage, the energy required by an electron is $\chi_r + \frac{{p_e}^2}{2m_e}$ where p_e and m_e are the momentum and mass of the electron and $\frac{{p_e}^2}{2m_e}$ is its K.E.

Let n_r and dn_{r+1} be the number densities of electrons in two ionization states r and (r + 1) and the electron at (r + 1) has the momentum in $(p_e, p_e + dp_e)$. Let the statistical weight of the free electron be $dg(p_e)$. Then,

$$\frac{dn_{r+1}}{n_r} = \frac{g_{r+1} dg(p_e)}{g_r} \exp\left(-\frac{\chi_r + {p_e}^2/2m_e}{kT}\right) \tag{1.30}$$

Then using (1.26)

$$\frac{dn_{r+1}}{n_r} = \frac{g_{r+1}}{g_r} \frac{8\pi {p_e}^2 dp_e}{n_e h^3} \exp\left(-\frac{\chi_r + {p_e}^2/(2m_e)}{kT}\right)$$

Summing over all momenta p_e

$$\frac{n_{r+1}}{n_r} = \frac{g_{r+1}}{g_r}\frac{8\pi}{n_e h^3}e^{-\chi_r/kT}\int_0^\infty {p_e}^2 \exp\left(\frac{-{p_e}^2}{2m_e kT}\right) dp_e$$

Now, for $y > 0$, $\int_0^\infty x^2 e^{-y^2x^2}dx = \frac{\sqrt{\pi}}{4y^3}$

So, using the above result,

$$\frac{n_{r+1}}{n_r}.n_e = \frac{g_{r+1}}{g_r}.2.\frac{(2\pi m_e kT)^{3/2}}{h^3}e^{-\chi_r/kT} \tag{1.31}$$

Here, n_{r+1}, n_r indicate ground states of excitation. More generally if $n_{r,k}$ is the number density of atoms in the r the stage of ionization and k the stage of excitation and $g_{r,k}$ is its corresponding statistical weight then, proceeding as before,

$$\frac{n_{r,s}}{n_{r,0}} = \frac{g_{r,s}}{g_{r,0}}e^{-\psi_{r,s}/kT}, n_r = \sum_{s=0}^{\infty} n_{r,s}$$

and

$$\frac{g_{r,0}}{n_{r,0}}n_r = g_{r,0}\sum_{s=0}^{\infty}\frac{n_{r,s}}{n_{r,0}} = g_{r,0} + g_{r,1}e^{-\psi_{r,1}/kT} + \ldots = u_r(T)$$

Hence,

$$n_r = u_r\frac{n_{r,0}}{g_{r,0}}$$

and

$$n_{r+1} = u_{r+1}\frac{n_{r+1,0}}{g_{r+1,0}}$$

$$\text{so, } \frac{n_{r+1}}{n_r} = \frac{u_{r+1}}{u_r}\frac{n_{r+1,0}g_{r,0}}{n_{r,0}g_{r+1,0}}$$
$$= \frac{u_{r+1}}{u_r}\frac{1}{n_e}\frac{2(2\pi m_e kT)^{3/2}}{h^3}e^{-\chi_r/kT} \quad (using\ (1.31))$$

i.e.

$$\frac{n_{r+1}}{n_r}n_e = \frac{u_{r+1}}{u_r}2\frac{(2\pi m_e kT)^{3/2}}{h^3}e^{-\chi_r/kT}$$

Since,

$$P_e = n_e kT$$

$$\frac{n_{r+1}}{n_r}P_e = \frac{u_{r+1}}{u_r}2\frac{(2\pi m_e)^{3/2}}{h^3}e^{-\chi_r/kT}(kT)^{5/2}$$

$$\text{i.e. } ln\frac{n_{r+1}}{n_r} = \frac{-5040.4}{T}\chi_r + 2.5lnT + ln\left(\frac{2u_{r+1}}{u_r}\right) - 0.48 - lnP_e \quad (1.32)$$

The above equation is the well-known generalized from of **Saha's ionization equation.**

Observations and Interpretation

It is clear from Eq. (1.32) that degree of ionization is inversely proportional to the electron pressure (P_e) and it is same for all elements. On the other hand degree of ionization increases with temperature T but it is not related to temperature T in the same way as with P_e as T is locked with ionization potential χ_r which is different for different element. If temperature remains constant, ionization will be greater for smaller χ_r and vice versa. Now in giants and supergiants P_e is several order lower than in dwarfs. So atoms with smaller χ_r will have higher degree of ionization in giants than in dwarfs and vice versa. Thus neutral metallic lines Ca I ($\chi_r = 6.09$ eV), Sr I ($\chi_r = 5.67$ eV) are stronger in dwarfs and ionized metallic lines Ca II ($\chi_r = 11.82$ eV), Sr II ($\chi_r = 1{,}098$ eV) are stronger in the spectra of giants. As degree of ionization increases with temperature and ionization potential of hydrogen atom is quite high, viz. 10.15 eV so in the atmosphere of cool stars a small number of H atoms are excited. So Balmer series of lines are weak in the spectra of cool stars. In the spectra of stars having comparatively higher temperature, these lines are available but in the spectra of hottest stars these lines again disappear as H atoms are ionized. So in the spectra of hottest stars ionized line of H and excited lines of *He* are available as *He* atom has the highest i.p (viz. 24.54 eV).

Hence it is clear from the above discussion that the spectra of different stars are mainly a temperature sequence combined with the property of ionization potential of various elements and the density of the atmosphere manifested by electron pressure. These features can be quantitatively explained with the help of Saha's Ionization Theory.

1.6 Celestial Co-ordinate Systems

For studying various astronomical objects we have introduced various co-ordinate systems. If at any point of the earth a straight line is drawn through an observer, it will meet the celestial sphere (the spherical dome of sky) at two points, Z, $Z^{'}$ (Fig. 1.11), called "Zenith" and "Nadir". The circular plane through the observer and perpendicular to this line cuts the sphere along a great circle, called "horizon" (viz. NS). Similarly if a line is drawn through the observer parallel to earth's rotational axis, then this line will intersect the sphere at two points, called north celestial pole (P) and south celestial pole (P') and the corresponding plane perpendicular to this line cuts the sphere along a great circle, called the equator (viz. QR). Due to annual motion of Earth around Sun, the Sun appears to move along a great circle, called "the ecliptic" and this intersects the equator and meridian (the circle along which plane of the paper cuts the celestial sphere) at four points, "First point of Aries" (γ, at 21st March), Vernal Equinox (at 22nd June), "First point of Libra" (Ω, at 23rd September) and Autumnal Equinox (22nd December). The shape of our Galaxy also provides co-ordinate systems. The central plane through the Galactic disc cuts the sphere along galactic equator. A line perpendicular to this plane cuts the sphere at two points, north galactic pole (NGP) and south galactic pole (SGP). Super galactic plane, is the plane of Local Supergalaxy. Corresponding reference points for this plane are north supergalactic pole (NSP) and south supergalactic pole (SSP). Then the various co-ordinate systems are defined as:

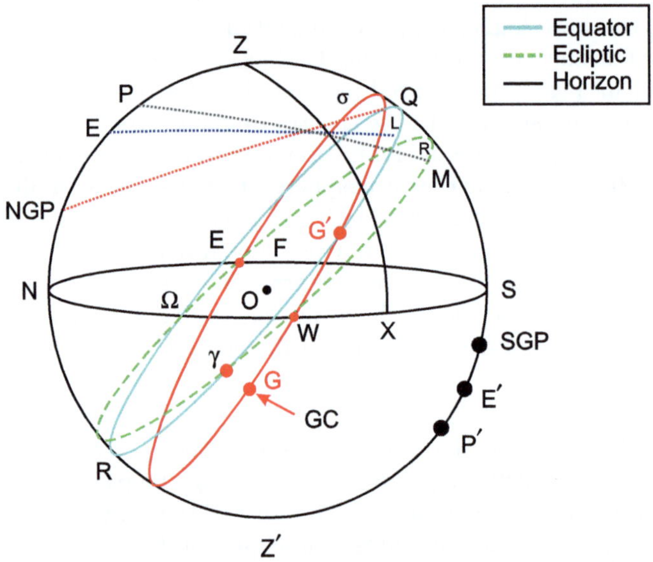

Figure 1.11 Various co-ordinate systems used in astronomy

1. Azimuth–Altitude: This is the co-ordinate systems with respect to $Z - Z'$ and horizon $NWSE$. The co-ordinates of a star σ are azimuth and altitudes are the angular distances corresponding to the arcs, NX and σX made at the observer O.
2. Right ascension (α)–declination (δ): This is the co-ordinate systems with respect to $P - P'$ and celestial equator. The co-ordinates of a star or any object δ are (α, δ), which are the angular distances by the arcs $\gamma M(\alpha)$ and $\sigma M(\delta)$ of great circles made at O.

3. Celestial Latitude (b) and Longitude (l): This is the co-ordinate system with respect to E and E' and ecliptic. The co-ordinates of a star σ or any object are the angular distances made by the arcs γR (l) and σR (b) of great circles made at O.
4. Galactic Latitude and Longitude: This is the co-ordinate system with respect to the galactic equator and NGP and SGP. The co-ordinates of any star δ or object are the angular distances of the arcs GL and σL of great circles (G being the Galactic centre). The equatorial co-ordinates of NGP are $\alpha = 12\,\text{h}\,49\,\text{min}$ and $\delta = 27.4°$ (α is measured in hour angle). The angle between celestial equator and galactic equator is roughly 63.5°.
5. Super Galactic Latitude (b) and Longitude (l): This is the co-ordinate system with respect to NSP, SSP and supergalactic equator. The co-ordinates of any star σ or any objects are the angular distances of the arcs $G'L$ and σL of great circles. The NSP has the galactic co-ordinates $l = 47.37°$ and $b = 6.32°$ and zero point (origin) (G') in the supergalactic plane has the co-ordinates ($l = 137.37°$ $b = 0°$).

Co-ordinate Transformation

Following Groningen Image Processing System (home page: Gipsy) let the co-ordinates (α, β) will have to be changed to (α', β'), where α is the longitude and β is the latitude (say). The unit vector along the direction (α, β) has co-ordinates

$$\begin{aligned} r_1 &= \cos\alpha\cos\beta \\ r_2 &= \sin\alpha\cos\beta \\ r_3 &= \sin\beta \end{aligned}$$

Then, if $\vec{s} = T\vec{r}$ where $\vec{s} = (s_1, s_2, s_3)$, then

$$\begin{aligned} \alpha' &= \tan^{-1}(s_2/s_1) \\ \beta' &= \sin^{-1}(s_3) \end{aligned}$$

$\alpha^{'}$ are chosen in appropriate quadrant.
The matrices T for useful co-ordinate transformations are (Cartesy: skyco. c, Gipsy source code) given in Appendix at the end of this chapter.

Projected Separation Between Two Stars in the Sky

If two points are very close together on the sky, within a degree or less then the approximate angular separation between them is given by

$$d = \{(\alpha_2 - \alpha_1)^2 \cos^2\left(\frac{\delta_1 + \delta_2}{2}\right) + (\delta_2 - \delta_1)^2\}^{1/2}$$

where $(\alpha_1, \delta_1)m$ and (α_2, δ_2) are the RA and DEC of the two celestial objects and they are measured in degrees. Then the distance is the projected distance measured in degrees.

For two objects separated by arbitrary distance the general formula is

$$\cos\gamma = \cos(90 - \delta_1)\cos(90 - \delta_2) + \sin(90 - \delta_1)\sin(90 - \delta_2)\cos(\alpha_1 - \alpha_2).$$

where γ is the arc length measured in degrees.

1.7 Hertzsprung–Russel Diagram

In early era of twentieth century (1913) E. Hertsprung of Denmark and H.N. Russel of USA independently found that spectral characteristics or colours of various stars are closely associated with their luminosity or absolute magnitudes and can be classified into groups O, B, A, F, G, K, M following a temperature sequence of approximately 40,000 to 2,400 K. Each spectral class is divided into ten subclasses, e.g. O1, O2, ..., O10, B1, B2 ... etc. The plot of absolute magnitude versus spectral classes of the stars show some definite trends (Fig. 1.12).

(1) Almost 90 % of the stars lie along a narrow band extending from left top to right bottom of the diagram known as the "main sequence" (MS, viz. branch V). The extent of absolute magnitude is from −7/−8 to +15. Sun belongs to this branch having the position (+4.8, G2) in the diagram.

(2) A short branch upwards the MS extending from F to M and −1.0 to +1.0, in absolute magnitude is known as "giants" (viz. branch III).

(3) Above the "giants" lies the group of stars extending up to a magnitude −3.0. This is called "bright giants" (viz. branch II).

(4) At the top of the diagram there are stars belonging to highly luminous absolute magnitude range −3.0 to −8.0, known as "super giants" (viz. Ia, Ib).

(5) At the lower left, far below the MS, there is a group of very faint stars belonging to middle B to G with absolute magnitude varying from +10.0 to +15.0. This group is called "white dwarfs" (viz. branch VII).

(6) Most of the variable stars, e.g. Cepheids, RR Lyrae, stars, occupy a large region belonging to giants and supergiants.

(7) Between MS and "white dwarfs" branches, there lies a group of stars known as subdwarfs (viz. branch VI).

(8) There are two other groups of dwarf stars very recently classified (1993) as L and T dwarfs. L dwarfs are objects having spectrum in

the range 6,400–9,000 Å and mainly show neutral alkali lines, e.g. NaI, KI, RbI, CsI, LiI, oxide bands (TiO, VO), hydride bands (CrH, FeH, CaOH), etc. T dwarfs mainly contain strong H_2O, lines, neutral alkalines but no hydrides. For late T, H_2O, NaI and KI have highest strength. About 403 L dwarfs and 62 T dwarfs are so far discovered. With the advent of theories of stellar evolution it is found that different trends in the H–R diagram are nothing but manifestations of different phases of stellar evolution from its birth towards death, which are described in the later part of this chapter.

1.8 Stellar Atmosphere

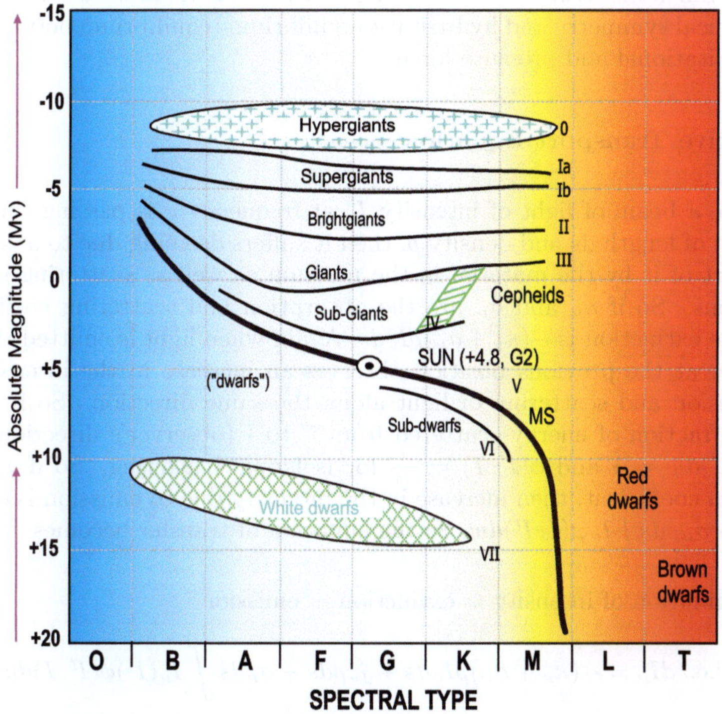

Figure 1.12 The Hertzsprung–Russel diagram (H–R)

The layers of stars producing various kinds of absorption and emission features in the corresponding spectra constitute the stellar atmosphere. The surface temperatures of stars generally vary from 3,000 to 40,000 K. At this high temperature the materials are more or less in gaseous form. As the stars continuously emit radiation, it is expected there is a central engine as the source of energy. So a study of the surface characteristics, together with various modelling helps us to delineate a quantitative picture of physical and chemical conditions inside stellar atmosphere.

For this the basic assumptions are

(1) The atmosphere is considered as stratified plane parallel layers as the atmosphere is very thin compared to the radii of stars.

(2) Since the atmosphere is more or less gaseous, convection or radiation is the only heat transfer mechanism, operative therein.

(3) At the surface of star the flux of radiation $F = \sigma {T_e}^4$ where T_e is the surface temperature and σ is the Stefan–Boltzmann constant. But below the surface, as the temperature increases thermodynamic equilibrium is not maintained but at each depth the flux is isotropic, i.e. we assume thermodynamic equilibrium at local temperature T(z) and call it as local thermodynamic equilibrium (LTE).

(4) Since the structure of star is more or less spherical, we assume a spherical symmetry and hydrostatic equilibrium (equilibrium between gravitational and pressure forces).

Radiative Transport Equation

Suppose a beam of light of intensity I_ν at frequency ν is passing through a medium of length ds and density ρ, then it suffers deeming due to absorption of a part of it by the material in the medium as well as scattering in other directions. So if κ_ν and σ_ν are the absorption and scattering coefficients, then the extinction is $-(\kappa_\nu + \sigma_\nu)\rho I_\nu ds$. Again when light is emitted then the observer at the previous position observes an increase in the intensity due to emission and scattering of light along the same direction. So if $\phi(I', I)$ be the fraction of energy scattered from I' to I (observer's direction), then $\int \phi(I', I)dw = 1$ and $\phi(I', I) = \frac{1}{4\pi}$ for isotropic scattering. So if j_ν is the emission coefficient, then increase in the intensity due to emission is equal to $j_\nu \rho ds + \sigma_\nu \rho ds \int I_\nu \varphi(I, I')dw$. So the equation of transfer becomes

Diminution of intensity = extinction + emission

$$\text{i.e. } dI_\nu = -(\kappa_\nu + \sigma_\nu)\rho I_\nu ds + j_\nu \rho ds + \sigma_\nu ds \int I_\nu(I')\phi(I', I)dw$$

From Fig. 1.13, if we take z axis along the normal and θ is the inclination between the normal direction and observer's direction, the $ds = -dz \sec\theta$. Putting, $\mu = \cos\theta$ the above equation reduces to

$$\mu \frac{dI_\nu}{d\tau_\nu} = I_\nu - S_\nu$$

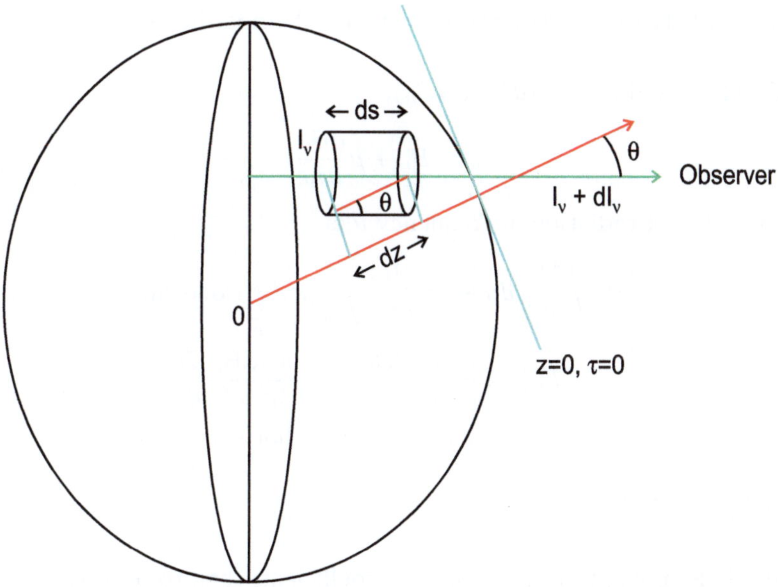

Figure 1.13 Spherical surface of a star viewed as a circular disc by the observer

where τ_ν is defined to be the optical depth as

$$d\tau_\nu = (\kappa_\nu + \sigma_\nu)\rho dz \text{ and } S_\nu = \frac{j_\nu + \sigma_\nu \int I_\nu(I')\phi(I', I)d\omega}{k_\nu + \sigma_\nu}$$

is called the source function.

For LTE, $j_\nu = k_\nu I_\nu = k_\nu B_\nu$ and for isotropic scattering,

$$\sigma_\nu \int I_\nu(I')\phi(I', I)d\omega = \frac{\sigma_\nu}{4\pi} \int I_\nu(I')d\omega = \sigma_\nu J_\nu$$

where J_ν is the mean intensity.

Then,

$$S_\nu = \frac{\kappa_\nu B_\nu + \sigma_\nu J_\nu}{\kappa_\nu + \sigma_\nu}$$

For hot star, $\sigma_\nu = 0$, so, $S_\nu = B_\nu$ and the transfer equation finally reduces to

$$\mu \frac{dI_\nu}{d\tau_\nu} = I_\nu - B_\nu \tag{1.33}$$

Solution of Transfer Equation in Grey Atmosphere

In LTE (1.33) reduces to ($dI_\nu \approx dB_\nu$),

$$I_\nu = B_\nu + \mu \frac{dB_\nu}{d\tau_\nu}.$$

Then the flux of radiation at frequency ν is

$$F_\nu(\tau_\nu) = 2\pi B_\nu \int_{-1}^{+1} \mu d\mu + 2\pi \frac{dB_\nu}{d\tau_\nu} \int_{-1}^{1} \mu^2 d\mu \text{ (following Eq. (1.5))}$$

$$i.e.\ F_\nu(\tau_\nu) = \frac{4\pi}{3}\frac{dB_\nu}{d\tau_\nu} = \frac{4\pi}{3\kappa_\nu\rho}\frac{dB_\nu}{dT}\frac{dT}{dz}$$

Since $d\tau_\nu = (\kappa_\nu + \sigma_\nu)\rho dz = \kappa_\nu \rho dz, \sigma_\nu = 0$ for hot star.

So if $\frac{1}{\kappa}\frac{dB}{dT} = \int_0^\infty \frac{1}{\kappa_\nu}\frac{dB_\nu}{dT} d\nu$

where κ is Rosseland mean absorption coefficient, then total flux,

$$F = \int_0^\infty F_\nu d\nu = \frac{4\pi}{3\rho}\frac{dT}{dz}\int_0^\infty \frac{1}{\kappa_\nu}\frac{dB_\nu}{dT} d\nu$$

and we have

$$F = \frac{4\pi}{3\rho\kappa}\frac{dB}{dT}\frac{dT}{dz} = \frac{4\pi}{3\rho\kappa}\frac{dB}{dz}$$

where $d\tau = \rho\kappa dz$, τ is the mean optical depth.

So,

$$\frac{4\pi}{3}\frac{dB}{d\tau} = F = Constant,$$

$$or, B = \frac{3F}{4\pi}(\tau + t)$$

t being the constant of integration.

Solving (1.33) as a linear equation,

$$I_\nu = \int_{\tau_\nu}^\infty B_\nu e^{-({\tau_\nu}' - \tau_\nu)/\mu} \frac{d{\tau_\nu}'}{\mu}$$

So, the intensity at the surface of a star ($\tau = 0, B_\nu = B$, for grey atmosphere) we have

$$\begin{aligned} I(\tau_\nu = 0) &= \int_0^\infty \frac{3F}{4\pi}({\tau_\nu}' + t) e^{-{\tau_\nu}'/\mu} \frac{d\tau_\nu'}{\mu} \\ &= \frac{3F}{4\pi}(\mu + t) \end{aligned}$$

So

$$F = 2\pi \int_0^1 I(\tau_\nu = 0)\mu d\mu = \frac{3F}{2}\left(\frac{t}{2} + \frac{1}{3}\right)$$

which gives $t = \frac{2}{3}$, so, $B = \frac{3F}{4\pi}(\tau + \frac{2}{3})$.

At the surface, $F = \sigma {T_e}^4$ and $\pi B = \sigma T^4$.

So, substituting B and F in the above relation,

$$T^4 = \frac{3}{4}{T_e}^4\left(\tau + \frac{2}{3}\right). \tag{1.34}$$

It is called the Milne–Eddington solution for grey atmosphere.

Convection and Radiation in Stellar Atmosphere

Let us consider a bubble of the atmosphere of a star becomes overheated. Let $T^{'}, P^{'}, \rho^{'}$ and T_s, P_s, ρ_s be the temperature, pressure and density of the volume element and surrounding atmosphere, respectively. Since the bubble is lighter it starts moving and after travelling a distance −dz the temperature of the bubble is $T' - (\frac{dT'}{dz})dz$ and that of surrounding is $T_s - (\frac{dT_s}{dz})dz$.

For convection to continue,

$$T' - \left(\frac{dT'}{dz}\right)dz > T_s - \left(\frac{dT_s}{dz}\right)dz.$$

We assume P = P' (otherwise the bubble will burst) and adiabatic condition within the bubble. Then, $dP = dP' = \frac{dP}{dz}dz$ and we have

$$\frac{1}{T}\frac{dT_s}{dP_s} > \frac{1}{T'}\frac{dT'}{dP'}$$

$$i.e. \nabla_s = \left(\frac{d\log T_s}{d\log P_s}\right) > \left(\frac{d\log T'}{d\log P'}\right) = \nabla_{ad}$$

If some part of the flux is carried by convection, then

$$\nabla_{ad} < \nabla_{rad}$$

Since only a part of the total flux is left for radiative transfer, we have

$$\nabla_{ad} < \nabla_{rad}$$

This is called Schwarzchild criterion for convection.

Now, from (1.34), $(\frac{1}{T}\frac{dT}{d\tau})_{rad} = \frac{3}{16}\frac{{T_e}^4}{T^4}$

For hydrostatic equilibrium,

$$dP = g\rho dz \text{ and } d\tau = \kappa\rho dz$$

$$\text{so, } \frac{dP}{d\tau} = \frac{g}{\kappa} \text{ i.e. } d\tau = \frac{\kappa}{g}dP$$

So,

$$\nabla_{rad} = \left(\frac{d\log T}{d\log P}\right) = \frac{3\kappa P{T_e}^4}{16gT^4}$$

Now, $F = \sigma {T_e}^4, \sigma = \frac{ac}{4}$ where

$$g = \frac{GM}{R^2}, L = 4\pi R^2 F$$

So,

$$\nabla_{rad} = \frac{3\kappa PL}{16\pi acGMT^4}$$

For adiabatic expansion, PV^γ = Const and PV = RT where $R = C_p - C_v$, then,

$$\nabla_{ad} = \left(\frac{d\log T}{d\log P}\right)_{ad} = \frac{\gamma - 1}{\gamma}$$

Basic Equations of Stellar Atmosphere

The basic equations governing the stellar atmospheres are (1) the equation of continuity, (2) the equation of hydrostatic equilibrium, (3) equation of thermal equilibrium and (4) equation of energy transfer through either convection or radiation.

Equation of Continuity

Let dM(r) present the mass of a star within a spherical shell radius r and r + dr and $\rho(r)$ be the density then,

$$\frac{dM(r)}{dr} = 4\pi r^2 \rho(r) \tag{1.35}$$

Equation of Hydrostatic Equilibrium

Let P(r + dr) and P(r) be the pressures exerted by the gaseous material over the two faces of an infinitesimal cylinder of thickness dr then the net pressure force along the increasing direction of r is $(P(r) - P(r+dr))$ (since the pressures on the curved surfaces in the upward and downward directions balance each other). If this net pressure force balances the gravitational pull of the material in the shell,

$$P(r) - P(r+dr) = \frac{GM(r)\rho(r)}{r^2}$$

i.e.

$$\frac{dP(r)}{dr} = -\frac{GM(r)\rho(r)}{r^2} \tag{1.36}$$

Equation of Thermal Equilibrium

If dL(r) is the change in the luminosity within the shell and $\in$ is the energy produced in the shell per unit time, then

$$dL(r) = 4\pi r^2 \rho(r). \in (r)dr$$

i.e.

$$\frac{dL(r)}{dr} = 4\pi r^2 \rho(r) \in (r) \tag{1.37}$$

Equation of Energy Transfer

For radiative equilibrium,

$$\nabla_{rad} = \left(\frac{d\log T}{d\log P}\right)_{rad} = \frac{3\kappa PL}{16\pi acGMT^4}$$

$$\text{i.e.} \left(\frac{dT}{dP}\right)_{rad} = \frac{3\kappa L}{16\pi acGMT^3}$$

Now, multiplying this relation by (1.36)

$$\left(\frac{dT}{dr}\right)_{rad} = -\frac{3\kappa(r)L(r)\rho(r)}{16\pi acr^2T^3(r)} \tag{1.38}$$

Similarly for convective equilibrium

$$\left(\frac{dT}{dr}\right)_{ad} = -\frac{\gamma - 1}{\gamma}\frac{T(r)}{P(r)}\frac{GM(r)\rho(r)}{r^2} \tag{1.39}$$

Various Models of Stellar Structure

Polytropic Models

In the above discussion there are three unknowns P, ρ and T and two independent equations, viz. equation of continuity and hydrostatic equation. So we are in need of a third equation for studying stellar structure which is served by the equation of state.

Initially when the energy generation mechanism and mode of energy transport were not very well known the stars were expected to be in convective equilibrium and hence an adiabatic equation of state is used to study stellar atmosphere. That is, if P and ρ are the pressure and density inside a star, then $P \propto \rho^{\gamma}$ where $\gamma = \frac{C_p}{C_v}$. C_p, C_v being the specific heats at constant pressure and volume, respectively, and γ is taken as $\gamma = 1 + \frac{1}{n}$. n is called the polytropic index and it describes various conditions of the atmosphere, e.g. when n = 0, ρ becomes constant, i.e. the atmosphere is of uniform density, when $n \rightarrow \infty$ it leads $P \propto \rho$, i.e. the atmosphere is of constant temperature (iso-thermal). Later it will be seen that when n = 1.5, it corresponds an atmosphere in convective equilibrium. If one substitutes $\rho = \rho_c y^{n+1}, P = P_c y^{n+1}$ where P_c and ρ_c are the central pressure and density of the star, respectively, then from Eqs. (1.35) and (1.36) introducing new variable y we have

$$\frac{1}{x^2}\frac{d}{dx}\left\{x^2\frac{dy}{dx}\right\} + y^n = 0 \qquad (1.40)$$

where $r = \alpha x, \alpha = \sqrt{P_c(n+1)/(4\pi G\rho_c^2)}$.

It is interesting to note that the above differential equation, known as Lane–Emden equation, has analytic solutions for n = 0, 1, 5, respectively, only and for other values of n, the solution is to be studied numerically.

The boundary conditions for this second order differential equation are at $x = 0$ (at the centre of the star), y = 1 (since at the centre $\rho = \rho_c$ so $y^n = 1$ giving y = 1) and seek a solution where one expects $\frac{dy}{dx} = 0$ as well as for many cases, e.g. for a uniform density sphere.

Prolem 2 *Solve Lane–Emden equation for n = 0.*

The corresponding form of Lane–Emden equation is (viz. Eq. (1.40))

$$\frac{1}{x^2}\frac{d}{dx}\left\{x^2\frac{dy}{dx}\right\} + 1 = 0$$

$$\text{i.e. } x^2\frac{dy}{dx} = -\frac{x^3}{3} + A \text{ (constant)}$$

At $x = 0$, $\frac{dy}{dx} = 0$ leads to zero value for the constant and

$$\frac{dy}{dx} = -\frac{x}{3}$$

$$\text{leading, } y = \frac{-x^2}{6} + B \text{ (constant)}$$

Since at $x = 0$, y = 1 this leads to B = 1, i.e. $y = 1 - \frac{x^2}{6}$.

Problem 3 *Solve Lane–Emden equation for n = 1.*

The corresponding form of Lane–Emden equation is

$$\frac{d}{dx}\left\{x^2\frac{dy}{dx}\right\} + x^2 y = 0$$

$$\begin{aligned}
\text{Let us substitute, } xy &= \mathcal{X} \\
\text{i.e. } y &= \frac{\mathcal{X}}{x} \\
\text{i.e. } \frac{dy}{dx} &= \frac{x\mathcal{X}' - \mathcal{X}}{x^2} \left(\mathcal{X}' \equiv \frac{d\mathcal{X}}{dx}\right)
\end{aligned}$$

So, the above equation reduces to

$$\mathcal{X}'' + \mathcal{X} = 0$$

$$\begin{aligned}
\text{This has a solution, } \mathcal{X} &= A\cos x + B\sin x \\
\text{i.e. } y &= \frac{A\cos x}{x} + \frac{B\sin x}{x}
\end{aligned}$$

Since, y = 1 as $x \to 0$. So, A = 0, B = 1, since, $\underset{x \to 0}{Lt} \frac{\cos x}{x} = 0$ and $\underset{x \to 0}{Lt} \frac{\sin x}{x} = 1$ so the solution is $y = \frac{\sin x}{x}$, the second boundary condition is explicitly satisfied after L' Hospital's rule.

Problem 4 *Show that for n = 1, the mass of a star is proportional to the central density.*

Now for n=1 the equation of state takes the form, $P = K\rho^2$, K is a constant. Hence we can write $P_c = K\rho_c^2$ for central pressure and central density. Let M and R be the mass and radius of a star. Now since, $r = \alpha x$,

$R = \alpha x_1 = \alpha\pi$ for n=1 (since, $x = x_1$, is at the boundary, so y = 0 there which leads to $\frac{sinx}{x} = 0$, i.e. $x_1 = \pi$) . So, $R^2 = \alpha^2\pi^2 = \frac{P_c(n+1).\pi^2}{4\pi G\rho_c^2} = \frac{K\pi}{2G}$.

Thus for n=1 the radius of a star is independent of the central density.

Now, M = $[-\frac{3}{x}\frac{dy}{dx}]_{x_1}\rho_c\frac{4}{3}\pi R^3$.

Then it can be shown that for n = 1, $M = (\frac{2\pi K^3}{G^3})^{1/2}.\rho_c$, i.e. for n = 1 mass of a star is proportional to the central density.

Problem 5 *Show that for n = 5, the radius of a star is infinite but the mass is finite.*

The corresponding form of Lane–Emden equation is

$$\frac{1}{x}\frac{d}{dx}\left\{x^2\frac{dy}{dx}\right\} + y^5 = 0$$

Substituting $x = \frac{1}{z} = e^{-t}, y = (\frac{z}{2})^{1/2}u = (\frac{1}{2}e^t)^{1/2}u$ the above equation reduces to $\frac{d^2u}{dt^2} = \frac{1}{4}u(1 - z^4)$

$$\text{which has the solution, } u = \pm\left[\frac{12Fe^{-2t}}{(1 + Fe^{-2t})^2}\right]^{1/4}, \text{ F is constant}$$

$$\text{i.e. } y = \left\{\frac{3F}{(1 + Fx^2)^2}\right\}^{1/4}$$

As, at $x = 0$, y = 1, it gives F = 1/3.

So, $y = (1 + \frac{1}{3}x^2)^{-1/2}$

Now for y = 0, x → ∞ i.e. for n = 5 the radius of a star is infinite. Also, for n = 5, $P_c = K\rho_c^{6/5}$. Then it can be shown that (similarly as for n = 1), M = $\frac{18\sqrt{2}K^{3/2}}{\sqrt{4\pi}\ G^{3/2}\rho_c^{1/5}}$, i.e. the mass of the star is finite.

Density and Pressure Profile Inside Star

If M, R be the mass and radius of a star, then integrating (1.35) with respect to r

$$M = 4\pi\int_0^R r^2\rho dr = 4\pi\alpha^3\int_0^{x_1}\rho_c y^n x^2 dx.$$

where $x = x_1$ at y = 0, i.e. P = 0 which represents the boundary surface of the star. Thus,

$$\begin{aligned} M &= -4\pi\alpha^3\rho_c \int_0^{x_1} \frac{d}{dx}\left\{x^2\frac{dy}{dx}\right\} dx. \text{ (using Eq. (1.40))} \\ &= 4\pi\alpha^3\rho_c\left(-x^2\frac{dy}{dx}\right)_{x=x_1=R/\alpha} \\ &= \frac{4\pi R^3\rho_c}{3}\left(-\frac{3}{x}\frac{dy}{dx}\right)_{x=x_1} \end{aligned}$$

If $\bar{\rho}$ be the mean density of the star, then

$$M = \frac{4}{3}\pi R^3\bar{\rho}$$

$$\begin{aligned} \text{Then, } \frac{M}{\frac{4}{3}\pi R^3} &= \bar{\rho} = \rho_c\left(-\frac{3}{x}\frac{dy}{dx}\right)_{x=x_1} \\ \text{i.e. } \rho_c &= \frac{3M}{4\pi R^3}\Big/\left(-\frac{3}{x}\frac{dy}{dx}\right)_{x=x_1} \end{aligned}$$

Now, at r = R, $x = x_1$. So, from the definition of α,

$$\frac{P_c(n+1)}{4\pi G{\rho_c}^2} = \alpha^2 = \frac{R^2}{{x_1}^2}$$

Using x_1 in terms of the other,

$$P_c = \frac{GM^2}{R^4}\Big/\left\{4\pi(n+1)\left(\frac{dy}{dx}\right)^2_{x=x_1}\right\}$$

Substituting P_c in $P = P_c y^{n+1}$ and $\rho = \rho_c y^n$ we can find the pressure and density profiles as a function of y which is a function of x, hence r for a given star of mass M and radius R.

Mass Radius Relation in a Polytropic Star

We have assumed in the derivation of Lane–Emden equation,

$r = \alpha x, \quad \alpha = \sqrt{\frac{P_c(n+1)}{4\pi G\rho_c^2}}$ where $P_c = K{\rho_c}^{1+\frac{1}{n}}$

Now, it is found $M = 4\pi\alpha^3\rho_c(-x^2\frac{dy}{dx})_{x=x_1}$

Then substituting α and P_c,

$$M = 4\pi[\frac{(n+1)K}{4\pi G}]^{\frac{3}{2}} \rho_c^{\frac{(3-n)}{2n}} (-x^2 \frac{dy}{dx})_{x=x_1}$$

Solving for ρ_c,

$$\rho_c = \{(\frac{M}{4\pi})[\frac{4\pi G}{(n+1)K}]^{\frac{3}{2}}[-x_1^2(\frac{dy}{dx})_{x=x_1}]^{-1}\}^{\frac{2n}{3-n}}$$

Now, $R = \alpha x_1 = [\frac{(n+1)K}{4\pi G}]^{\frac{1}{2}} \rho_c^{\frac{(1-n)}{2n}} x_1$

Substituting ρ_c in the above equation,

$$R = (4\pi)^{\frac{1}{n-3}}[\frac{(n+1)K}{G}]^{\frac{n}{3-n}}[-x_1^2(\frac{dy}{dx})_{x=x_1}]^{\frac{n-1}{3-n}} M^{\frac{1-n}{3-n}}$$

Hence $R \propto M^{\frac{1-n}{3-n}}$

For $n = \frac{3}{2}, R \propto M^{-\frac{1}{3}}$

Thus for a white dwarf or a fully convective star (as we will see later that in both cases $n = \frac{3}{2}$) undergoing mass transfer, stellar radius is inversely proportional to mass.

Homologous Model

The equations governing stellar structure (viz. Eqs. (1.34)–(1.38)) can be directly integrated if the sources of energy and the form of opacity within a star are identified.

The huge amount of energy emitted by a star, e.g. Sun (say) is 4×10^{33}ergs s^{-1}. This huge energy cannot be supplied by ordinary chemical reaction or the energy emitted by gravitational contraction. The source should be much intense than those. So along the line of above speculation, Bethe in 1939, for the first time suggested thermonuclear reactions. These are known as carbon–nitrogen cycle or C–N–O cycle. Later George Gamow suggested another thermonuclear reactions, known as, proton–proton reactions or p–p chain.

C–N–O cycle

$$C^{12} + H^1 = N^{13} + \gamma$$

$$N^{13} = C^{13} + \beta^+ + \nu$$

$$C^{13} + H^1 = N^{14} + \gamma$$

$N^{14} + H^1 = O^{15} + \gamma$

$O^{15} = N^{15} + \beta^+ + \nu$

$N^{15} + H^1 = O^{16} = C^{12} + He^4$

p–p chain

$H^1 + H^1 = H^2 + \beta^+ + \nu$

$H^2 + H^1 = {H_e}^3 + \gamma$

${H_e}^3 + {H_e}^3 = H^4 + H^1 + H^1$

Rates of Thermonuclear Reactions

Let us consider the reaction A + a = B + b where small letters correspond to the light particles. If $\sigma(a,b)$ is the cross section for the above reaction and N_a(v)dv be the number of particles per unit volume within the velocity range v and v + dv, then the number of encounters of these particles with A particles per second is $\sigma(a,b)N_A N_a(v)vdv$. Then the rate of reaction

$$R_c = N_A \int_0^\infty \sigma(a,b)N_a(v)vdv$$

Assuming a Maxwellian velocity distribution form it can be shown that $R_c \infty N_A N_a T^n$ where N_a is the total number of light particles per unit volume and $n = \frac{\tau-2}{3}$ where $\tau = (\frac{27B^2}{4kT})^{1/3}, B = \frac{4\pi^2}{h}\sqrt{2\mu_a} Z_A Z_a e^2, Z_A, Z_a$ being the charges of A and a and $\mu_a = \frac{M_a M_A}{M_a + M_A}, M_a, M_A$ are atomic weights of a and A, respectively.

Now, for p–p chain, the first reaction gives $Z_A = Z_a = 1, \mu = \frac{1}{2}, T \sim 12 \times 10^6, \tau = 14.8, n = 4.3$. So, $R_c \infty N_A N_a T^{4.3}$. Now, $N_A = N_a = (\frac{\rho X}{m_H})\,\text{cm}^{-3}$ where X is the fraction of H atom in the star, m_H is the mass of H atom, ρ is the density of the star. Then, $R_{pp} = 4 \times 10^{-6} \rho X^2 (\frac{T}{10^6})^{4.3}\,\text{erg}\,\text{s}^{-1}$. Similarly for CNO cycle, for the fourth reaction in CNO cycle, $Z_A = 7, Z_a = 1, \mu_a = \frac{14}{15}, T \sim 20 \times 10^6\,\text{K}, n = 18.5, N_a = \frac{\rho X}{m_H}, N_A = \frac{\rho A_{CN}}{13 m_H}$ where A_{CN} is the fraction of cosmic mixture of carbon and nitrogen $\sim 5.3 \times 10^{-3} X$.

$$\text{Then, } R_{CNO} = 11 \times 10^{-22} \rho X A_{CN} (T/10^6)^{18.5}$$

$$\text{So, } \frac{R_{CNO}}{R_{pp}} = 2.75 \times 10^{-16} \frac{A_{CN}}{X} (T/10^6)^{14.2}$$

It has been seen that for massive stars R_{CNO}/R_{total} is close to 1, i.e. there CNO cycle dominates and for less massive stars R_{pp}/R_{total} is close to 1, i.e. the reverse situation occurs.

Stellar Opacity

Opacity is the diminution of stellar radiation as it passes from the centre towards the envelope of the star. The opacity might happen due to various processes, e.g. (1) scattering of radiation by electrons in all frequencies (κ^e), (2) absorption of photons by electrons (κ^{ff}) and (3) absorption of photons by electrons within an atom or ion which then become free (κ^{bf}). The corresponding form of opacity, $\kappa = \kappa_0 \rho^\alpha T^\beta$ where for electron scattering, $\kappa_0 = 0.34, \alpha = \beta = 0$, for bound-free scattering $\kappa_0 = 1.5 \times 10^{23}, \alpha = 1, \beta = -3.5$ and for free-free scattering $\kappa_0 = 6.3 \times 10^{22}, \alpha = 1, \beta = -3.5$, respectively. The above relation is known as **Kramer's law of opacity.**

Now we concentrate on the nature of transport mechanism operating near the centre of a star. If we assume that the stellar material consists of monatomic gas, then $\gamma = 5/3$. So $\bigtriangledown_{ad} = \frac{\gamma-1}{\gamma} = 0.4$. Hence for convection process we must have $\bigtriangledown_{rad} > \bigtriangledown_{ad} = 0.4$.

It has been shown that $\bigtriangledown_{rad} = \frac{3\kappa P}{16\pi acGT^4}\frac{L}{M}$ (viz. Eq. (1.38)). From Eq. (1.37) near central region, $L(r) = \frac{4}{3}\pi r^3 \rho_c R_c$ where R_c is the rate of energy generation and $M = \frac{4}{3}\pi r^3 \rho_c$. Substituting, $\bigtriangledown_{rad}$ at the centre $= \frac{3\kappa_c P_c R_c}{16\pi acG{T_c}^4}$.

Putting $\kappa_c = \kappa_0 \rho^\alpha T^{-\beta}, R_c = R_0 \rho T^n$ we can see that $\bigtriangledown_{rad}$ at the centre is greater than 0.4 for massive stars where CNO cycle is operative and vice versa for less massive stars. So we can at once say that **for massive stars the stellar core is convective, stellar envelope is radiative and CNO cycle is the dominating source of nuclear energy whereas for less massive stars we have radiative cores, convective envelope with p–p cycle taking place at the centre.**

Homologous Model

Introducing the dimensionless variables $x = \frac{r}{R}, q = \frac{M(r)}{M}, f = \frac{L(r)}{L}, p = P/\left(\frac{GM^2}{4\pi R^4}\right), t = T/\left(\frac{\mu GM}{RR}\right)$ and $\sigma = \rho/\left(\frac{M}{4\pi R^3}\right)$ and substituting $\kappa = \kappa_0 Z^\lambda \rho^\alpha$ and $\in = \in_0 \rho T^n$ [in place of R_c we have used $\in$ to keep parity with reaction rate equation] Eqs. (1.35)–(1.38) reduce to

$$\begin{aligned}
\frac{dq}{dx} &= \frac{p}{t}x^2 \text{ (Equation of continuity)}\\
\frac{dp}{dx} &= -\frac{p}{t}\frac{q}{x^2} \text{ (Hydrostatic equation)}\\
\frac{df}{dx} &= Dp^2t^{n-2}x^2 \text{ (Equation of energy production)}
\end{aligned}$$

$$\frac{dt}{dx} = -C\frac{p^{\alpha+1}f}{x^2t^{\alpha+\beta+4}} \text{ (Radiative equilibrium)}$$

$$\text{and } \frac{dt}{dx} = -\frac{2}{5}\frac{q}{x^2} \text{ (Convective equilibrium)}$$

where C, D are constants, with boundary conditions at x = 0, q = 0, f = 0 and at x = 1, q = 1, f = 1, t = 1, p = 0. The above set of equations are easily solvable under given boundary conditions.

The empirical relations observed for stars are

$$\begin{aligned}(L/L_\odot) &= (M/M_\odot)^{3.5}\\ (R/R_\odot) &= (M/M_\odot)^{0.75}\\ (T_e/T_{e,\odot}) &= (M/M_\odot)^{0.5}\\ \text{and } (L/L_\odot) &= (T_e/T_{e,\odot})^{6.9}\end{aligned}$$

It is found that the profiles found, solving equations following homologous model, do not match with the above profiles if the stellar atmosphere is assumed to be either in full convective or in full radiative equilibrium. So the stellar atmosphere is a combination of the two modes of energy transport, radiation as well as convection for massive and solar type stars as discussed in the previous part of stellar opacity.

1.9 Stellar Evolution and Connection with H–R Diagram

As a star passes through various evolutionary phases, from its birth to death its position in the H–R diagram continuously changes and the evolutionary track depends on the mass of the star.

Pre Main Sequence Contraction

Let us consider a system of particles under gravitational attraction and some external forces. Then, if I be the moment of inertia, then

$$\frac{1}{2}\frac{d^2I}{dt^2} = 2T + \Omega - 3PV$$

where T, Ω, P, V are the kinetic energy, potential energy, external pressure and volume of the system. If the system be in equilibrium, $\frac{d^2I}{dt^2} = 0$ and in absence of any external pressure the above relation reduces to $2T + \Omega = 0$, which is known as **Virial Theorem** (Goldstein et al. 2001). So, when $\frac{d^2I}{dt^2} < 0$, $\frac{dI}{dt}$ will decrease over time, hence $I = \sum_i m_i {r_i}^2$ will also decrease over time, i.e. the size of the system will decrease and we say that the system is gravitationally unstable. Now, a star is speculated to begin its life from the gravitational contraction of a big cloud of interstellar matter. The contraction continues until the central temperature is as high as 10^7 K. At this temperature the nuclear reaction starts and hydrogen is converted to helium via fusion reaction (viz. p–p/CNO cycles). Once the nuclear reaction starts, the star achieves a quasi hydrostatic equilibrium. If the energy produced in the nuclear reaction is less than the radiated energy, the star contracts, its central temperature increases along with an increase in energy production. If the energy produced is greater than the energy released, then the star expands which decreases its central temperature. In this way the star attains a quasi equilibrium state by a valve mechanism and we see the star on the main sequence. The time scale for pre main sequence contraction, τ_g, is found as follows. For a perfect gas of mass M,

$$T = \frac{3}{2}(\gamma - 1)U, \text{ where U is the internal energy.}$$

For monatomic gas (as the spherical contracting cloud of gas mostly comprises of hydrogen gas in atomic stage), $\gamma = 5/3$. So, T = U. Hence, total energy $E = U + \Omega = \frac{\Omega}{2}$. If L be the luminosity of the contracting cloud, then

$$L = -\frac{dE}{dt} = -\frac{1}{2}\frac{d\Omega}{dt}$$

where $\Omega = -\frac{3}{5-n}GM^2 = -\frac{3}{5}\frac{GM^2}{R}$ for monatomic gas of uniform density (n = 0).

$$\text{Then, } L = -\frac{3}{2(5-n)}\frac{GM^2}{R^2}\frac{dR}{dt}$$

$$\text{i.e. } dt = -\frac{3}{2(5-n)}\frac{GM^2}{LR^2}dR$$

For radiative equilibrium, n = 3 then

$\tau_g = \tau_{gr} = \frac{3}{4}\frac{GM^2}{LR}$ (here the contraction is slow enough due to slow transfer of energy to keep L nearly constant).
The tracks for contraction through radiation are shown in Fig. 1.14.

For convective equilibrium, n = 1.5 (solving the equation of transfer for full convective equilibrium for homologous model) and it is an efficient mechanism for quick transport of energy making luminosity of the star to decrease rapidly and keeping surface temperature almost constant. Then L is a function of R only and,

$$\tau_g = \tau_{gc} = \frac{GM^2}{7LR}$$

Comparing the above two time scales, $\tau_{gc} << \tau_{gr}$. During this pre main sequence contraction, radiative tracks are called **Heney tracks** (1955) and convective tracks are called **Hayashi tracks** (1961).

Post Main Sequence Evolution

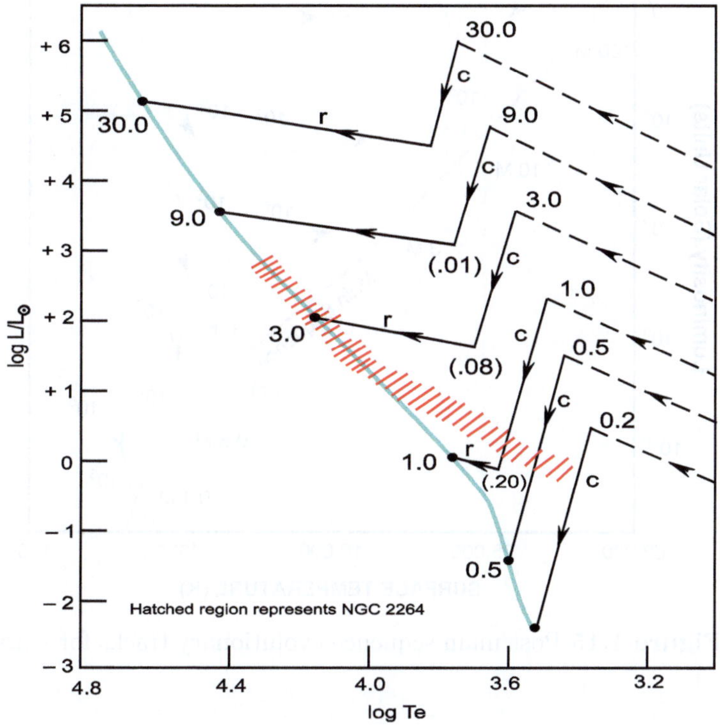

Figure 1.14 Pre main sequence evolutionary tracks for stars. r stands for radiative track and c stands for convective track (courtesy: Abhyankar 2001)

When hydrogen gas in the core gets exhausted producing a helium core, the nuclear reaction propagates in a thin shell of hydrogen of low density surrounding the core. So we have a helium core, a hydrogen burning shell and an envelope surrounding the shell.

We have seen that for a massive star the envelope is fully radiative and the energy generated in the shell balances the energy radiated keeping the luminosity almost constant. So the star moves horizontally along a track of constant luminosity to the right (Fig. 1.15). But the time for this phase is so short, that almost no stars are seen in this phase on the H–R diagram and is called **"Hertsprung Gap"**.

On the contrary, for the less massive stars the envelope is fully convective so it transfers the energy produced in the shell rapidly outwards which in turn increases it luminosity.

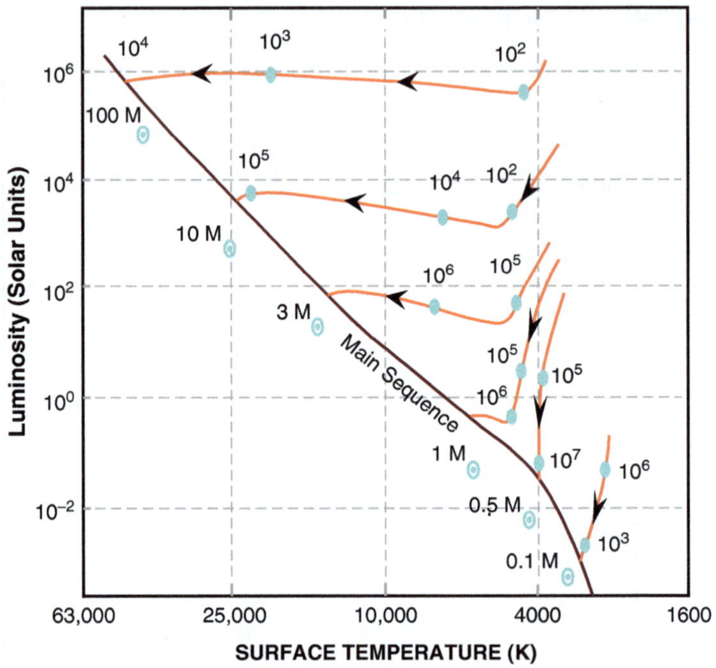

Figure 1.15 Post main sequence evolutionary tracks for stars

Also due to the contraction of the core, the central temperature rises which generates a pressure force and the envelope expands. So as a result of expansion of the envelope the surface temperature falls but the luminosity is high due to larger size of the envelope. As a result the stars move to the top right corner of the H–R diagram and are called "Red Giants". During the contraction of the helium core the central temperature rises and as it reaches 10^8 K Heluim burning takes place via the reaction $3He^4 \rightarrow C^{12}$ but the process depends upon the mass of stars. For massive stars $(M > 3M_\odot)$, most of the time the core remains non-degenerate throughout. The degeneracy is discussed during the final stage of stellar evolution. So, the star undergoes a steady state of helium burning and passes through the left and undergoes a pulsation phase indicated as "Cepheid variable strip".

As helium burning gets exhausted the reaction again propagates to the shell and the star expands like the previous situation of hydrogen burning shell and follows path to the right again and the process continues successively producing oxygen, neon, silicon at the centre via the reactions $C^{12} + C^{12} \rightarrow Ne^{20} + He^4, C^{12} + He^4 \rightarrow O^{16}, O^{16} + O^{16} \rightarrow Si^{28} + He^4$, and the following events occur. Stars of masses between $0.5M_\odot - 3M_\odot$, become white dwarfs with a degenerate C–O core. For stars of masses between $3M_\odot - 8M_\odot$, they finally produce C–O–Mg core white dwarfs. For stars of masses larger than $8M_\odot$, nuclear burning at the core finally produces iron which is an endothermic reaction which produces instability and as a result the core contracts and the envelope is thrown with a big explosion called **Supernova** and the remnants core contracts to become a **Neutron star** and or **Black hole (BH)** if the star is as massive as $20M_\odot$.

For low mass stars ($M < 3M_\odot$) the core becomes degenerate and the degenerate electron pressure halts the core contraction. The stars become **White dwarf** and occupy the bottom left position of the H–R diagram.

Final Stages of Stellar Evolution

We have envisaged in the previous sections that a star starts its life from the onset of gravitational instability taking place in a large big cloud, comprising mostly of hydrogen gas and undergoes various phases through main sequence, red giant, super giant and ends its life as a supernovae, white dwarf, neutron star or black hole depending upon its mass.

In the following part we will describe in detail about the compact objects, e.g. white dwarf, neutron star and a black hole.

White Dwarf

Degenerate State

Let us consider a cloud of electrons in the volume element dv. Then the number of electrons in the spherical shell of momentum and velocity ranges (p, p + dp) and (v, v+dv), according to Boltzmann distribution is

$$n(p)dpdv = n_e \frac{4\pi p^2}{(2\pi m_e kT)^{3/2}} \exp\left(-\frac{p^2}{2m_e kT}\right) dpdv$$

The maximum of this distribution occurs at $p_{max} = \left(\frac{2m_e}{kT}\right)^{1/2}$. Here n_e is the number density of free electrons, T is the temperature, k is the Boltzmann constant and m_e is the mass of the electron. The distribution functions are shown in Fig. 1.16 for different T and constant n_e. Now, since electrons are fermions they must follow Pauli's exclusion principle. So each quantum cell in six-dimensional phase space cannot contain more than two electrons. Now, volume of such a cell is $dp_x dp_y dp_z dx dy dz = h^3$ where h is Planck's constant. Therefore in the momentum shell the number of such cells $= \frac{4\pi p^2 dpdv}{h^3}$. Hence, number of electrons in this shell $= \frac{8\pi p^2 dpdv}{h^3}$. Thus, $n(p)dpdv \leq \frac{8\pi p^2 dpdv}{h^3}$, the corresponding curve (dashed one) is shown in Fig. 1.16. It is clear from

Fig. 1.16 that for low temperature, at a constant n_e, the Boltzmann distribution is in contradiction of quantum mechanics. This happens again if T remains constant and n_e is too high, so there is a need to include quantum mechanical idea when temperature is too low or electron density is too high and call this state as **degenerate** state.

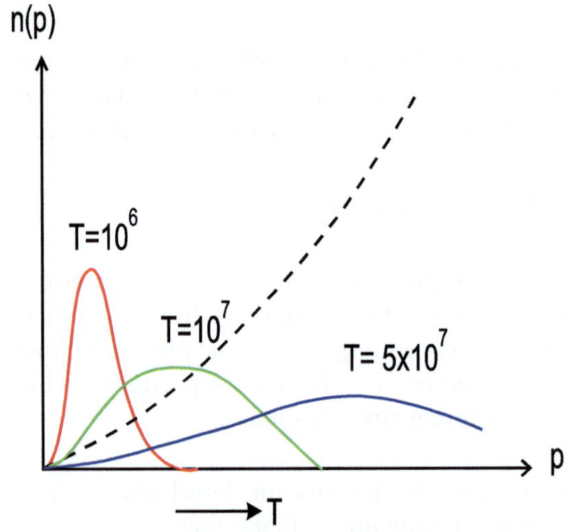

Figure 1.16 Boltzmann distributions of electrons at three temperatures (*thin lines*) and that (*dotted line*) corresponding to Pauli exclusion principle

Let us consider electron gas at lowest energy (i.e. temperature almost zero). This degenerate state without violating Pauli exclusion principle is the state in which all the electrons up to a certain momentum p_F (say) occupy two states in a phase-space quantum cell.

$$\begin{aligned}\text{Then, } n(p) &= \frac{8\pi p^2}{h^3} \text{ for } p \leq p_F \\ &= 0 \text{ for } p > p_F\end{aligned}$$

Then the total number of electrons in the volume dV is given by

$$n_e(V) \quad = dV \int_0^{p_F} \frac{8\pi p^2 dp}{h^3} = \frac{8\pi p_F^3}{3h^3} dV$$

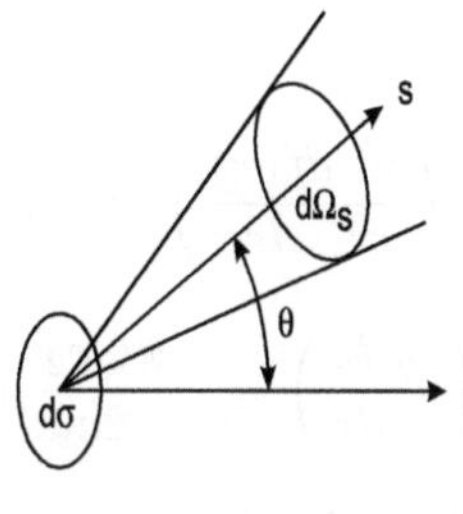

Figure 1.17 Surface element $d\sigma$ with normal vector $\hat{n}$ and unit vector s along the axis of solid angle $d\Omega_s$

Let us consider the flux of momentum of electron through a unit surface per second which is its pressure. Let us consider the flux of electrons through an elementary surface $d\sigma$, entering into a solid angle $d\Omega_s$ along any arbitrary direction s. Then the pressure of electrons over all directions s of a hemisphere and over all absolute values of p is (Fig. 1.17)

$$P = \frac{1}{4\pi}\int_{\phi=0}^{2\pi}\int_{\theta=0}^{\pi}\int_{p=0}^{\infty} n(p)v(p)p\cos^2\theta dp(\sin\theta d\theta d\phi)$$

where $d\Omega_s = \sin\theta d\theta d\phi$ and v(p) is the velocity of electrons in (p, p + dp) and $p = m_e v(p)/\sqrt{(1-v(p)^2)/c^2}$

Then,

$$P = \frac{8\pi}{3h^3}\int_0^{p_F} p^3 v(p)dp$$

Using the above relation involving p and v(p)

$$P = \frac{8\pi c^5 {m_e}^4}{3h^3}\int_0^x \frac{\xi^4 d\xi}{(1+\xi^2)^{1/2}} = \frac{\pi {m_e}^4 c^5}{3h^3} f(x) \tag{1.41}$$

where $\xi = p/(m_e c)$ and $x = p_F/(m_e c), f(x) = 8\int_0^x \frac{\xi^4 d\xi}{\sqrt{1+\xi^2}}$.
So, n_e (number density of electrons)$= \frac{\rho}{\mu_e m_H} = \frac{8\pi}{3h^3} m_e^3 c^3 x^3$

Then these can be written as

$$P = c_1 f(x), \rho = c_2 x^3 \text{ and } x = p_F/(m_e c)$$

where ρ is the density and c_1, c_2 are constants.

Then from (1.35) and (1.36),

$$\frac{c_1}{c_2}\frac{1}{r^2}\frac{d}{dr}\left(\frac{r^2}{x^3}\frac{df(x)}{dr}\right) = -4\pi G c_2 x^3$$

$$\text{i.e. } \frac{1}{r^2}\frac{d}{dr}\left(r^2\frac{dz}{dr}\right) = -\frac{\pi G {c_2}^2}{2c_1}(z^2-1)^{3/2}$$

where $z^2 = x^2 + 1$ and $\frac{1}{x^3}\frac{df(x)}{dr} = 8\frac{dz}{dr}$.

Replacing r,z by dimensionless variables, $x^{'}$ and y,

$$x^{'} = \frac{r}{\alpha}, \alpha = \sqrt{\frac{2c_1}{\pi G}}\frac{1}{c_2 z_0}$$

$y = \frac{z}{z_0}$, where z_0 is the central value of z, we have

$$\frac{1}{x'^2}\frac{d}{dx'}\left(x'^2\frac{dy}{dx'}\right) + \left(y^2 - \frac{1}{{z_0}^2}\right)^{3/2} = 0 \qquad (1.42)$$

This is **Chandrasekhar's differential equation** for white dwarf. The boundary conditions are at r = 0, $x^{'} = 0$ y = 1, $\frac{dy}{dx'} = 0$ and at r = R, x= 0, z = 1, $y = \frac{1}{z_0}, x^{'} = x_1(say)$.

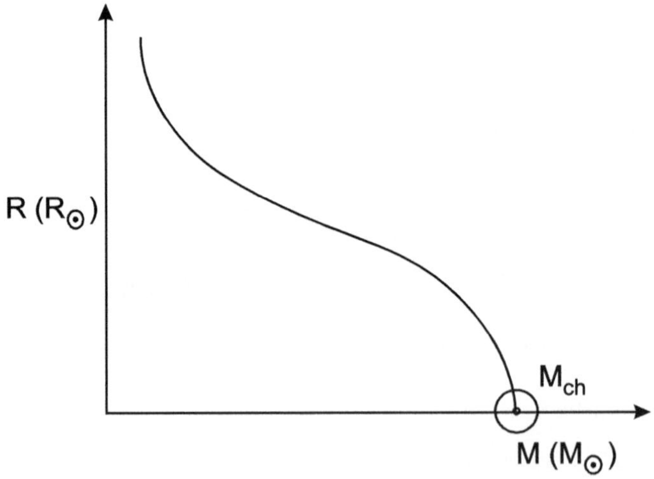

Figure 1.18 Mass radius relation for white dwarf

Then mass of the white dwarf is

$$M(R) = \int_0^R 4\pi r^2 \rho dr = 4\pi\alpha^3 c_2 {z_0}^3 \int_0^{x_1} {x'}^2 \left(y^2 - \frac{1}{{z_0}^2}\right)^{3/2} dx'$$
$$= \frac{4\pi}{{c_2}^2}\left(\frac{2c_1}{\pi G}\right)^{3/2}\left(-{x'}^2\frac{dy}{dx'}\right)_{x_1} \quad (1.43)$$

Solving numerically, $\frac{dy}{dx'}$ from (1.42) and substituting in (1.43) we get mass vs radius relation for different values of z_0 from $\propto$ to 1, i.e. from $x = \propto$ (fully relativistic) to x = 0 (non-relativistic), and shown in Fig. 1.18. It is clear from Fig. 1.18 that the mass of white dwarf cannot exceed the **Chandrasekhar mass** given by $M_{ch} = (\frac{2}{\mu_e})^2 \times 1.459 M_{\odot}$ where μ_e is the mean molecular weight of the electrons. For He white dwarf star, $\mu_e = 2$ so, $M_{ch} = 1.459 M_{\odot}$.

Neutron Star

Neutron stars are the compact objects where the gravitational force is balanced by degenerate neutron stars, either relativistic or non-relativistic or partially relativistic. Since the neutrons are fermions, the equation of state is same as that of a white dwarf, with m_e replaced by m_n (viz. Eq. (1.41)) and μ_e replaced by $\mu_n = 1$. For extreme relativistic state the equation of state for white dwarf reduces to $P \sim {\rho_0}^{4/3}(\xi >> 1)$ and $P \sim {\rho_0}^{5/3}(\xi << 1)$ where ρ_0 is the rest mass density. But in case of neutron gas one has to consider the density, $\rho = \rho_0 + u/c^2$ where u/c^2 is the energy density. In case of white dwarf $\rho_0 >> u/c^2$, so $\rho \sim \rho_0$. In case of neutron stars, for non-relativistic neutrons, $\rho_0 >> u/c^2$ so, $\rho \sim \rho_0$. So equation of state for non-relativistic neutrons is $P \sim \rho^{5/3}$ (stiffer equation of state). For relativistic neutrons, $\rho_0 << u/c^2$ so, $\rho \sim u/c^2$. Again P = u/3, i.e. $P = \rho c^2/3$. So the equation of state is $P \sim \rho$ (softer equation of state, i.e. gas is more compressible).

Relativistic Hydrostatic Equation of State

For compact objects, Einstein field equation is

$$R_{ik} - \frac{1}{2} g_{ik} R = \frac{k}{c^2} T_{ik}$$

where R_{ik} is the Ricci tensor, g_{ik} is the metric tensor, R is the Riemann curvature, T_{ik} is the energy momentum tensor. The corresponding metric for spherically symmetric static mass distribution is

$$ds^2 = e^{\nu} c^2 dt^2 - e^{\lambda} dr^2 - r^2(d\theta^2 + \sin^2\theta d\phi^2)$$

where $\nu = \nu(r), \lambda = \lambda(r)$. T_{ik} has only nonzero components $T_{00} = \rho c^2$, $T_{11} = T_{22} = T_{33} = P$.

Then the field equations are

$$\frac{\kappa P}{c^2} = e^{-\lambda}\left(\frac{\nu'}{r}+\frac{1}{r^2}\right)-\frac{1}{r^2} \quad (1.44)$$

$$\frac{\kappa P}{c^2} = \frac{1}{2}e^{-\lambda}\left(\nu" + \frac{1}{2}\nu'^2 + \frac{\nu'-\lambda'}{r} - \frac{\nu'\lambda'}{2}\right) \quad (1.45)$$

$$\kappa\rho = e^{-\lambda}\left(\frac{\lambda'}{r}-\frac{1}{r^2}\right)+\frac{1}{r^2} \quad (1.46)$$

where prime denotes differentiation with respect to r.

Multiplying (1.46) by $4\pi r^2$ and using $m = \int_0^r 4\pi r^2 \rho dr$, we get after integrating (1.46)

$$\kappa m = 4\pi r(1 - e^{-\lambda})$$

Here mass does not include not only the rest mass but also the total energy divided by c^2. Differentiating (1.44) with respect to r and eliminating $\lambda, \lambda^{'}, \nu^{'}, \nu^{''}$ using the other two equations we get **Tolman–Oppenheimer–Volkoff (TOV)** equation which is the relativistic form of hydrostatic equilibrium

$$\frac{dP}{dr} = -\frac{Gm}{r^2}\rho\left(1+\frac{P}{\rho_c^2}\right)\left(1+\frac{4\pi r^3 P}{mc^2}\right)\left(1-\frac{2Gm}{rc^2}\right)^{-1}$$

This equation along with equation of state gives the mechanical structure for a chosen value of central density ρ_c at r = 0. When $\rho = p = 0$, the corresponding r = R and m = m(R) give the size and total mass of the star. For several choices of ρ_c, one gets several $m = m(\rho_c), R = R(\rho_c)$ and by elimination of ρ_c one gets the mass radius relation R = R(m) for a particular equation of state. In Fig. 1.19, the mass radius relation is plotted for a few choices of equation of states, from softer to stiffer equations of state (suffix 1–6). For a stiffer equation of state, the matter is less compressible and hence one expects for a given M larger R and a smaller ρ_c. For a given ρ_c, one can put more mass on top until $\rho = 0$. This lowers the gravitational pull inside the star and hence maximum mass m_{max} is higher. For a soft equation of state, the reverse occurs, hence m_{max} is lower. For relativistic degenerate neutron star, $P \sim \rho$ (softer), hence $m_{max} \sim 0.72 M_{\odot}$ which is the solution obtained by Oppenheimer and Volkov (1939) after the discovery of neutron star in 1934. For other cases m_{max} generally varies from 1 to $3M_{\odot}$. But particle physicists have not become able to correctly formulate yet the exact form of equation of state so m_{max} is still not unique unlike white dwarfs.

Black Holes

In the previous sections of stellar evolution we have described that a massive star undergoes a big explosion due to an instability in the core arising out of the endothermic reaction producing iron and the remnant is a neutron star or black hole depending upon its mass. These black holes are called "stellar mass black holes" (BH). So black holes represent the ultimate fate of a massive star ($M > 20M_{\odot}$). Black holes are compact objects which can be described in general relativity (GR) by a metric, known as, Schwarzchild metric in four-dimensional space time. The metric is

$$ds^2 = \left(1 - \frac{2GM}{rc^2}\right) c^2dt^2 - \left(1 - \frac{2GM}{rc^2}\right)^{-1} dr^2 - r^2d\theta^2 - r^2 \sin^2\theta d\phi^2$$

The value $\frac{2GM}{c^2} = r_s$ is called the Schwarzchild radius. It has a great physical importance.

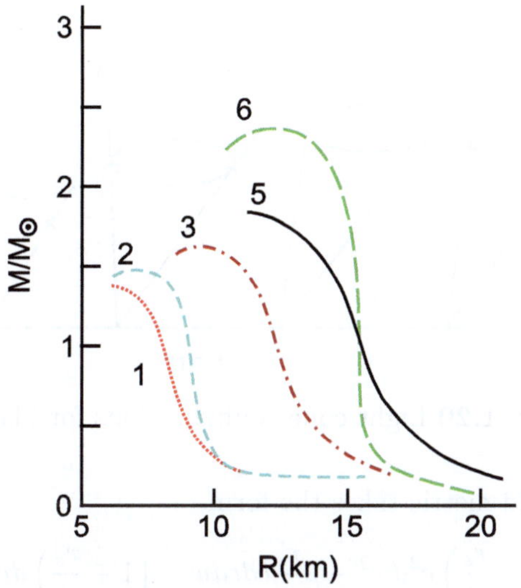

Figure 1.19 Mass radius relation for neutron star

Now, for a stationary observer at infinity (i.e. $dr = d\theta = d\phi = 0$ and $r \to \infty$), $ds^2 = c^2dt^2$. So, if τ is the proper time (the time measured by an observer carrying a standard clock), then $d\tau = \frac{ds}{c} = dt$ in this particular case. Let us consider two stationary observers, one at (r, θ, ϕ) and the other at infinity ($r \to \infty$), then,

$$\frac{d\tau}{d\tau_{\infty}} = \left(1 - \frac{r_s}{r}\right)^{1/2}$$

Let two consecutive signals are emitted from the point of first observer. Then the corresponding frequency, often called, rest frequency, $\nu_0 = \frac{1}{d\tau}$. The other observer at infinity receives the signals at interval, $d\tau_{\infty}$, at a frequency, often called observer's frequency, $\nu = \frac{1}{d\tau_{\infty}}$. So, redshift due to the gravitational field of the compact object under consideration in observer's frame at infinity is $z = \frac{\nu_0 - \nu}{\nu} = \left(1 - \frac{r_s}{r}\right)^{-1/2} - 1$ and which shows as $r \to r_s, z \to \infty$, i.e. no light is received from the point $r \le r_s$. The corresponding situation can be geometrically described as follows. The trajectory of a photon is described by null geodesic, $ds^2 = 0$. If $ds^2 > 0$, it is called "time like" and if $ds^2 < 0$, it is called "space like". The material particles, following "causality principle" are always time like. The null geodesics are called "light cones". The Schwarzchild metric has a singularity at $r = r_s$, but it is not a physical singularity and can be removed by the following substitution,

$$w = t + \frac{r_s}{c} \ln\left|\frac{r}{r_s} - 1\right|$$

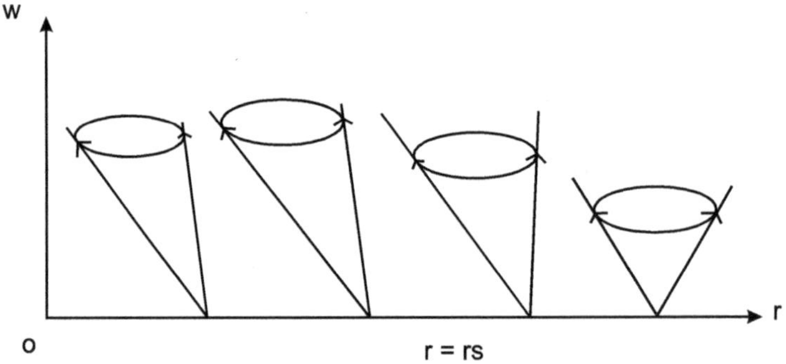

Figure 1.20 Light cone configurations for black hole

Then, Schwarzchild metric takes the form,

$$ds^2 = \left(1 - \frac{r_s}{r}\right) c^2 dw^2 - 2\frac{r_s}{r} cdrdw - \left(1 + \frac{r_s}{r}\right) dr^2 - r^2 d\theta^2 - r^2 \sin^2\theta d\phi^2$$

Then null geodesic for this metric yields (along radial boundaries of the light cone, i.e. $d\theta = d\phi = 0$),

$$\left(1 - \frac{r_s}{r}\right)\left(\frac{dw}{dr}\right)^2 - \frac{2r_s}{cr} - \frac{1}{c^2}\left(1 + \frac{r_s}{r}\right) = 0$$

which has the solutions

$$\left(\frac{dw}{dr}\right) = -\frac{1}{c} \text{ and } \left(\frac{dw}{dr}\right) = \frac{1}{c}\frac{1 + r_s/r}{1 - r_s/r}$$

The first slope is constant but the second slope changes sign from negative to positive for $r < r_s$ to $r > r_s$. For $r < r_s$, both the slopes are negative, i.e. the light cone shrinks and all material particles are drawn towards the centre

(Fig. 1.20), i.e. no static solution is possible for $r < r_s$, since it requires a motion vertically upwards, i.e. outside the light cone. That is why the configuration $r \leq r_s$ is called a BH.

1.10 Variable Stars

Variable star is that whose brightness changes over time. There are two broad groups of variable stars: (1) extrinsic variables and (2) intrinsic variables. Extrinsic variables are the variables whose brightness varies due to some external agent, e.g. obscuration of light by another star or dust, etc. Eclipsing binary stars or multiple stars, visual binaries and spectroscopic binaries belong to this category. On the contrary when brightness varies due to some internal physical process within the star, it is called an intrinsic variable. Table 1.2 gives a brief classification of variable stars.

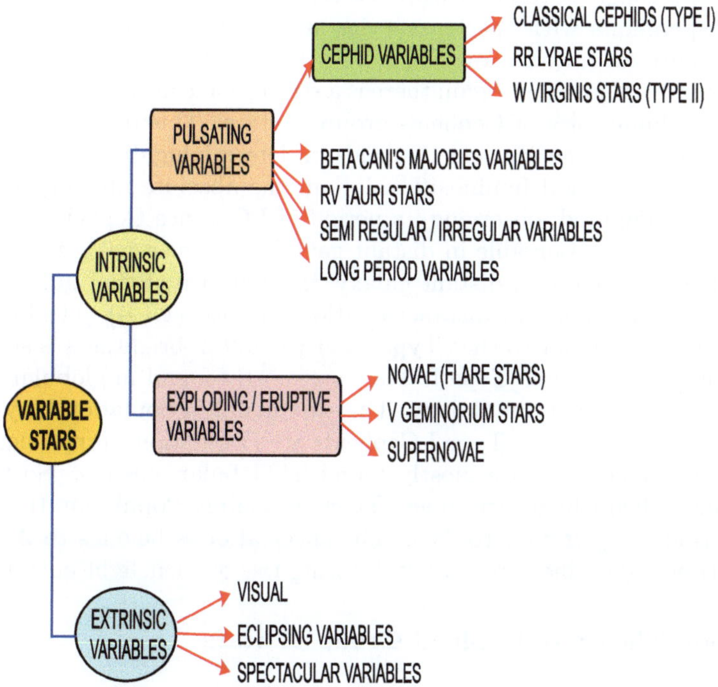

Table 1.2 Classification of variable stars.

Nomenclature of Variable Stars

In a particular constellation the variables are designated by double letters starting from A and ending with Z followed by the IAU approved abbreviations of the corresponding constellation, e.g. AA, ..., AZ, BB, ..., BZ,

..., ZZ and then Aql for the constellation Aquilae, etc. When all these are exhausted the variable is designated by the letter V followed by the number denoting the number of variables in a constellation followed by the abbreviated name of that constellation, e.g. V 1359 Aql is the 1359th variable star of the constellation Aquila.

The Cepheid Group Stars

Classical Cepheids (Type I), W-Virginis stars (Type II) and RR Lyrae stars are the members belonging to this group. The light curves (LC), defined as the magnitude versus time period, are shown in Fig. 1.21. The rise of the curves are rapid for Type I Cepheids of periods 2–6 and 7–8 days but the curve is more or less symmetric for type I Cepheids of period 10 days. For longer period again the rise is rapid compared to the fall in brightness. The LC is strongly correlated with the velocity versus time curve. They are yellow supergiants with spectra varying from F to K and highly luminous with absolute magnitudes varying from -1.5 to -6. So they are 10^4 times more luminous than Sun. Again there is a strong correlation between the time period and luminosity of Cepheids group of stars. This particular property has a great advantage of using Cepheids in determining distance to external galaxies. From period luminosity relation, the absolute magnitude M of a Cepheid is computed, observing its period of LC. Since Cepheids are bright objects, they are observable in distant galaxies. So measuring their median apparent magnitude in a distant galaxy, the distance of that galaxy can be measured using magnitude distance relation (viz. Eq. (1.19)). The LC of type II Cepheids are similar to that Type I but the fall in brightness is stiffer and also fainter compared to Type I. They are mostly found in globular clusters and near the centre of our Galaxy. They are mainly Population II type stars (metal deficient) whereas Type I Cepheids are Population I type (metal rich) stars. RR Lyrae stars are mostly found in globular clusters. So they are sometimes called Cluster variables. They are mainly Population II stars and their periods vary from 2 to 24 h. The spectral class belongs to A to F. H and CaII emission lines are observed during rise of their light curves.

Pulsation Theory of Cepheid Group of Stars

We have seen in the previous sections of stellar structure that under hydrostatic equilibrium, pressure, density and size of a star are independent of time. Let us consider the various equations under a small radial perturbation. Then,

$$P = P_0 + P_1, P_1 = P_0 p e^{iwt}$$

$$r = r_0 + r_1, r_1 = r_0 q e^{iwt}$$

$$\rho = \rho_0 + \rho_1, \rho_1 = \rho_0 s e^{iwt}$$

where the quantities with suffix "0" are functions of M(r). Now the equation of continuity and equation of motion can be written as

$$\frac{\partial r}{\partial M(r)} = \frac{1}{4\pi r^2 \rho(r)}$$

$$\text{and } \frac{\partial P}{\partial M(r)} = -\frac{GM(r)}{4\pi r^4} - \frac{1}{4\pi r^2}\frac{\partial^2 r}{\partial t^2}$$

Then substituting the expressions for P, r and ρ in equation of motion and using $g_0 = \frac{GM(r)}{r_0^2}$, linearizing and using the fact that P_0, r_0 obey the hydrostatic equation $\frac{\partial P_0}{\partial M} = -\frac{GM}{4\pi r_0^4}$ we find

$$\frac{P_0}{\rho_0}\frac{\partial p}{\partial r_0} = w^2 r_0 q + g_0(p + 4s)$$

Similarly equation of continuity leads to $r_0 \frac{\partial q}{\partial r_0} = -3q - s$.

If we assume the perturbation is adiabatic,

$$p = \gamma s$$

then from the above equations, (assuming γ constant) differentiating with respect to r_0,

$$\frac{\partial q}{\partial r_0} + r_0 \frac{\partial^2 q}{\partial r_0^2} = -3\frac{\partial q}{\partial r_0} - \frac{1}{\gamma}\frac{\partial p}{\partial r_0}$$

Then eliminating $\frac{\partial p}{\partial r_0}$, p and s we get

$$\frac{\partial^2 q}{\partial r_0^2} + \left(\frac{4}{r_0} - \frac{\rho_0 g_0}{P_0}\right)\frac{\partial q}{\partial r_0} + \frac{\rho_0}{\gamma P_0}\left[w^2 + (4 - 3\gamma)\frac{g_0}{r_0}\right] = 0 \qquad (1.47)$$

Let ρ be the constant and $\rho = \rho_0$. Then

$$r_0 = \left(\frac{3M}{4\pi\rho_0}\right)^{1/3}, g_0 = \frac{GM}{r_0^2} = \frac{4\pi}{3}G r_0 \rho_0$$

Then hydrostatic equation in equilibrium gives

$$P_0(r_0) = \frac{2\pi}{3}G\rho_0^2(R_0^2 - r_0^2)$$

where R_0 is the radius of the star at equilibrium.

We introduce a dimensionless variable, $\eta = r_0/R_0$ and put

$$B = \frac{3w^2}{2\pi G\rho_0\gamma} + \frac{2(4-3\gamma)}{\gamma}$$

Then, from Eq. (1.47) we get

$$\frac{d^2q}{d\eta^2} + \left(\frac{4}{\eta} - \frac{2\eta}{1-\eta^2}\right)\frac{dq}{d\eta} + \frac{B}{1-\eta^2}q = 0$$

The simplest solution of the above equation is $q = Constant = q_0$ provided, B = 0. Then for B = 0, $w^2 = w_0^2 = \frac{4\pi}{3}G\rho_0(3\gamma-4)$. This is the period density relation for pulsating stars obtained by A. Ritter in 1879. So if T = period of the pulsating star,

$$T = \left[\frac{3\pi}{(3\gamma-4)G\rho_0}\right]^{1/2}$$

and it shows that $T^2\rho_0 = Constant.$

The above model of radial adiabatic pulsation theory is over simplified and cannot explain the phase leg between size and luminosity over a quarter of a period, observed for variable stars. The most satisfactory answer was due to Martin Schwarzchild in 1938. He argued that a star as a whole does not pulsate. The interior of a star pulsates and it sends compressional waves to the outer layers. So a star is not brightest when it is smallest but when the compressional waves are moving fastest, i.e. when velocity of approach is maximum. This is what is observed for light curve and velocity curve for Cepheid variables.

Other Variable Stars

RV Tauri variables are mostly red giants or yellow super giants of spectral class G to K having periods between 30 and 150 days. The LC has alternately large and small maxima. The cause of two maxima is speculated to be due to pulsation together with some other physical process.

Long period variables or Mira variables have periods ranging from 100 days to 1,000 days. Median absolute magnitude has the range from +2 to −2. Their spectra belong to M class. They are red giant stars. They are Population II type stars.

Beta canis majoris variables are blue giants having spectral class B and absolute magnitude in the range −2 to −4. Their periods range from 6 to 8 h. Due to small periods they are hardly observed.

U Geminorium stars belong to eruptive variable class. Their light variation is sudden. It rises to maximum in 2, 3 days and declines in another 10–20 days. The variation in magnitude ranges from 2 to 6 magnitudes. The spectral class ranges from A to F.

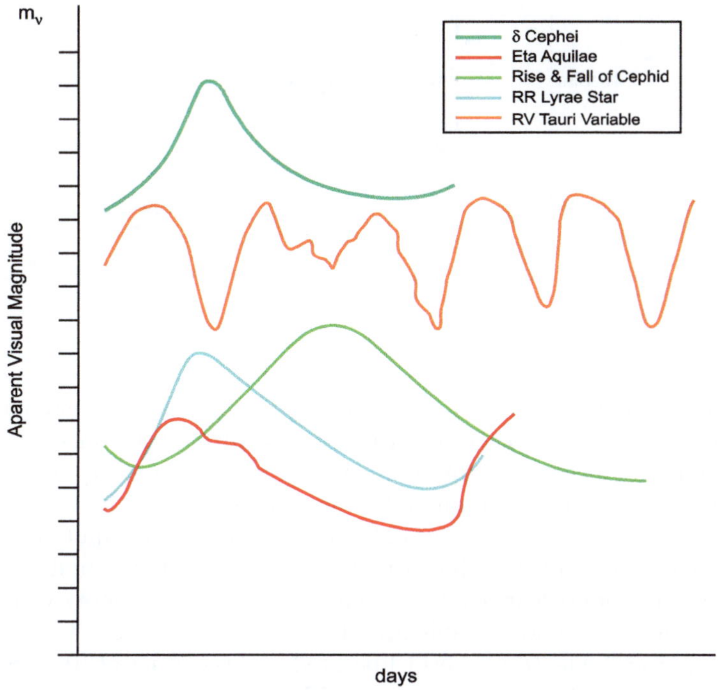

Figure 1.21 Light curves of Cepheids group of stars, RR-Lyrae and RV Tauri stars

Novae and Supernovae belong to the group of exploding variables. Novae are generally faint subdwarfs which suddenly flare up to very high luminosity and occasionally undergo explosion. Figure 1.22 describes in detail the light curve of a nova during various stages of light variation. Like Cepheids, novae are also used as distance indicators. The distances of nearby novae are mostly determined by the angular rate of expansion of the nebulae (big cold gas cloud) around them which are used for calibration.

There are many theories regarding the nova outburst but the proper explanation is due to Martin Schwarzchild. During the stellar evolution hydrogen converts into helium through fusion process and this liberates huge amount of energy. This energy generates shock waves which propagates into stellar surface shooting materials outwards. This phenomenon is observed as novae.

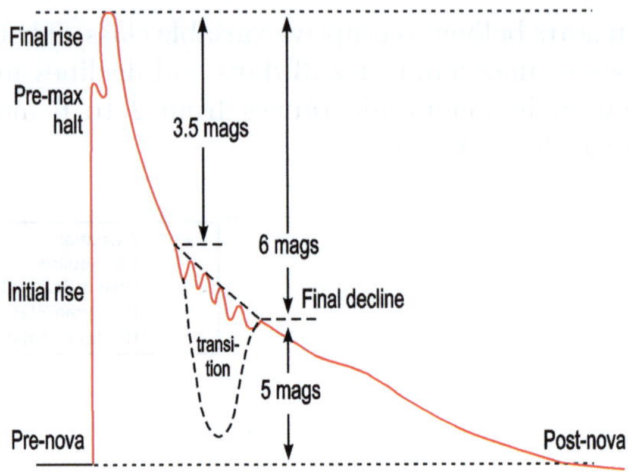

Figure 1.22 Various stages of light variation of a nova (courtesy: Basu et al. 2010)

Supernovae are the results of a big explosion occurring at the stellar core of very massive stars when there is an instability during the transformation of silicon to iron, which is endothermic. At this stage iron group of metals are suddenly transformed to helium and some neutrons are set free. This change requires huge amount of energy and this is supplied by rapid contraction. The core can hardly bear such contraction. The outer layers undergo, at the same time, thermonuclear reactions which liberates huge energy which is not radiated quickly to hold a stable structure. As a result the star undergoes violent explosion. The total energy liberated is of the order of 10^{50} ergs which corresponds to an absolute magnitude -17.

Extrinsic Variable Stars

We have already discussed that the light variation occurs in extrinsic binary system due to obscuration of light from one by another partner.

Visual Binaries

When the stars in a binary system are widely separated, so that a telescope can resolve them, they are called Visual binaries. The brighter star in a binary is called primary and the fainter one is called secondary. Sirius, 61 Cygni are examples of such binaries.

Spectroscopic Binaries

If the stars in a binary system be so close that they cannot be resolved visually with the aid of a telescope, but their spectra contain single and double lines, it is called a Spectroscopic binary. This happens because when the two stars revolve about their common centre of gravity and one is approaching towards the observer while the other is receding from the observer, then light from the approaching one is shifted towards the violet end whereas that from the receding one is shifted towards the red end. So, two separate line patterns in opposite phases are observed. When both move perpendicular to the line of sight, no such shift is observed giving a single pattern. So the spectrum of the pair reveals alternately single and double lines, provided they are sufficiently close and are of almost equal brightness.

Eclipsing Binary Stars

When the stars in a spectroscopic binary are such that the orbit is lined up with the line of sight or makes a very small angle, the variation of light occurs periodically. The light curve of such system shows two minima in a total period (Fig. 1.23). When the orbital plane is perpendicular to line of sight there is no eclipse. When the plane is along the line of sight, the eclipse is total or annular. For other positions the eclipse is partial.

Modelling of Light Curves of Eclipsing Binaries

In the following section it is described how light curves can be used for computing the physical characteristics of the companion stars using a simple model of circular orbits in absence of effects like limb darkening, tidal distortion, etc. At first a theoretical light curve is generated assuming the

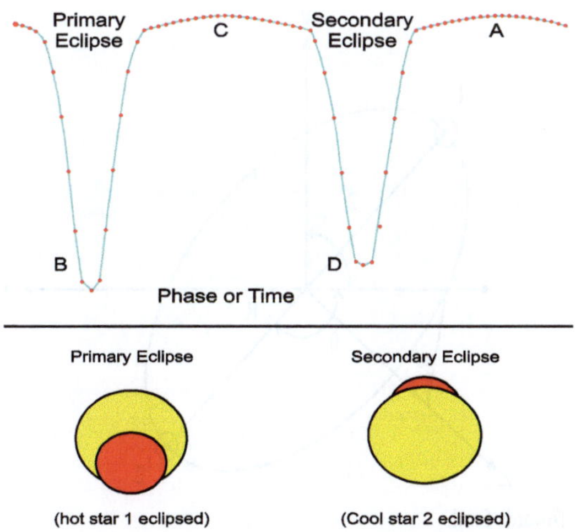

Figure 1.23 Light curve of a typical eclipsing binary star

physical parameters for the stars. Then it is compared with the observed light curve. Then by adjusting the values of the parameters the theoretical curve is matched with the observed one. The assumed parameters are the estimated parameters for the companion stars with some error of precession. The present model is due to Dan Bruton. Let $M_1, M_2, L_1, L_2, R_1, R_2, i$ be the masses, luminosities, radii and orbital inclination of star1 and star2, respectively (Fig. 1.24). Let R be the distance between the centres of the stars. Let θ be the azimuthal angle. Let (x_1, y_1, z_1) and (x_2, y_2, z_2) be the cartesian co-ordinates of the two stars, then

$$x_1 = -\frac{x}{1+(1/q)}, y_1 = -\frac{y}{1+(1/q)}, z_1 = -\frac{z}{1+(1/q)}$$

and

$$x_2 = \frac{x}{1+q}, y_2 = \frac{y}{1+q}, z_2 = \frac{z}{1+q}$$

where $x = R\ \sin\theta,\ y = R\ \cos i\ \cos\theta,\ z = R\sin i\ \cos\theta$ and $q = M_2/M_1$ (Goldstein et al. 2001).

Let F_1, F_2 be the brightness of the two stars. Then,

$$F_1 = \frac{L_1}{4\pi {R_1}^2}, F_2 = \frac{L_2}{4\pi {R_2}^2}$$

Then the brightness to an observer is $F = K(F_1A_1 + F_2A_2)$ where A_1, A_2 are the areas of the star discs seen by an observer and K is a constant that can be determined from the area of the observer's detector and the distance between earth and the binary system.

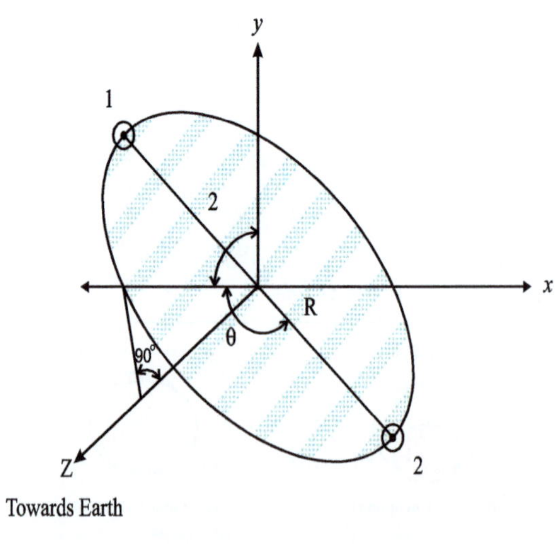

Figure 1.24 Co-ordinates of the centre of mass of an eclipsing binary system

A_1 and A_2 can be found by considering the geometry of the eclipse. The distance between the two stars to an observer is $\zeta = \sqrt{(x_2 - x_1)^2 + (y_2 - y_1)^2}$. So for $\zeta > R_1 + R_2$, there is no eclipse, for $(R_1 + R_2) > \zeta > \sqrt{R_1^2 - R_2}$, there is shallow eclipse and for $\sqrt{R_1^2 - R_2^2} > \zeta > R_1 - R_2$, there is deep eclipse and for $\zeta < (R_1 - R_2)$, there is annular eclipse. The corresponding positions are shown in Fig. 1.25 and the corresponding observed areas are listed in Table 1.3.

	$z_1 > z_2$		$z_1 < z_2$	
State	A_1	A_2	A_1	A_2
No Eclipse	πR_1^2	πR_2^2	πR_1^2	πR_2^2
Shallow eclipse	πR_1^2	$\pi R_2^2 - \triangle A_1 - \triangle A_2$	$\pi R_1^2 - \triangle A_1 - \triangle A_2$	πR_2^2
Deep eclipse	πR_1^2	$\pi R_2^2 - \triangle A_2 - \triangle A_1$	$\pi R_1^2 - \triangle A_2 - \triangle A_1$	πR_2^2
Annular or total eclipse	πR_1^2	$\pi R_2^2 - \pi R_1^2$	0	πR_2^2

Table 1.3 Areas of the stellar discs during various stages

To calculate $\triangle A_1$ and $\triangle A_2$ from Fig. 1.25,

$$\triangle A_1 = \frac{1}{2} R_1^2(\theta_1 - \sin\theta_1), \triangle A_2 = \frac{1}{2} R_2^2(\theta_2 - \sin\theta_2)$$

θ_1, θ_2 can be found from $R_2^2 = R_1^2 + \zeta^2 - 2R_1\zeta\cos(\theta_1/2)$ and $R_1^2 = R_2^2 + \zeta^2 - 2R_2\zeta\cos(\theta_2/2)$

Now, $\theta = 2\pi \times$ phase (ϕ) and phase = (Time since primary eclipse) / (orbital period).

So, intensity I can be plotted against ϕ, which gives the theoretically predicted light curve.

So the best choice of L_1, L_2, R_1, R_2, i, q give the best light curve with minimum error.

The software package Binary maker 3.0 predicts the physical parameters more accurately considering other effects also, given the observed light curve data.

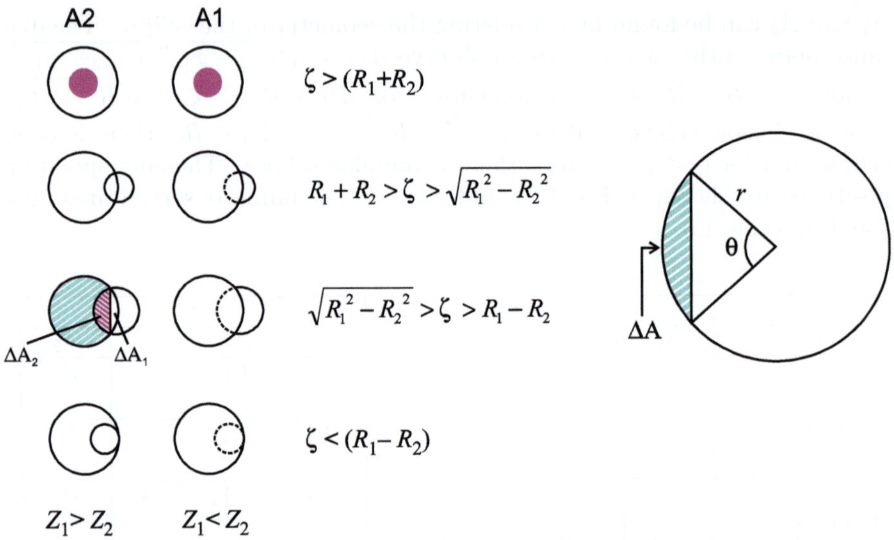

Figure 1.25 The various geometrical positions of the eclipse (courtesy: Dan Bruton: astro at sfasu.edu)

1.11 Stellar Populations

The solar system belongs to a vast ensemble of stars and gas, called our Galaxy (viz. Sect. 1.12). Our Galaxy contains almost 10^{11} stars. Most of these stars are found as isolated ones, called "field" star, or in association with another star or a triple system called "binary" or "multiple star" whereas a few (1 %) are associated with groups of various shapes and sizes. These are called "star clusters". There are primarily two types of star clusters: (1) galactic or open clusters and (2) globular clusters.

1.11.1 Galactic Clusters

Galactic clusters contain stars of the order of 10^2–10^4. They do not possess any particular shape. The stars belonging to this category are Population I objects, i.e. they are metal rich, take part in the Galactic rotation and contain bright stars belong to O and B classes. All the stars belonging to any particular star cluster have a common motion which is different from its surrounding objects. This is called, its **systematic velocity**. Also the stars in a cluster are at the same distance. So if the stars in a cluster are plotted for apparent magnitude versus colour that is equivalent to its H–R diagram, shifted parallel to match the zero age main sequence (ZAMS) which is equal to its distance modulus. So on the other way various galactic clusters whose distances are known by other methods can be used to calibrate the ZAMS.

The C–M diagram of various Galactic clusters are shown in Fig. 1.26. It is clear from the figure that (1) the main sequence is almost scatter free, (2) the turn of points of various clusters is different for different clusters, (3) the termination of the giant branches is different in different clusters and (4) the main sequence contains primarily hot blue stars. All these observations reveal that the stars in Galactic clusters have almost homogeneous chemical composition, the clusters are not "coeval" and they are young.

1.11.2 Globular Clusters

Globular clusters contain stars of the order of 10^4–10^6 and they are mainly Population II objects, i.e. metal deficient. They possess a spherical structure and are concentrated near the Galactic bulge and halo. Globular clusters far away from the Galactic centre do not take part in the Galactic rotation. If the stars in globular clusters are plotted in H–R diagram, the following features are observed (Fig. 1.27). (1) The main sequence has large scatter, (2) the clusters are more-or-less "coeval", and (3) contain cool stars belonging to classes F–M. So it can be concluded that the chemical composition of stars largely vary in globular clusters. In fact more or less all giant elliptical galaxies have globular clusters which show bimodality in their integrated metallicities. The reason of such bimodality is not clear yet. There are various theories regarding this bimodality exist among which "merger model" (Ashman and Zepf 1992) is somewhat popular. According to this theory, elliptical galaxies have been formed by the merger of two progenitor gas rich spiral galaxies. The globular clusters which are comparatively metal poor are the clusters of the progenitor spirals and the metal rich globular clusters have been formed by the dissipative (wet merger) merger of the associated gas of the spiral galaxies in the next episode of star formation. But there are other theories of galaxy formation besides merger model, e.g. monolithic collapse model, multiphase dissipational collapse model, dissipationless merger model (dry merger) and accretion and in situ hierarchical merging (viz. Sect. 1.12). Sometimes the formation of galaxies has been suggested by various statistical models in multivariate set-up (Chattopadhyay et al. 2009 and the references therein). Since globular clusters contain cool stars so they are older. The ages of globular clusters have been determined by using simple stellar population (SSP) model and they are comparable to the age of their host galaxies ($\sim Gyr$), i.e. they are the robust counterpart overcoming the complicated process of galaxy formation. Hence they can be considered as the fossil records of galaxy formation.

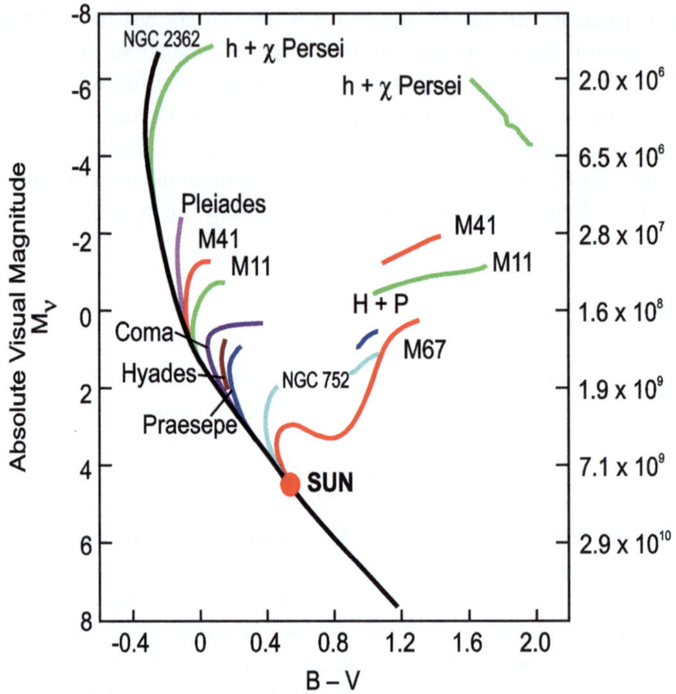

Figure 1.26 The C–M diagram of Galactic clusters

Fundamental Plane

In 1995, Djorgovsky found a scaling relation among the core radius (r_c), central velocity dispersion (σ_e), central surface brightness ($\mu_v(0)$) of globular clusters in our Galaxy. The relation found is

$$\mu_v(0) = (-4.9 \pm 0.2)(\log \sigma - 0.45 \log r_c) + (20.45 \pm 0.2)$$

which corresponds to $r_c \sim \sigma^{2.0\pm0.15} I_0^{-1.1\pm0.1}$

where I_0 is the luminosity density and

$$\log I_0 = 0.4(M_{v,\odot} - M_v) - \log(2\pi) - 2\log r_c$$

$$\text{and } \mu_v(0) = M_{V,\odot} + 20.652 - 2.5 \log I_0$$

This is similar to the Virial Theorem,

$$r_c \sim \sigma^2 I_0^{-1} (M/L)^{-1}$$

for a constant mass to light ratio within the measurement errors. This means that the globular clusters are dynamically stable system, with respect to core parameters and for a universal mass to light ratio.

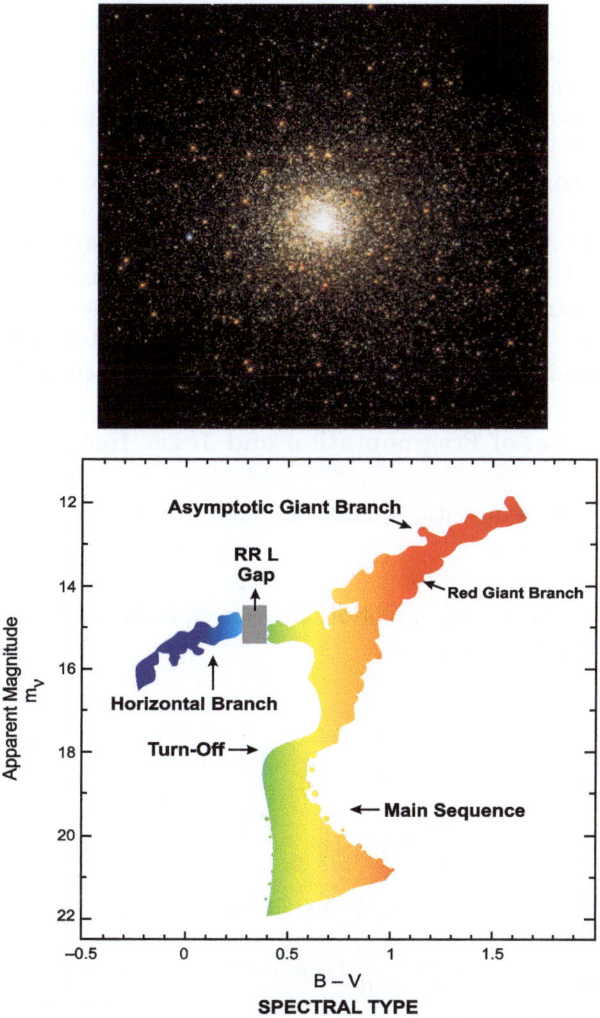

Figure 1.27 H–R diagram of globular clusters (bottom) and image of a typical globular cluster (top)

For effective radius parameters the FP is $r_e \sim \sigma^{1.45\pm0.2} I_e^{0.85\pm0.1}$. This is close to the FP of giant elliptical galaxies $R{\sim}\sigma^{1.4\pm0.2}\ I^{-0.8\pm0.1}$.

1.11.3 Fragmentation of Molecular Clouds and Initial Mass Function (IMF)

Star formation remains a tantalizing problem in modern astronomy. Most of the astronomers now believe that star formation is triggered by the gravitational collapse of the dense molecular clouds. Molecular clouds are large gaseous cold (10–100 K) clouds ($10^2 M_\odot$–$10^6 M_\odot$) of size varying from 10 to 100 pc. It primarily consists of H_2 and He and contains traces of molecules such as CO, NH_3, H_2CO, CS, and N_2H^+. Molecular hydrogen is difficult to detect by infrared or radio observations so cloud properties are demonstrated by strong rotational or vibrational emission lines of CO in cm, mm and sub mm ranges. The core is traced by NH_4. These molecular clouds are thought to be regions of recent star formation as many recently formed open clusters, e.g. NGC 3293, NGC 2264 are found to be embedded in such clouds. Many theories have been developed to demonstrate the existence of a second generation of fragmentation, i.e. the fragmentation of these molecular clouds in star clusters. In this context a hierarchy of fragmentation scenario of an infinite gas cloud to galaxies and stars has been developed by Hoyle (1953).

A Simple Model of Fragmentation and Jeans Instability

Suppose a system is in static equilibrium having density ρ_0 pressure p_0, gravitational potential ϕ_0.

Now equations of continuity, motion, Poisson equation and equation of state (adiabatic) of such a system are

$$\frac{\partial \rho}{\partial t} + \vec{\bigtriangledown}.(\rho \vec{v}) = 0$$

$$\frac{\partial \vec{v}}{\partial t} + (\vec{v}.\vec{\bigtriangledown})\rho = -\frac{1}{\rho}\vec{\bigtriangledown} p + \vec{F}$$

$$\bigtriangledown^2 \phi = 4\pi G \rho$$

$$\frac{d}{dt}(p/\rho^\gamma) = 0$$

where $\vec{F}$ is the external force.

Now for static equilibrium,

$\vec{F} = -\vec{\bigtriangledown}\phi_0, \vec{\bigtriangledown} p_0 = -\rho\vec{\bigtriangledown}\phi_0, \bigtriangledown^2\phi_0 = 4\pi G\rho_0$ (since $\vec{v_0} = \vec{0}$), $p_0 =$ constant. $\rho^\gamma = \frac{kT}{\mu m_H}.\rho_0^\gamma$, where k is Boltzmann constant, μ is the mean molecular weight and m_H is the mass of hydrogen atom.

If there are small perturbations of the form

$$\begin{aligned} \phi &= \phi_0 + \phi_1 \\ \rho &= \rho_0 + \rho_1 \\ p &= p_0 + p_1 \end{aligned}$$

$$\vec{v} = \vec{0} + \vec{v_1}$$

then linearizing the above set of equations (assuming the perturbations are small enough) we have

$$\frac{\partial \rho_1}{\partial t} + \rho_0 \vec{\nabla}_0.\vec{v_1} = 0 \text{ (equation of continuity)}$$

$$\rho_0 \frac{\partial \vec{v_1}}{\partial t} = -\vec{\nabla} p_1 + \rho_0 \vec{\nabla} \phi_1 \text{ (equation of motion)}$$

$$\nabla^2 \phi_1 = 4\pi G \rho_1 \text{ (poisson equation)}$$

$$\text{and } p_1 = {c_s}^2 \rho_1 \text{ (equation of state)}$$

where c_s is the isothermal speed of sound and $c_s^2 = \frac{dp}{d\rho}$.

Then the above four equations can be combined to

$$\frac{\partial^2 \rho_1}{\partial t^2} = {c_s}^2 \nabla^2 \rho_1 + 4\pi G \rho_0 \rho_1$$

This is a wave equation which admits a plane wave solution,

$$\rho_1 = \rho_0 e^{i(\vec{k}.\vec{x} - wt)}$$

where $\vec{k}$ is the wave number and $|\vec{k}| = \frac{2\pi}{\lambda} = k$, λ being the wavelength of the perturbation. Then substituting ρ_1, in the latter equation we have the dispersion relation

$$w^2 = {c_s}^2 k^2 - 4\pi G \rho_0 = {c_s}^2 (k^2 - {k_J}^2)$$

where ${c_s}^2 {k_J}^2 = 4\pi G \rho_0$ (say).

If $k < k_J, w^2 < 0$, the disturbance grows exponentially with time. $\lambda_J = \frac{2\pi}{k_J}$ is defined as the Jeans (1902) critical wavelength. Its value is

$$\lambda_J = \frac{2\pi}{k_J} = \left(\frac{\pi k T}{\mu m_H G \rho_0} \right)^{1/2}$$

The **Jeans mass**, $M_J = \frac{\pi}{6} \rho_0 \lambda_J^3 = 10^{23} \left(\frac{T}{\mu} \right)^{3/2} {\rho_0}^{-1/2}$ g.
So a cloud mass $M > M_J$ will be gravitationally unstable and will undergo isothermal collapse.

It is to be noted that the unperturbed conditions assumed by Jeans are not consistent. Because for a constant pressure $p_0, \vec{\nabla} p_0 = -\rho \vec{\nabla} \phi_0$ leads to a constant ϕ_0 but a constant ϕ_0 leads to $\rho_0 = 0$ from $\nabla^2 \phi_0 = 4\pi G \rho_0$. So for a proper stability analysis one needs a proper equilibrium configuration (Spitzer 1978). Still Jeans analysis is important for its simplicity as the correct stability analysis yields qualitatively similar results.

Initial Mass Function

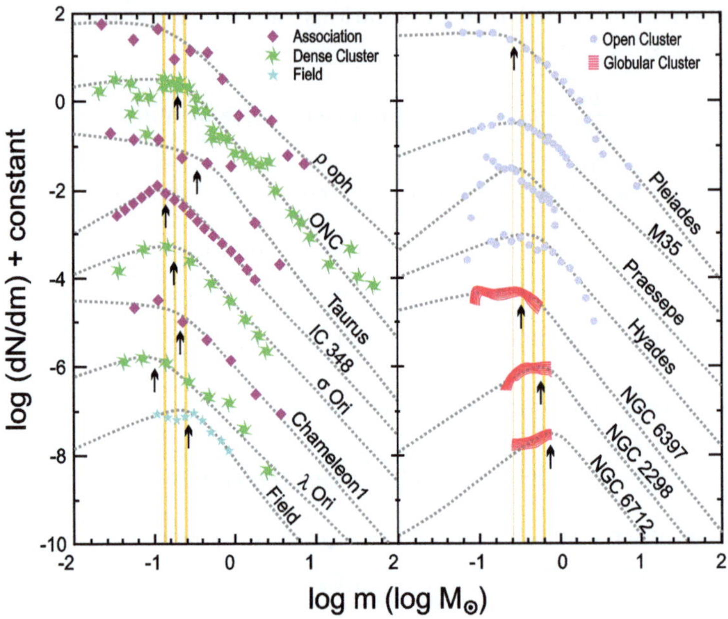

Figure 1.28 Mass spectrums of various star clusters and stellar associations (courtesy Bastian et al. 2010)

Now as the cloud collapses its density increases, so under isothermal collapse (i.e. constant temperature), Jeans mass decreases, i.e. the cloud fragments. In this way a hierarchical fragmentation scenario sets in until the cloud becomes opaque to trap the radiation and switches over from isothermal to adiabatic phase and the fragmentation stops. The mass spectrum developed at this stage is what is called "IMF" and is defined by Salpeter (1955) as a power law of the form,

$\xi = \frac{dN}{d\log m} \propto m^{\Gamma}$ where m is the mass of a star and N is the number of stars in the logarithmic mass range $\log m$ and $(\log m + d\log m)$. It is found by Salpeter that $\Gamma \sim -1.35$ for $0.4M_{\odot} \leq m \leq 10M_{\odot}$. The IMF in linear mass units takes the form $\frac{dN}{dm} \propto m^{-\alpha}$ so that $\Gamma = 1 - \alpha$. Later a segmented power law was derived by Kroupa et al. (1993) that gave $\Gamma \sim -1.35$ above a few solar masses with a shallower power law ($\Gamma \sim 0$ to -0.25) for low mass stars. The turn over occurs at $0.3M_{\odot}$, often known as the characteristic mass (m_c). The mass spectrums for various star clusters, galactic or globular are shown in Fig. 1.28.

Integrated Galaxial IMF

Study of IMF is very important regarding the link between stellar and galactic evolution and hence the insight on theories of star formation. It is questionable whether the IMF is universal or not. Under the assumption of an invariant canonical stellar IMF in star clusters and an invariant embedded cluster mass function (ECMF), i.e. the mass function of the star clusters, the integrated galaxial initial mass function (IGIMF) is defined as (Weidner and Kroupa 2005)

$$\xi_{IGIMF} = \int_{M_{ecl,min}}^{M_{ecl,max(SFR(t))}} \xi(m \leq m_{max})\xi_{ecl}(M_{ecl})dM_{ecl}$$

where ξ is the stellar IMF, ξ_{ecl} is the star cluster ECMF, m_{max} is the maximum mass of a star in a particular star cluster and $M_{ecl,min}$ and $M_{ecl,max}$ are minimum and maximum masses of star cluster and $M_{ecl,max}$ is a function of the star formation rate (SFR) which varies over time. If the ECMF is taken as

$$\xi_{ecl} \infty M_{ecl}^{-\beta}$$

then assuming $\beta \sim 2.2$ (Lada and Lada 2003; Hunter et al. 2003) it is seen that IGIMF (α_{IGIMF}) is steeper than $IMF(\alpha)$ and for $\beta > 2$, α_{IGIMF} largely differs from α for increasing values of α (Fig. 1.29). Now for $\beta > 2$ the number of white dwarfs falls rapidly and the fall of supernova II even stronger. The effect is more pronounced for steeper α. So depression of SN II slows down the chemical enrichment, hence formation of stars particularly in galaxies with low SFR (e.g. dwarf galaxies). As a result the IGIMF steepens.

The scenario changes completely for a slightly lower $\beta(\sim 2.0)$ and the sampling procedure for the maximum mass $M_{ecl,max}$ of the star cluster. In Fig. 1.30 it is seen that IMF and IGIMF are almost identical for low $\beta(\sim 2.0)$ and stochastic sampling of $M_{ecl,max}$. So at this point it is very important to have an accurate estimation on $\beta, M_{ecl,max}$ and knowledge of birthplace of high mass stars using kinematic conditions.

1.12 Galaxies

A vast ensemble of stars and gaseous matter pervaded by magnetic field, cosmic rays together with unseen matter is known as a galaxy. The galaxy to which our solar system belongs is called "Milky Way" or Galaxy. Its shape is like a flattened spheroid with certain ring like structures, called spiral arms. Sun lies along an arm (Orion) at a distance 8.2 kpc from the centre of the Galaxy. The arm structure is confined in a thin disc like structure, called the Galactic disc. The arms generate from a dense region, called nuclear bulge

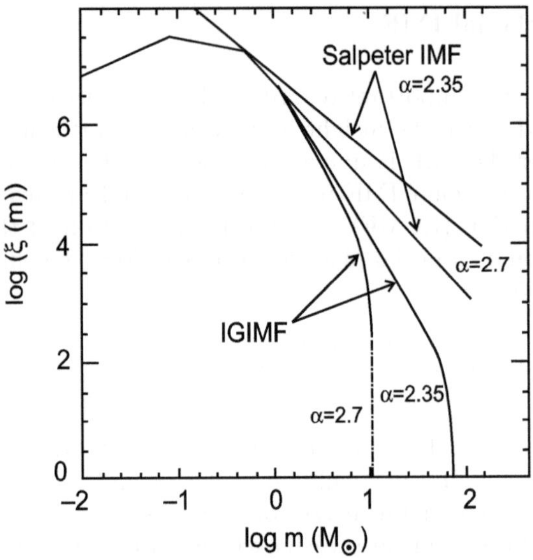

Figure 1.29 Trends of IGIMF and IMF for different values of β (courtesy Kroupa 2004)

in which the densest part of the Galaxy, called nucleus, is embedded. The nucleus, bulge, disc and spiral arms primarily contain Population I objects. The disc rotates around the Galactic centre with a high velocity whose type is different from that of a solid body rotation. This is known as "differential galactic rotation". In solid body rotation, all the objects situated at different distances from the centre of rotation have the same angular velocity ω so, if v is the linear velocity at a distance r from the centre since $v = \omega r$, so $v \propto r$. In case of "differential galactic rotation" objects away from the centre move more slowly. In Keplerian motion objects move around a central massive objects of mass M such that $v^2 = GM/r$, i.e. the linear velocity falls off as the inverse square root of the radius since M is constant. Differential galactic rotation is different from Keplerian motion (applicable to planetary orbits) in a sense that here mass M as we will see is not constant and it occurs for gaseous bodies such as Sun and planets with atmosphere.
In Fig. 1.31 the profiles of three types of motion have been shown. Although the primary constituents are confined in a thin disc, the Population II objects e.g., globular clusters, high velocity stars like subdwarfs, Cepheids (RR Lyraes and Type II Cepheids) define a more or less spherical structure enclosing the disc. This is known as Galactic halo. Observations indicate that neutral hydrogen gas clouds extend in the halo to at least 1 kpc above the Galactic plane. The mass of our Galaxy has been estimated by various authors (Schmidt 1956, 1965; Lohmann 1953; Bucerius 1934) to be of the order of $10^{11} M_{\odot}$. But later observations of V.C. Rubin and her co-workers indicate that galaxies are much more massive and are of much greater extension. The

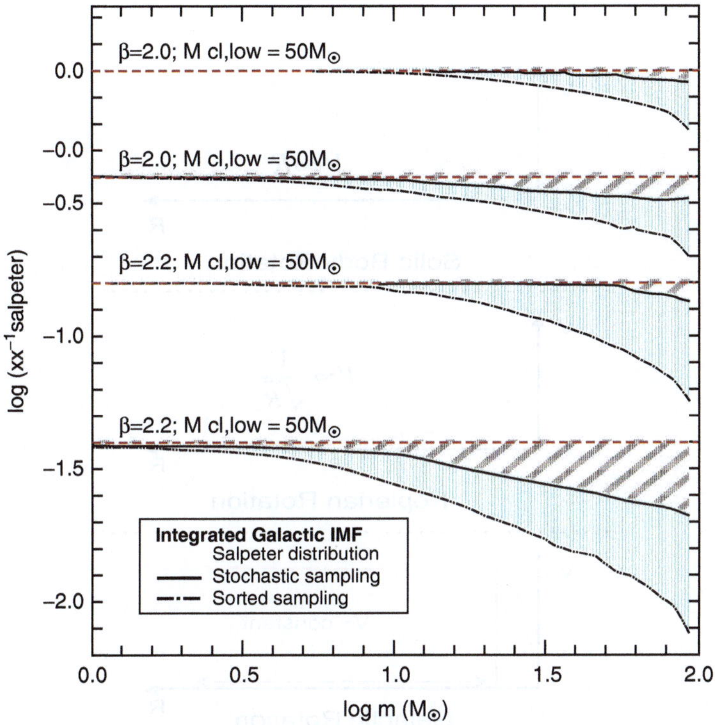

Figure 1.30 Trends of IGIMF and IMF under various sampling procedures (courtesy Bastian et al. 2010)

mass, however, exists in unobserved form. For Galactic rotation we have $v^2/r = GM/r^2$. If $M \propto r$, then v = constant and this is what is seen for galaxies (Fig. 1.31). So a thrice compound model of the Galaxy (nuclear region, disc and halo) considered by several authors yields a mass of the order of $10^{12} M_\odot$.

Oort's Constants

We have discussed that differential galactic rotation introduces some kind of shearing motion in the Galactic plane. If this shear can be determined quantitatively, the local force law can be estimated. Oort's constants are the well-known constants which give measure of this local shear and the force law. In this part we give a derivation of these constants. Let us assume circular motion of gaseous material around Galactic centre. Then at a given galactic longitude the radial velocity will be maximum at a position where the line of sight is closest to the Galactic centre. This point is the tangential point T corresponding to the Galactic longitude l (say) (viz. Fig. 1.32).

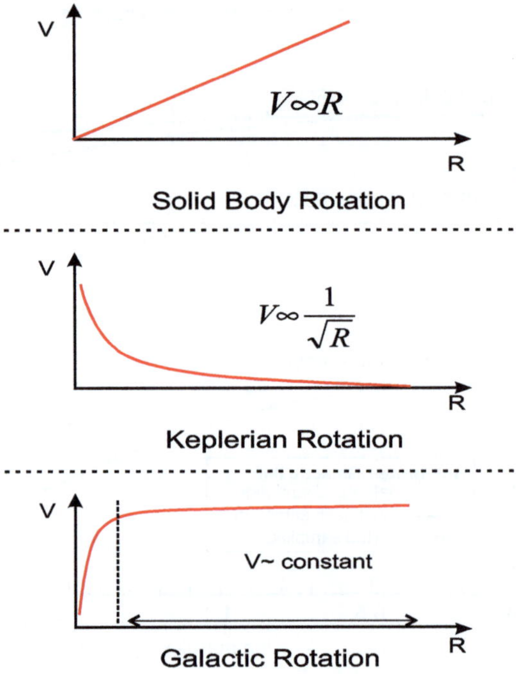

Figure 1.31 Schematics of various types of rotations

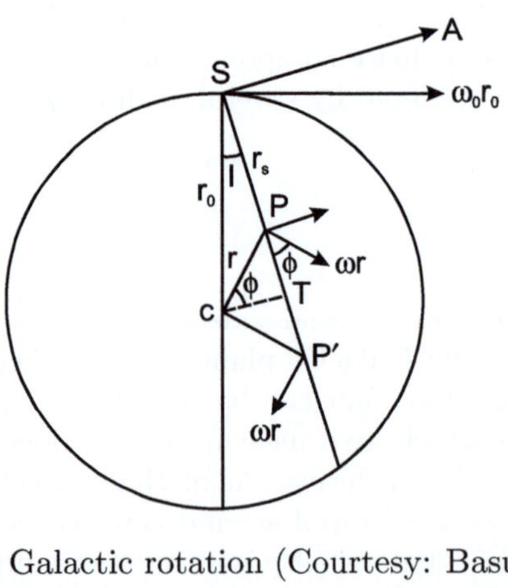

Figure 1.32 Galactic rotation (Courtesy: Basu et al. 2010)

Let r_{min} be the distance of T from Galactic centre C (say). Then $CT = r_{min} = r_0 \sin l$, where r_0 is the distance of sun S from C. Then, $ST = r_0 \cos l$. Let P be any other point on the Galactic plane and CP = r and $SP = r_s$. Then from the triangle SCP,

$$\frac{\sin l}{r} = \frac{\sin(90+\phi)}{r_0} \text{ i.e. } \frac{\sin l}{r} = \frac{\cos\phi}{r_0}$$

Then the radial velocity of any material at P relative to that near Sun is $v_r = r_0(w(r) - w_0)\sin l$ where ω and w_0 are the angular velocities at P at a distance r and at S at a distance r_0 from C. If P has the latitude b, then

$$v_r = r_0(w(r) - w_0)\sin l \cos b$$

We assume here, b = 0. The expanding $\omega(r)$ by Taylor series in neighbourhood or r_0,

$$w(r) = w_0 + (r - r_0)\left(\frac{dw}{dr}\right)_{r_0} + O(r - r_0)^2$$

Neglecting smaller terms,

$v_r = r_0 \sin l (r - r_0)\left(\frac{dw}{dr}\right)_{r_0}$, to first order of approximation. The first Oort's constant A is defined as

$$A = -\frac{1}{2}r_0\left(\frac{dw}{dr}\right)_{r_0} = -\frac{1}{2}r_0\left[\frac{d}{dr}\left(\frac{v}{r}\right)\right]_{r_0} = \frac{1}{2}\left(\frac{v}{r} - \frac{dv}{dr}\right)_{r_0}$$

It represents the rate of local shear. Then,

$$v_r = 2A(r_0 - r)\sin l$$

Also, from triangle SCP, $r^2 = {r_0}^2 + {r_s}^2 - 2r_0 r_s \cos l$

$$\text{So, } \frac{r}{r_0} = 1 - \frac{r_s}{r_0}\cos l + O\left(\frac{r_s}{r_0}\right)^2 \text{ i.e.}$$

$$r_0 - r = r_s \cos l \text{ (neglecting smaller terms)}$$

$$\text{Then, } v_r = A r_s \sin 2l$$

The transverse velocity of the material at P with respect to that at S is given by

$$v_T = v\sin\phi - v_0 \cos l$$

Then using some algebraic manipulations and assuming $r_s/r_0 << 1$,

$$v_T = -r_s\left[r_0\left(\frac{dw}{dr}\right)_{r_0}\cos^2 l + w_0\right]$$

$$\text{Then, } v_T = r_s(A\cos 2l + A - w)$$

Introducing another constant $B = A - w_0$,

$$v_T = r_s(A\cos 2l + B)$$

Now, $B = A - w_0 = \frac{1}{2}(\frac{v}{r} - \frac{dv}{dr})_{r_0} - \frac{v_0}{r_0} = -\frac{1}{2}(\frac{v}{r} + \frac{dv}{dr})_0$

So $(\frac{dv}{dr})_{r_0} = -w_0 - 2B = -(A+B)$

Let $F = \frac{v^2}{r}$ be the force law at a distance r from C for the material at P. Then,

$$\left(\frac{dF}{dr}\right)_{r_0} = \left(\frac{2v}{r}\frac{dv}{dr} - \frac{v^2}{r^2}\right)_{r_0} = 2w_0\left(\frac{dv}{dr}\right)_{r_0} - w_0^2$$

$$\text{i.e.} \left(\frac{dF}{dr}\right)_{r_0} = -(A-B)(3A+B)$$

If the force law is of the form $F = Dr^n$, then

$$\left(\frac{d\ln F}{d\ln r}\right)_{r_0} = n = -\left(\frac{3A+B}{A-B}\right)$$

Thus the value of Oort's constants give local force law of the Galaxy. Observations suggest that n varies from +1 to −2 from close to the centre to outer region. Working values of A and B are $14.4 \pm 1.2\,\mathrm{km\,s^{-1}\,kpc^{-1}}, -12 \pm 2.8\,\mathrm{km\,s^{-1}\,kpc^{-1}}$ which have been accepted in IAU, 1985. The other rotation parameters are $r_0 = 8.5 \pm 1.1\,\mathrm{kpc}, v_0 = 222 \pm 20\,\mathrm{km\,s^{-1}}, w_0 = 26.4 \pm 1.9\,\mathrm{km\,s^{-1}\,kpc^{-1}}$.

External Galaxies

In the previous section we have discussed various features of our Galaxy. But actually there are many such galaxies in the Universe. The galaxies belong to groups and many more to clusters. The clusters in general contain a large numbers of galaxies of different types and of various sizes. The nearest cluster is Local Group which contains our Galaxy, Andromeda Galaxy (M31), Large Magellanic Cloud (LMC), Small Magellanic Cloud (SMC) and many dwarf galaxies with a total of 30 in number. The next ones are The Local Volume (LV), Virgo and Coma clusters at distances of 10 Mpc, 16 Mpc and ... Mpc, respectively. The galaxies are first classified by E.P. Hubble in early 1920s on the basis of their morphological structures. His scheme consists of three regular classes, viz. ellipticals, spiral/barred spirals and irregulars. It is represented by his famous tuning fork diagram (Fig. 1.33). The elliptical galaxies have spheroidal structures having no spiral arms and are subclassified as E0 to E7 depending on the degree of flattening. Spiral galaxies are like our Galaxy and they are subclassified as Sa, Sb, Sc on the degree of tightness of the spiral arms. If the spiral galaxies have bars at their centre, they are termed barred spiral and denoted by SB, Irregular galaxies do not posses any particular shape. Our Galaxy is a spiral galaxy whereas LMC, SMC are irregular galaxies. Recently there are several catalogues of galaxies, e.g. Messier Catalogue, abbreviated by M followed by a number, New General Catalogue abbreviated by NGC followed by a number, etc.

Scaling Relations for Galaxies

Virial Theorem

As we have discussed in Sect. 1.9 that a gas cloud or any other isolated

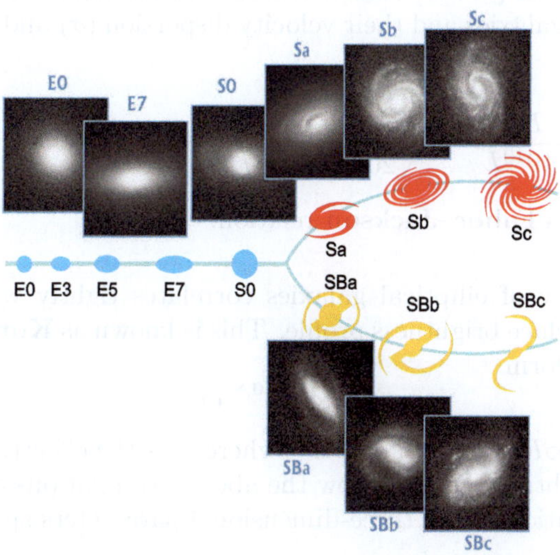

Figure 1.33 Hubble's Tuning fork diagram of galaxy classification

dynamical system in equilibrium will satisfy the Virial theorem, $2T + \Omega = 0$, where T, Ω are the kinetic energy and potential energy of the system and $T = \sum_i \frac{1}{2} m_i {v_i}^2$ or $2T = M < v^2 >= M\sigma^2$, where M is the total mass of the system and σ^2 is the mean square velocity. Also,

$$\Omega = \sum_{i>j} -\frac{Gm_i m_j}{|\vec{r_i} - \vec{r_j}|}$$

$$\text{i.e. } \Omega = -\frac{GM^2}{r_g}$$

where r_g is the separations of the stars and hence must represent a characteristic extent of the system. For a spheroidal system $\frac{1}{r_g} = \frac{\alpha}{R_e}$ where R_e is the effective radius and α is some constant of order unity. Then the Virial theorem takes the form,

$$M\sigma^2 = \frac{\alpha G M^2}{R_e}$$

$$\text{i.e. } M = \frac{1}{\alpha} R_e \sigma^2 G^{-1} = C G^{-1} R_e \sigma^2$$

which gives an estimate of the gravitational mass of the system if R_e and σ can be estimated. This mass is called **Virial** mass of the system.

Fundamental Plane

Faber and Jackson (1976) studied the correlation between the luminosities (L) of elliptical galaxies and their velocity dispersion (σ) and found a relation of the type,

$$\frac{L}{2 \times 10^{10} L_\odot} \simeq \left(\frac{\sigma}{200\,\text{km}\,\text{s}^{-1}}\right)^4, 50 \leq \sigma \leq 500\,\text{km}\,\text{s}^{-1}$$

This is known as **Faber–Jackson** relation.

Again luminosity of elliptical galaxies correlates tightly with the effective radius of the surface brightness profile. This is known as **Kormendy** relation which is of the form

$$L \propto {R_e}^{0.8} \text{ or}$$

equivalently $I_e \propto {R_e}^{-1.2}$ (as $L \propto I_e {R_e}^2$ where I_e is the effective luminosity, i.e. the surface brightness at R_e). Now the above two relations can be combined into a third relation in the three-dimensional parameters space R_e, I_e and σ as,

$$R_e \propto \sigma^{1.4} {I_e}^{-0.83}$$

or $R_e \propto \sigma^{5/4} {I_e}^{-5/6}$ or in terms of luminosity, $L \propto {I_e}^{-2/3} \sigma^{5/2}$ which is known as the **Fundamental Plane** (FP) of elliptical galaxies. The Kormendy and the Faber–Jackson relations are the projections of this FP to the respective two-dimensional slices. Note that we can deduce L if we can measure σ and I_e for the corresponding galaxy. Then if we can measure apparent magnitude hence apparent luminosity we can use the usual inverse square law to deduce the distance of that galaxy.

Hubble's Law

In 1929 Edwin Hubble studied the recessional velocities of galaxies and their distances and published a paper in the "Proceedings of the National Academy of Sciences" which is the trendsetter of modern observational cosmology. He concluded that the speed of recession of a galaxy is proportional to its distance from us, i.e. if v is the recessional velocity and $v << c$ and D is its distance, then $v \propto D$, i.e. $v = H_0 D$. Here H_0 is called the Hubble's constant. Again from Doppler's law we know that $\frac{v}{c} = \frac{\triangle\lambda}{\lambda} = z(\text{say})$, i.e. $v = cz$ where z is the redshift. Here $z << 1$ (maximum z = 0.003). Here it is to be noted that for high redshift the distance of a galaxy becomes function of redshift and is given as

$$D_1 = r_1 S(to)(1+z)$$

where S(to) is the scale factor at present epoch and r_1 is the distance of the galaxy from us when the light left from the galaxy at any epoch t_1(say) to reach us at the present epoch to (say) and then

$$H(t) = \frac{\dot{S(t)}}{S(t)} \text{ where } H(to) = H_0$$

With this value of D_1 the bolometric flux (flux integrated over all wavelengths)

$$F_{bol} = \frac{L_{bol}}{4\pi {D_1}^2} = \frac{L_{bol}}{4\pi {r_1}^2 S^2(to)(1+z)^2}$$

where L_{bol} is the bolometric luminosity.

This reduces to $m_{bol} - M_{bol} = 5\log D_1 - 5$

where M_{bol} and m_{bol} are the absolute and apparent bolometric magnitudes of the galaxy. Now, when the magnitudes are measured using a particular filter (λ_0, say) then an astronomer has to apply corresponding correction to redshift so that

$$m(\lambda_0) - M(\lambda_0) = 5\log D_1 - 5 + BC$$

where BC is called the K-correction.

K-Space Parameters

Now, mass to luminosity ratio following Virial theorem and FP is

$$\frac{M}{L} \propto \frac{\sigma^2 R_e}{\sigma^{5/2} {I_e}^{-2/3}} \propto \sigma^{-1/2} R_e \left(\sigma^{-3/2} {R_e}^{6/5}\right)^{-2/3} \propto \sigma^{1/2} {R_e}^{1/5}$$

As ellipticals lie close to the FP so by combining parameters it is possible to find projection which gives exact edge on view of the plane. Thus defining,

$$\begin{aligned} k_1 &= \frac{1}{\sqrt{2}} \log\left(\frac{M}{c_2}\right) = \frac{1}{\sqrt{2}} \log(\sigma^2 R_e), \text{ where } M = c_2 \sigma^2 R_e \\ k_2 &= \frac{1}{\sqrt{6}} \log\left(\frac{c_1}{c_2}\frac{M}{L} {I_e}^3\right) = \frac{1}{\sqrt{6}} \log\left(\frac{\sigma^2 {I_e}^2}{R_e}\right) \text{ where, } L = c_1 I_e {R_e}^2 \end{aligned}$$

$$\text{and } k_3 = \frac{1}{\sqrt{3}} \log\left(\frac{c_1}{c_2}\frac{M}{L}\right) = \frac{1}{\sqrt{3}} \log\left(\frac{\sigma^2}{I_e R_e}\right)$$

it is easy to see that k_1 is proportional to the virial mass M, k_3 is proportional to M/L and $k_1 - k_3$ plot gives the edge on view of the FP.

Galaxy Formation: Existing Theories

Classical formation of galaxies can be divided into five major categories: (1) the monolithic collapse model, (2) the major merger model, (3) the multi-phase dissipational collapse model, (4) the dissipationless merger model (dry merger) and (5) accretion and in situ hierarchical merging.

According to monolithic collapse model, elliptical galaxies start its life from the collapse of an isolated massive gas cloud at high redshift. So the colour distribution of its globular clusters is unimodal and their rotation is produced by the tidal force from satellite galaxies. Larson (1975), Carlberg (1984), Arimoto and Yoshii (1987), Peebles (1969) and many more have worked in support of the above theory. But in most of the giant elliptical galaxies, bimodality in metallicity cannot be explained by the above theory. So Ashman and Zepf (1992), Zepf et al. (2000) and many astrophysicists devised the "merger theory". According to merger model, elliptical galaxies are formed by merger of two or more disc galaxies. Younger globular clusters are formed out of the shocked gas in the disc while blue metal poor globular clusters come from halos of the merging galaxies (Bekki et al. 2002). As a result the colour distribution is bimodal. In this model, the kinematical properties of the globular clusters depend weakly on the orbital configuration of the merging galaxies, but metal rich globular clusters are generally located in the inner region of the galaxy and the metal poor ones in the outer regions.

In spite of over all success of the merger model it suffers from the following aspects, (1) in merger model the metal rich globular clusters are speculated to be produced in merger of gaseous discs. So a strong correlation is expected between mean metallicity of globular clusters and its specific frequency. But in practice no such correlation exists. (2) In many elliptical galaxies higher mean metallicity does not reflect the higher proportion of metal rich globular clusters, moreover (3) the blue peak of metallicity of globular clusters of NGC 3311, 3923 is redder than halo globular clusters of Milky Way. So these globular clusters are probably not the globular clusters of original spiral and (4) specific frequency (S_N) of metal poor globular clusters in M87 is larger than S_N of spiral galaxies. The multiphase dissipational collapse has been proposed by Forbes et al. (1997). According to this model the globular clusters form in distinct star formation episodes through dissipational collapse and there is tidal stripping of globular clusters from satellite galaxies. So we have blue metal poor globular clusters in initial stage and metal rich globular clusters form at a later stage. This produces a bimodality in metallicity. So blue accreted globular clusters have no rotation but red globular clusters show rotation depending on the degree of dissipation (Côté et al. 2001).

1.13 Quasars

A "quasar" or "quasi stellar object" is a star like object having large ultra violet flux of radiation accompanied by generally broad emission lines and absorption lines in some cases found at large redshift. Some of the quasars (10 %) are radio-loud.

In the optical region, the continuum spectrum of quasars can crudely be approximated by a power law of the form, $F(\nu) \propto \nu^{-\alpha}$, $0.5 \leq \alpha \leq 1$ in the colour range $-0.8 \leq U - B \leq -0.7$. So for identifying quasars one should identify all star like objects having U–B in the above range but the list is contaminated by white dwarfs. We have mentioned in Sect. 1.3.5 that redshift of light might occur due to Doppler effect (Doppler shift) or due to passing of light near a very compact object (gravitational redshift) or expansion of the universe (cosmological redshift). Now since the redshift of quasars are generally larger than those of normal galaxies and are similar to those of seyfert galaxies and active galactic nuclei (AGN) whose redshifts are believed to be cosmological, the redshifts of quasars are assumed to be of similar origin.

K-Correction for Quasars

Now observed flux at frequency ν_0 from a quasar at redshift z is related to its luminosity distance D_L and luminosity L is given by

$$F(\nu_0) = \frac{(1+z)L(\nu_0(1+z))}{4\pi {D_L}^2}$$

where D_L for a closed universe (curvature K = +1) is

$$D_L = (c/H_0 {q_0}^2)[q_0 z + (q_0 - 1)(\sqrt{1 + 2q_0 z} - 1]$$

H_0 is the value of Hubble's constant at present time t_o, q_0 is the deceleration parameters (i.e. the rate at which the universe is slowing down its expansion) at t_o.

Now if m is the apparent magnitude of the quasar, then

$$m = -2.5 \log F + constant \quad \text{(similar to Eq. (1.21))}$$

where the constant depends on the filter used in obtaining the flux. Then substituting $F(\nu_0)$ in the above relation and using (1.21) and $H_0 = h_{100} \times 100\,\mathrm{km\,s^{-1}Mpc^{-1}}$

$$m = M + 5\log\left\{\frac{1}{{q_0}^2}[q_0 z + (q_0 - 1)(\sqrt{1+2q_0 z} - 1)]\right\}$$

$$-2.5\log(1+z) + k(z) + 42.39 - 5\log h_{100}$$

Here, $k(z) = -2.5\log\left[\frac{L(\nu_0(1+z))}{L(\nu_0)}\right]$ is called the K-correction. This allows for the relevant correction so that absolute magnitude corresponds to zero redshift, i.e. the absolute magnitude is measured in rest frame.

Sample Completeness: V/V_m Method

The V/V_m test was first used by Schmidt (1968) to study the space distribution of a complete sample of radio quasars from 3cR catalogue.

Let F_m be the limiting flux of a survey. We define two columns, $V(r) = \frac{4}{3}\pi r^3$ and $V_m = \frac{4\pi}{3}{r_m}^3$ where r is the radial distance to a quasar and $r_m = \left(\frac{L}{4\pi F_m}\right)^{1/2}$ is the limiting distance at which flux of a quasar with luminosity L reduces to F_m. For $r > r_m$, the quasars do not belong to the sample under consideration. If the quasars are expected to be uniformly distributed, then V/V_m are uniformly distributed in [0, 1]. Then,

$$< V/V_m >= 0.5 \text{ and } \sigma(V/V_m) = \frac{1}{\sqrt{12N}}$$

where N is the number of objects in the sample. Also this is true for galaxies having distances $<< c/H_0$. When the parent population has a very large density which increases outwards, then $< V/V_m >> 1/2$.

Luminosity Function of Quasars

This is the distribution of quasars as a function of their luminosity. It is a segmented power law in the low and high luminosity regimes, fitted by various authors (Boyle et al. 1991; Warren et al. 1987). The luminosity function increases with redshift up to $z \simeq 2$ after which it slows down (Fig. 1.34) and there is a decline towards higher redshifts. This feature of luminosity evolution is tried to be explained by various accretion models on a supermassive black hole (Cavaliere and Padovani 1989; Small and Blandford 1992; Haehnelt and Rees 1993).

Active Galactic Nuclei

AGN are the nuclei of some violent activities with high degree of rapid variation emitting infrared, radio, ultraviolet and X-ray radiation of the electromagnetic spectrum. Models of active galaxies primarily rest on the existence of a supermassive blackhole at the centre of the galaxy and materials which accrete onto the blackhole release large amount of gravitational energy in the high energy zone. There are several types of AGN, e.g. seyfert 1 and 2, quasars and blazars. Most scientists believe that they are practically same viewed from different directions. For example, Seyfert galaxies have low redshift and have galactic envelope and quasars are star like and have no such envelope. Seyferts have low redshifts whereas quasars have high redshifts (viz. $z > 1$). So if a Seyfert is observed at high redshift its galactic envelope might be very faint to be observed and designated as a quasar. Seyfert 1 has broad emission lines while Seyfert 2 has narrow emission lines. Blazers are very bright in the radio band which results looking directly a jet which is emitting nonthermal radiation. So blazers are not optically luminous.

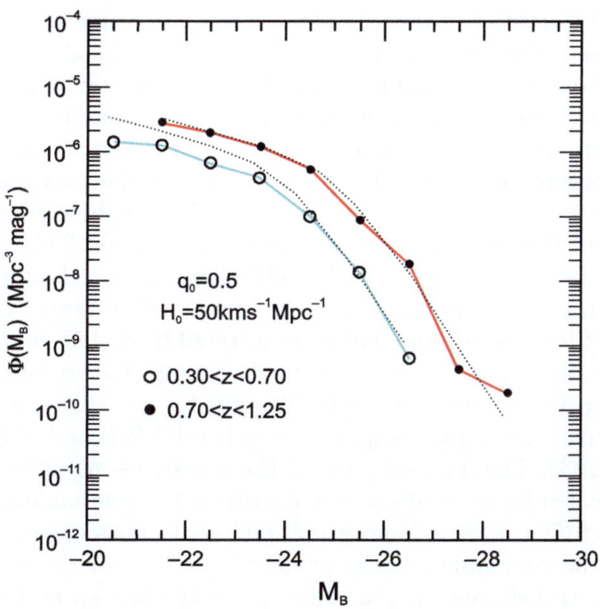

Figure 1.34 Luminosity functions of quasars (courtesy: Boyle et al. 1991)

1.14 Pulsars

Pulsars are highly magnetized rotating neutron stars emitting electromagnetic radiation, similar to that of a light house. The radiated beam is observed when it is pointing towards earth. The first pulsar was discovered in

1967 by Jocelyn Bell Barnell and Hewish while they were observing interplanetary scintillation. Subsequently Pacini and Gold established the theory that the pulsars are rotating neutron stars and not white dwarfs and the radiation was due to rotation and not due to oscillations with the discovery of very short period pulsars (33 ms) in Crab nebula. It was also found that there is an increase in the spinning period (P) of the pulsar due to the braking of the dipole magnetic field strength. Later a separate class of pulsars were discovered whose spinning periods are of the order of few milliseconds. They are called "millisecond pulsars" (MSP). In most cases they are found in "binary system" and they are comparatively older (10^8–10^9 years) compared to ordinary radio pulsars (10^7 years). The formation of a pulsar happens when the core of a massive star contracts as a result of supernova explosion. Since the core radius is very small, there is a sharp decrease of moment of inertia. As a result due to the conservation of angular momentum, it is accompanied with a very high rotational motion. This rapidly rotating system with a strong dipolar magnetic field acts as a very energetic source of electromagnetic radiation in a small cone. The magnetic axis is not aligned with the rotation axis and this miss alignment causes the "pulsed" nature of the radiation. The rotation gradually slows down as electromagnetic power is emitted and thus the birth of an ordinary radio pulsar leads to a death when the radiation becomes negligibly small. When a neutron star is in a binary system then the interaction between them occurs through transfer of mass from the least massive companion (m_e) to the primary (m_p). Binary and millisecond pulsars are generally speculated to be the precursors of X-ray binaries. There are two groups of X-ray binaries—high and low depending on the mass of the companion, $m_c > 10M_\odot$ and $m_c < 2M_\odot$, respectively. High mass X-ray binaries with $m_c > 10M_\odot$ are giants and supergiant stars. In this system the companion star has an extended envelope which in most cases fills its "Roche lobe" and matter is accreted from this companion to the primary. The high mass system is eccentric having short orbital period, e.g. 2.1 days (Cen X - 3) and 35 days (1223 - 62) for the strong X-ray sources. For less massive system the companion is generally WD and of later spectral type (A, F or G). The neutron stars of these systems are surrounded by an accretion disc, fed by the outflow of matter from the companion. The orbit is more or less circular having long orbital period. In both cases, mass transfer occurs from the companion to the primary. As a result the orbital angular momentum is transferred to spin of the primary through accretion and the spin period of the primary decreases to the order of milliseconds. Sometimes the orbit might collapse as a consequence of gravitational radiation losses. It is an NS–WD pair, the WD will be disrupted and a part of it mass will be accreted by the NS. As a result a rapidly rotating solitary pulsar will emerge (van den Heuvel and Bonsema 1984) (Fig. 1.35). The present pulsar catalog can be found at `www.atnf.csiro.an/people/pulsar/psrcat/`.

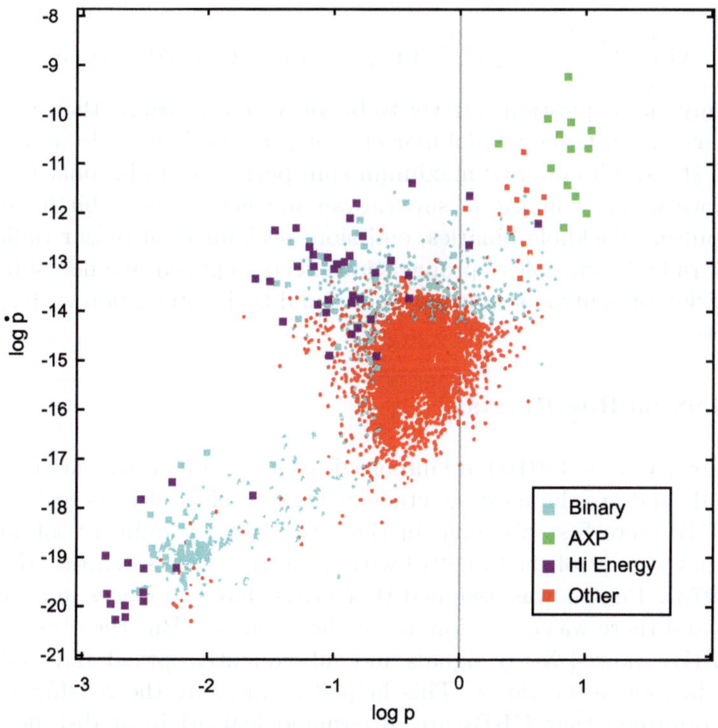

Figure 1.35 P–$\dot{P}$ diagram of ATNF pulsars

If ω and M is the angular velocity and mass of a pulsar then $\omega^2 r = GM/r^2$. So the spin period $P = \frac{2\pi}{w} = \left(\frac{3\pi}{G\rho}\right)^{1/2}$ where $M = \frac{4}{3}\pi r^3 \rho$ and ρ is the mean density. The characteristic age of a pulsar is $\tau_c = \frac{1}{2}\frac{P}{\dot{P}}$. The mass function of a binary pulsar is defined as $f(m_P, m_c, i) = \frac{4\pi^2}{G}\frac{a^3 \sin^3 i}{P_{orb}} = \frac{{m_c}^3 \sin i}{(m_P+m_c)^2}$ where a is the length of semi major axis of the elliptic orbit, P_{orb} is the orbital period, i is the inclination of the plane of the orbit to the observer. From spectroscopic Doppler shift measurement i cannot be measured but instead the analysis gives a value of $\sin i$. So if P_{orb} is measured then $f(m_P, m_c, i)$ is found easily from which an estimate of the mass range of the companion can be found for a value of m_P. The magnetic field strength of a pulsar is related to $P, \dot{P}$ as,

$$B = \left(\frac{3c^3}{8\pi^2}\frac{I}{R^6 \sin^2 \alpha} P\dot{P}\right)^{1/2}$$

where α is the angle between magnetic axis and rotation axis, I is the moment of inertia of the NS, R is the size of the NS and c is the speed of light. For a typical NS, $I = 10^{45}\,\mathrm{g\,cm^2}, R = 10\,\mathrm{km}$ and assuming $\alpha = 90°$,

$B = 10^{12}\,\mathrm{G}\left(\frac{\dot{P}}{10^{-15}}\right)^{1/2}\left(\frac{P}{s}\right)^{1/2}$. In spite of the great advancement in pulsar astronomy many questions are yet to be answered regarding their population characteristics in Galaxy, Globular clusters, magnetic field decay in solitary neutron stars, minimum or maximum spin periods of radio pulsars, correlation between core collapse in supernovae and neutron star birth rate properties, pulsar–blackhole binaries, emission mechanism of pulsar radio beam, shape of radio beam, role of propagation effects in pulsar magnetospheres and composition of neutron star atmospheres and their interaction with magnetic fields, etc.

1.15 Gamma Ray Bursts

Gamma ray bursts (GRBs) are intense flashes of gamma rays, lasting for tens of seconds and are the most spectacular astronomical objects to be explored yet. GRBs were first observed in the late 1960s by military satellites and the results were published in 1973 with data from Vela satellites (Klebesadel et al. 1973). First it was assumed that GRBs have no traces in other wavelengths and there was confusion about their sources. But the observations of some GRBs showed X-ray signals and subsequently optical and radio wavelengths in their after glows. This helped to measure the redshift distances which confirmed that GRBs are of cosmological origin at distances of the order of millions of light years away, similar to distant galaxies and quasars. At this distance the energy varies in the range 10^{15}–10^{54} ergs, larger than that of any other astronomical source. The major advancement in the exploration of GRB phenomena occurs after the launch of Compton Gamma-Ray Observatory (CGRO) and the Burst and Transient Experiment (BATSE) onboard CGRO. With that attempt almost 3,000 GRBs are known today. They are isotropically distributed with no dipole or quadruple components. This suggests a cosmological distribution. The spectra of GRBs are nonthermal with a power law pattern, $\sim\in^{-\alpha}$ where α is 1 at low energy and increases to 2–3 above photon energy ∼0.1–1 MeV. Depending on the duration GRBs can be classified into short (≤ 2 s) and long bursts (>2 s) though in recent multivariate approach an intermediate class (Chattopadhyay et al. 2007) has been suggested. Several models of GRBs have been suggested on the basis of its high energy. Primarily the high energy and casuality suggest that the source is confined in a region whose size is of the order of kilometres in a time scale of the order of seconds. So a fire ball of $(e^{\pm}, \gamma)$ forms, which expands relativistically.

The difficulty with this is that such a fire ball immediately converts the energy into kinetic energy of the baryons and produces a quasi-thermal spectrum instead of increasing the luminosity and this occurs within a time scale of the order of milliseconds. So a "fire ball shock model" is introduced which explains that as soon as the fire ball becomes transparent, a shockwave will occur in the outflow which reconverts the kinetic energy of expansion to non-thermal particle and radiation energy. In Fig. 1.35 the observed light curves of the afterglow of GRB 9702228 at various wavelengths have been compared with the predicted fire ball shock model. The GRB radiation, starting with γ-ray range during the burst subsequently radiates into X-rays, UV, optical, IR and radio in its afterglow. Though the mechanism of the bursts is somewhat explained the progenitors of GRB are not well identified so far. The current theories existing so far conjectured that a very small fraction of stars $\sim 10^{-6}$ give rise to GRBs. One category includes massive stars whose core collapses and it releases huge amount of energy when it merges with a companion, termed hypernova or collapser and another category is a neutron star–neutron star (NS) or a neutron star–black hole (BH) binary system which loses orbital angular momentum, releases gravitational energy during the merger. Both these systems finally end in a BH and the debris form a torus surrounding the BH. The accretion of debris material onto the BH undergoes a sudden release of gravitational energy comparable to the observed ones. The current effort is to devise a proper understanding of the progenitor scenarios, i.e. how the progenitors along with its environment can give rise to the observable bursts and afterglow characteristics.

Appendix

Transformation matrices T for various co-ordinate systems are as follows:

From (α, δ) 1950.0 to (α, δ) 2000.0.

$$T = \begin{pmatrix} 0.9999257453 & -0.0111761178 & -0.0048578157 \\ 0.0111761178 & 0.9999375449 & -0.0000271491 \\ 0.0048578157 & -0.0000271444 & 0.9999882004 \end{pmatrix}$$

From (α, δ) 2000.0 to galactic (l, b)

$$T = \begin{pmatrix} -0.0548808010 & -0.8734368042 & -0.4838349376 \\ 0.4941079543 & -0.4448322550 & 0.7469816560 \\ -0.867666568 & -0.1980717391 & 0.4559848231 \end{pmatrix}$$

From (α, δ) 2000.0 to supergalactic (l, b)

$$T = \begin{pmatrix} 0.3751891698 & 0.3408758302 & 0.8619957978 \\ -0.8982988298 & -0.957026824 & 0.4288358766 \\ 0.2286750954 & -0.9352243929 & 0.2703017493 \end{pmatrix}$$

From galactic (l, b) to (α, δ) 2000.0

$$T = \begin{pmatrix} -0.0548808010 & 0.49410795443 & -0.8676666568 \\ -0.8734368042 & -0.4448322550 & -0.1980717391 \\ -0.4838349376 & 0.7469816560 & 0.4559848231 \end{pmatrix}$$

From galactic (l, b) to supergalactic (l, b)

$$T = \begin{pmatrix} -0.7353878609 & 0.6776464374 & 0.0000000002 \\ -0.0745961752 & -0.0809524239 & 0.9939225904 \\ 0.6735281025 & 0.7309186075 & 0.1100812618 \end{pmatrix}$$

From supergalactic (l, b) to (α, δ) 2000.0

$$T = \begin{pmatrix} 0.3751891698 & -0.8982988298 & 0.2286750954 \\ 0.3408758302 & -0.0957026824 & -0.9352243929 \\ 0.8619957978 & 0.4288358766 & 0.2703017493 \end{pmatrix}$$

From supergalactic (l, b) to galactic (l, b)

$$T = \begin{pmatrix} -0.7353878609 & -0.0745961752 & 0.6735281025 \\ 0.6776464374 & -0.0809524239 & 0.7309186075 \\ 0.0000000002 & 0.9939225904 & 0.1100812618 \end{pmatrix}$$

Exercise

1. Find the mass of a galaxy if its velocity is known. [Hint. use Virial Theorem 2T + V = 0]
2. What is the wavelength chosen by a radio telescope and why?
3. What is the wavelength of emission line of HeII from a transition $n_i = 5$ to $n_f = 2$? [Hint. $\frac{1}{n_f^2} - \frac{1}{n_i^2} = \frac{91.16\,\text{nm}}{Z^2\lambda}$, Z being the atomic number of He]
4. A galaxy moves towards the observer with a velocity of 1,500 $\text{km}\,\text{s}^{-1}$. What will be the shift of first Balmer line (H_α) in the spectrum? Given $H_\alpha = 6,563\,\text{Å}$. [Hint. $\frac{\Delta\lambda}{\lambda} = \frac{v_r}{c}$]
5. Show that Lyman series limit lies between 911.25 and 1,215 Å. Given I.P of hydrogen atom is 13.6 eV. [Hint. $\frac{1}{\lambda(cm)} = 109740(\frac{1}{n_f^2} - \frac{1}{n_i^2})$]
6. Compute the fraction of calcium atoms in the first ionized state at T = 15,000 K, $P_e = 400\,\text{dynes}\,\text{cm}^{-2}$. For calcium $log\frac{2u_1}{u_0} = 0.18$.
7. Derive Wien's displacement law from Planck's law. How is it different from the case $h\nu = kT$—why?
8. What is the ratio of energies emitted by two stars of same size having surface temperatures at T = 15,000 K and 5,000 K, respectively?
9. How long a star can shine if its mass and luminosity are M = 100 $M_\odot$ and L = $10^6 L_\odot$, if initially it is composed of pure hydrogen? Given that the efficiency to convert H to He is 0.7 %. [Hint. Time = Total Energy/Luminosity]
10. Consider a cloud of molecular hydrogen with $T = 10\,\text{K}, n = 10^6\,\text{cm}^{-3}$. What is the minimum mass for gravitational collapse and compute the time scale for it. Given $m_H = 1.67 \times 10^{-24}\,\text{g}\,\text{cm}^{-3}$. [Hint. $M_J = 10^{23}(\frac{T}{\mu})^{\frac{3}{2}}\rho_0^{-\frac{1}{2}}$ g]
11. Calculate the time taken by light from a star with parallax p = $0''.75$ to reach Earth.

12. The apparent magnitude of a star is +3.5 and its parallax is $0''.025$. What are its absolute magnitude and luminosity?
13. Show that a massive has a much shorter life than Sun.
14. Using the principle of hydrostatic equilibrium find the central pressure of a star in terms of its mass and radius.
15. Compute the mass of an elliptical galaxy having a typical velocity $350\,\mathrm{km\,s^{-1}}$ and radius 10 kpc.

References

Abhyankar, K.D. 2001. *Astrophysics Stars and Galaxies* (Universities Press).

Arimoto, N., and Y. Yoshii. 1987. *Astronomy & Astrophysics* 173:23.

Ashman, K.M., and S.E. Zepf. 1992. *The Astrophysical Journal* 384:50.

Bastian, N., K.R. Covey, and M.R. Meyer. 2010. *Annual Review Astronomy & Astrophysics* 48:339.

Basu, B., T. Chattopadhyay, and S. Biswas. 2010. *An Introduction to Astrophysics* (Second Edition) (Prentice Hall of India).

Bekki, K., D.A. Forbes, M.A. Beasley, and W.J. Couch. 2002. *Monthly Notices of Royal Astronomical Society* 335:1176.

Bohr, N. 1913. *Philosophical Magazine* 26(151):1

Boyle, B.J., et al. 1991. *The Space distribution of quasars.* ASP Conference Series No. 21, ed. D. Crampton, 191. Provo: Brigham Young.

Bucerius, H. 1934. *Astronomische Nachrichten* 259:369.

Canterna, R. 1976. *Astronomical Journal* 81:228.

Carlberg, R.G. 1984. *The Astrophysical Journal* 286:40.

Cavaliere, A., and P. Padovani. 1989. *The Astrophysical Journal* 333:L33.

Chattopadhyay, A.K., T. Chattopadhyay, E. Davoust, S. Mondal, and M. Sharina. 2009. *The Astrophysical Journal* 705:1533.

Chattopadhyay, T., et al. 2007. *The Astrophysical Journal* 667:1017.

Côté, P., et al. 2001. *The Astrophysical Journal* 559:828.

Faber, S.M., and R.E. Jackson. 1976. *The Astrophysical Journal* 204:668.

Forbes, D.A., J.P. Brodie, and C.J. Grillmair. 1997. *Astronomical Journal* 113:1652.

Geisler, D. 1990. *Publication Astronomical Society of Pacific* 102:344.

Goldstein, H., C.P. Poole, and J.L. Safko. 2001. *Classical Mechanics* (Addison-Wesley Professional).

Haehnelt, M.G., and M.J. Rees. 1993. *Monthly Notices of Royal Astronomical Society* 263:168.

Hayashi, C. 1961. *Publication Astronomical Society of Japan* 13:450.

Heney, L.G., R. Le Leveier, and R.D. Levee. 1955. *Publication Astronomical Society of Pacific* 67:396.

Hoyle, F. 1953. *The Astrophysical Journal* 118:513.

Hunter, D.A., et al. 2003. *Astronomical Journal* 126:1836.

Jeans, J.H. 1902. *Philosophical Transactions of the Royal Society A* 199:1.

Johnson, H.L., and W.W. Morgan. 1953. *The Astrophysical Journal* 117:313.

Klebesadel, R.W., I.B. Strong, and R.A. Olson. 1973. *The Astrophysical Journal* 182:L85.

Kroupa, P., C.A. Tout, and G. Gilmore. 1993. *Monthly Notices of Royal Astronomical Society* 262:545.

Kroupa, P. 2004. *New Astronomy* 48:47.

Lada, C.J., and E.A. Lada. 2003. *Annual Review Astronomy & Astrophysics* 41:57.

Larson, R.B. 1975. *Monthly Notices of Royal Astronomical Society* 173:671.

Lohmann, W. 1953. *Zeitschrift für Astrophysik* 33:159.

Oppenheimer, J.R., and G.M. Volkoff. 1939. *Physical Review* 55:374

Peebles, P.J.E. 1969. *The Astrophysical Journal* 155:393.

Planck, M. 1901. *Annalen der Physik* 4:553.

Rutherford, E. 1911. *Philosophical Magazine* 21:669.

Saha, M.N. 1920. *Philosophical Magazine* 40:472.

Salpeter, E.E. 1955. *The Astrophysical Journal* 121:161.

Schmidt, M. 1956. *Bulletin Astronomical Institute of Netherlands* 13:15.

Schmidt, M. 1965. *Stars and stellar systems, galactic structure*, vol. 5, eds. A. Blaauw and M. Schmidt, 513. Chicago: University of Chicago Press.

Schmidt, M. 1968. *The Astrophysical Journal* 151:393.

Small, T.A., and R.D. Blandford. 1992. *Monthly Notices of Royal Astronomical Society* 259:725.

Spitzer, L. Jr. 1978. *Physical Processes in the Interstellar medium* (John wiley and Sons, New York).

van den Heuvel, E.P.J., and P. Bonsema. 1984. *Astronomy & Astrophysics* 139:L16.

Warren, S.J., P.C. Hewett, M.J. Irwin, R.G. McMahon, and M.T. Bridgeland. 1987. *Nature* 325:131.

Weidner, C., and P. Kroupa. 2005. *The Astrophysical Journal* 625:754.

Zepf, S.E., et al. 2000. *Astronomical Journal* 120:2928.

Chapter - 2

Introduction to Statistics

2.1 Introduction

Statistical tools and techniques play a major role in many areas of science like biology, economics, physics etc. In particular, a systematic development in statistical methodology for biological problems started long back and constituted a new area of biostatistics. Compared to biology, the rate of progress in the development of statistical techniques for physical science is very slow. The reason may be due to lack of interaction between Physicists and Statisticians. In areas like Astronomy and Astrophysics the application of sophisticated statistical analysis is a comparatively recent phenomenon. Astronomy is the science to study the different features of planets, stars, galaxies and the universe as a whole. Astrophysicists try to model observed astronomical properties by using laws governing physical process. The problem is to make inference for the underlying properties on the basis of observations related to a few external characteristics. During the past two decades the inter-disciplinary field of astrostatistics has newly emerged in order to study important astrophysical issues through appropriate statistical analysis. With the advancement of technology, at present several data archives have been prepared which contain tera bytes of astronomical data. Statistical analysis of these high dimensional large data sets is a challenging problem and a good solution can be obtained only through interaction among Astronomers, Statisticians and Computer Scientists.

The subject statistics (in singular sense) is concerned with the collection of data and with their analysis and interpretation. Data may be of two types, viz. real life and simulated. For situation where it is difficult or even impossible to collect direct observations, one can take help of simulated data generated from some conceived probability model. One can take help of simulation to study the formation of stars and galaxies.

In order to prepare the proper data set (simulated or real life), the final task is to identify the significant parameters (in statistics these are called variables) which are actually responsible for variation in the data. For astronomical objects, the commonly used variables (parameters) are luminosity, magnitude, mass, redshift, etc.

A.K. Chattopadhyay, T. Chattopadhyay, *Statistical Methods for Astronomical Data Analysis*, Springer Series in Astrostatistics 3,
DOI 10.1007/978-1-4939-1507-1_2

2.2 Variable

2.2.1 Discrete-Continuous

Variables are classified as discrete or continuous according to the nature of values they take. Theoretically a variable is said to be discrete if it takes values from a finite and usually small domain (set of values it takes). The values are usually, by nature, integers. In particular, a variable taking only two possible values can be defined as a binary variable. For example, number of misprints per page in a book, the presence or absence of bar in a galaxy, etc. are discrete variables.

Continuous variable can take infinite number of values from a large domain. For example, day temperature, luminosity, mass, etc. are continuous variables.

Practically variables are classified as discrete or continuous according to the number of values they can take. Due to precision limitation of measuring instruments actual measurement of a variable always occurs in a discrete manner. Variables which can take lots of values are usually identified as continuous whereas variables taking fewer values are identified as discrete. The actual distinction should be done on the basis of the nature of the variables which may not be known, in particular, for astronomical variables.

2.2.2 Qualitative–Quantitative

Variables for which the different states are not primarily determined by numbers are called qualitative whereas variables that do have numerical distance between any two values are called quantitative. For example, favourite type of music (classical, semi-classical, modern, etc.), colours of astronomical objects, etc. are qualitative and functional life of a computer CPU, distance of an astronomical object, etc. are quantitative.

For qualitative variables, there is a separate branch of statistical analysis known as categorical data analysis although this includes also ordered categories (ordinal variables).

Most of the statistical analysis are meant for quantitative variables. Among them techniques like correlation and regression, parametric and non-parametric inference, multivariate analysis, Monte Carlo simulation, Bayesian analysis, time series analysis, directional data analysis, etc. are very much important for the proper analysis of astronomical data.

2.2.3 Cause and Effects

The distinction between cause (explanatory or independent variable) and effect (response or dependent variable) is very much important for different types of regression analysis. For example, in order to predict wheat yield (effect), the independent variable can be chosen as minimum and maximum temperature, daily rain, evaporation from basin, sun hours, etc. But there may be situation where it is difficult to distinguish between the cause and effects as their inherent natures are not known. This is particularly important for astronomical variables (parameters). For early type galaxies there is linear relation among three parameters, viz. effective radius (r_e), central velocity dispersion (σ) and surface brightness average over effective radius ($< \mu_e >$) given by

$$\log r_e = a \log \sigma + b < \mu_e > +c$$

known as the fundamental plane. But here the dependent and independent variables are not known and astrophysicists try to predict any one of them on the basis of the other two. Since this is not possible by ordinary least square regression, some alternative methods have been developed. For two variables, methods like bisector regression or orthogonal regression may be used.

2.3 Frequency Distribution

In statistics frequency distribution is a tabular presentation of the original raw data in order to study the concentration of the values over different intervals of the total range. We initially subdivide the total range of the values into a number of intervals (this number should be properly chosen) and then compute the frequency (i.e. the number of observations belonging to) of each class. Such a representation will help us to study the following inherent features of the data.

2.3.1 Central Tendency

This feature is the most widely used among all other features of a frequency distribution. Central tendency relates to the way in which quantitative data tend to cluster around a central value. This feature always helps us to represent a total data by a single value as it is supposed to be a representative of the entire data set. So, a proper measure of central tendency should be so chosen that it reflect the average nature of the values. Some of the common measures of central tendency are mean, medium and mode. Among them arithmetic mean is the most popular as for most data it is a best representative. But when there are some unusual objects in the data sets, whose values are widely different from most of the values (known as outliers), median, being the middle most value, should be preferred over mean as the mean is very much affected by the outlier. For open ended data sets also median is a possible measure of central tendency. There are also a few situations where mode is the only possible measure of central tendency. If somebody is interested to know average preference of people for a car size, then mode is the only option as car sizes are of some fixed values and mode is that size which is preferred by the maximum number of persons. For astronomical data usually mean and median are the two possible measures of central tendency.

2.3.2 Dispersion

Statistical dispersion determines the variability or spread in a variable (parameter). This is also known as variation. Dispersion is a very important concept as it indicates the amount of scatter present in a data set. Generally a good data set is expected to have less scatter and more central tendency. The most common and useful measure of dispersion is the variance (or standard deviation $= +\sqrt{variance}$). Some other possible measures are the mean deviation and range. General form of the measures variance and mean deviation are given by

$$\text{Mean square deviation about A} = \frac{1}{n}\sum_{i=1}^{n}(x_i - A)^2 \tag{2.1}$$

$$\text{and Mean deviation about A} = \frac{1}{n}\sum_{i=1}^{n}|x_i - A| \tag{2.2}$$

respectively, where $(x_1, x_2, \ldots, x_n)$ is a set of n values of the variable (parameter) and A is any measure of central tendency. In particular when A is equal to the arithmetic mean $\bar{x} = \frac{1}{n}\sum_{i=1}^{n} x_i$, the Mean square deviation reduces to variance and the mean deviation about A reduces to mean deviation about mean. Standard deviation of a sample estimate corresponding to a population parameter is used to measure the amount of error involved in that estimate as a good estimator.

2.3.3 Skewness

Skewness describes the degree of departure of a frequency distribution from symmetry. A distribution which is not symmetrical is called asymmetrical or skewed. Positive skewness implies that the longer tail of the distribution is towards the higher values and negative skewness compounds to the situation where the longer tail is towards the lower values of the parameter (variable) under study. For a unimodal symmetric distribution, the values of mean, median and mode are equal. If the distribution is positively skewed mean > median > mode and for a negatively skewed distribution mean < median < mode. Hence one can consider the following coefficients as possible values of skewness.

$$sk1 = \frac{(mean - mode)}{sd} \tag{2.3}$$

$$sk2 = \frac{3(mean - median)}{sd} \tag{2.4}$$

$$sk3 = \frac{(Q_3 - median) - (median - Q_1)}{Q_3 - Q_1} \tag{2.5}$$

where Q_1 and Q_3 are the first and third quartiles of the distribution.

Two other possible measures are
$\beta_1 = \mu_3^2/\mu_2^3$ and

$\gamma_1 = \sqrt(\beta_1)$
where μ_2, μ_3 and μ_4 are second, third and fourth population central moments.

All odd ordered central moments are zero for a symmetric distribution, positive for a positively skewed distribution and negative for a negatively skewed distribution. It may also be noted that variance is the second order central moment.

2.3.4 Kurtosis

Another method of describing a frequency distribution is to specify its degree of peakedness or kurtosis. This is measured by

$\beta_2 = \mu_4/\mu_2^2$ or

$\gamma_2 = \beta_2 - 3$

2.4 Exploratory Data Analysis

Distinguished statistician J.W. Tukey described exploratory data analysis (EDA) as the future of data analysis. EDA is an approach for data analysis that uses a number of techniques (mostly graphical) to explain the underlying structure of a data set, extract important variables, detect outliers and many other features. Unlike confirmatory analysis, EDA does not consider any assumption regarding the underlying model but allows the data to reveal its underlying structure and model. Under EDA, the graphical techniques include either plotting of raw data or plotting of simple statistics. Some of these plots can be described as follows.

2.4.1 Histogram

Histogram (Fig. 2.1) is a graphical display of tabulated frequencies. It is a graphical version of a table which shows the proportion of cases fall into each of several specified categories. The categories are usually specified as non overlapping intervals of some variable. The bars must be adjacent. The matter of selecting the number of categories (or bins) to be used in the histogram is not trivial. The choice of the width of a bin is also subjective. Although a histogram can be very useful for examining the distribution of a variable, the graph can differ significantly depending on the number of bins used. As an alternative one may use nonparametric density estimation which is an attempt to estimate the probability density function of a variable based on the sample. It can also be considered as a way of averaging and smoothing the histogram. According to this method, the kernel (weight function) density estimator, corresponding to a random sample $x_1, x_2, \ldots, x_n$ with a density function f is given by

$$f(x,h) = \frac{1}{nh}\sum_{i=1}^{n} k\left(\frac{x - X_i}{h}\right) \tag{2.6}$$

with kernel k and bandwidth h. Both the kernel function and bandwidth must be specified by the methodologist. For convenient mathematical properties, the standard normal density is often is used as the kernel density function k. The bandwidth of the kernel is a free parameter and exhibits a strong influence on the resulting estimate. Several attempts have been made to find the optimum value of the bandwidth.

2.4.2 Box Plot

A box plot (Fig. 2.2), also known as box-and-whisker plot is a method of displaying data invented by John Tukey. Under this method, draw a box with ends at the first (Q1) and third quartiles (Q3). The width of the box may be arbitrary. Draw the median as a horizontal line in the box. Then extend the whiskers to the furthest points that are not outliers. These two points denoted by $L1$ and $U1$ (inner fence in Fig. 2.2) can be calculated by using the relations $L1 = Q_1 - 1.5Q$ and $U2 = Q_3 + 1.5Q$, where $Q = (Q_3 - Q_1) = \text{IQR}$ (in Fig. 2.2). Any point below $L1$ and above $U1$ may be treated as an outlier. Such a plot also helps to know about the skewness of the distribution by depending on the position of the median line. Under some modification, one

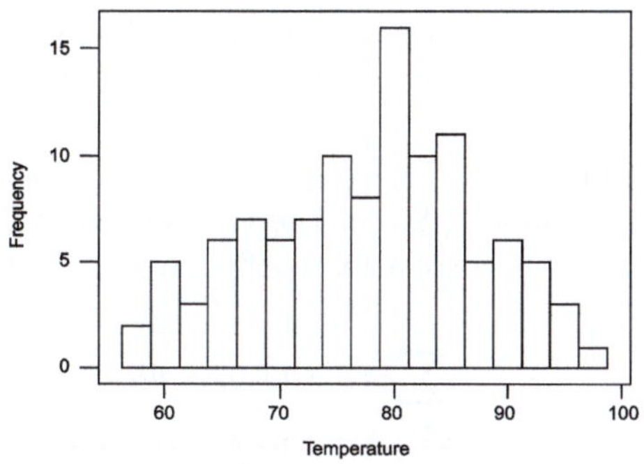

Figure 2.1 Histogram of temperature values corresponding to 111 selected points for an environmental pollution survey

may also further extend the whiskers to the points $L2$ and $U2$ (outer fence in Fig. 2.2) given by $L2 = Q_1 - 3Q$ and $U2 = Q_3 + 3Q$. Points in between $L1$ and $L2$ or $U1$ and $U2$ may be considered as suspected outliers and points above $U2$ and below $L2$ may be considered as confirmed outliers.

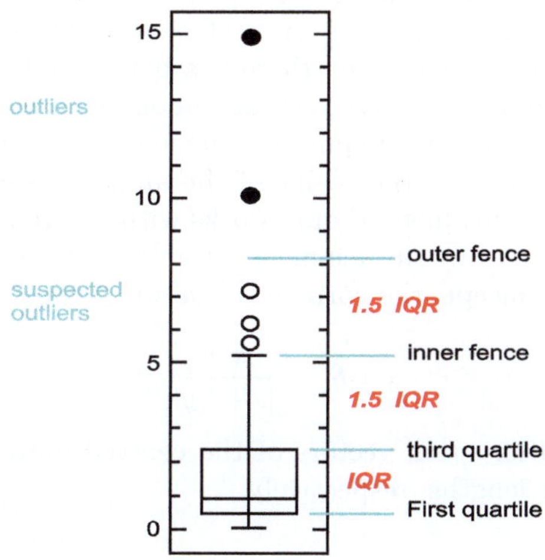

Figure 2.2 Upper part of a Box plot

2.5 Correlation

Correlation is a statistical technique which can show whether and how strongly pairs of variables are related. For example, height and weight of a group of persons, central velocity dispersion and surface brightness average over effective radius of galaxies, etc. are correlated variables. Correlations works for data in which numbers are meaningful. It cannot be used for categorical data such as colour and gender. By correlation we usually mean linear correlation. So any such measure does not reflect any non linear relationship.

Given n pairs values $(x_i y_i)$ $i = 1, 2, \ldots n$, corresponding to two jointly distributed random variables (X, Y), the correlation coefficient (also called Pearson's product moment correlation coefficient) is given by

$$r = \frac{\sum_{i=1}^{n}(x_i - \bar{x})(y_i - \bar{y})}{\sqrt{\sum_{i=1}^{n}(x_i - \bar{x})^2 \sum_{i=1}^{n}(y_i - \bar{y})^2}} \tag{2.7}$$

It ranges from -1.0 and $+1.0$. The closer is r to $+1$ or -1, the more closely the two variables are related. A zero value of r indicated that there is no linear correlation. The square of the coefficient (i.e. r^2) is equal to the percent of the variation of one variable that is related to the variation in the other. An r-value of 0.6 means 36 % of the variation is related.

While working with correlations one must be careful about spurious correlation. Over the last few years sales of mobile phones and half cooked foods have both risen strongly and there is a high correlation between them but one cannot assume that buying mobile phones causes people to buy half cooked foods. These are called spurious correlations. The correlation coefficient can also be viewed as the cosine of the angle between the two vectors of observations. But this method only works with centred data, i.e. the data which have been shifted by the sample mean so as to have an average of zero. According to this concept, the correlation between two variables x and y is given by

$$\cos\theta = \frac{x.y}{||x||\ ||y||}$$

x and y are the observation vectors of the centred data and $||x||$ and $||y||$ correspond to their lengths, respectively.

2.5.1 Scatter Plot

The simplest mode of a diagrammatic representation of bivariate data is the use of scatter plot (or XY plot). Taking two perpendicular axes of co-ordinates, one for x and other for y, each pair of values is plotted as a point on graph paper. The whole set of points taken together constitutes the scatter diagram. This method is not very suitable when the number of individuals is very large. If it is found that as one variable increases, the other also increases on the average, the two variables are said to be positively correlated. On the other hand, as one variable increases, the other may decrease on the average, then the two variables are said to be negatively correlated. A third situation corresponds to the case where if one variable increases or decreases, the other remains constant on the average. Such a situation may be interpreted as zero correlation (or no linear correlation).

2.6 Regression

Regression analysis is the statistical methodology for predicting values of one or more response (dependent) variables from a single or collection of predictor (independent) variables. It can also be needed for assessing the effects of the predictor variables on the responses. The name "regression" has been introduced by F. Galton. While correlation investigates the interrelation between pairs of variables, regression corresponds to the effect of the independent variable on the dependent variable. In the simplest situation with one dependent (y) and another independent (x) variable, the relationship between y and x has to be expressed in a mathematical form. If it is possible to assume a linear relationship, the approximate relation may be represented by

$$y = a + bx \tag{2.8}$$

where the constants a and b have to be estimated from the observed data. Given the n paired values of x and y denoted by (x_i, y_i) $i = 1, \ldots, n$, the above line gives an estimate of y_i as

$$Y_i = a + bx_i \text{ (say)} \tag{2.9}$$

The difference $(y_i - Y_i)$ is the error of estimate for the ith pair. The values of a and b should be such that these errors of estimate are as small as possible. For this, for estimating a and b from the observed data, the least square method is used which consists in minimizing the sum of squares of the errors of estimation. Hence the problem is to choose a and b in such a way as to minimize

$$\begin{aligned} s^2 &= \sum_i (y_i - Y_i)^2 \\ &= \sum_i (y_i - a - bx_i)^2 \end{aligned} \tag{2.10}$$

One has to minimize s^2 with respect to a and b to get the estimates of a and b.

The estimated values of a and b can be obtained by solving the simultaneous normal equations

$$\frac{\partial s^2}{\partial a} = 0 \text{ and } \frac{\partial s^2}{\partial b} = 0$$

which lead to the equations.

$$\sum_i y_i = na + b\sum_i x_i \tag{2.11}$$

$$\text{and } \sum_i x_i y_i = a\sum_i x_i + b\sum_i {x_i}^2 \tag{2.12}$$

The final estimated values are given by

$$b = \frac{n\sum x_i y_i - (\sum x_i)(\sum y_i)}{n\sum_i {x_i}^2 - \left(\sum_i x_i\right)^2} \tag{2.13}$$

$$= r\frac{s_y}{s_x} \tag{2.14}$$

$$\text{and } a = \bar{y} - b\bar{x} \tag{2.15}$$

As a result, the desired predicted formula is given by

$$y = \bar{y} + r\frac{s_y}{s_x}(x - \bar{x}) \tag{2.16}$$

and is known as the regression equation of y on x. Regression, in general, is the problem of estimating a conditional expected value.

Linear regression is called linear because the relation of the response (dependent) to the explanatory variable(s) is assumed to be a linear function of some unknown constant parameters like a and b.

The model

$$y_i = a + bx_i + c{x_i}^2 \tag{2.17}$$

is again a linear regression model even though the graph is not a straight line.

In non linear regression the observational data are modelled by a function which is a non linear combination of model parameters and depends on one or more dependent variables. For example,

$$y_i = \frac{ax_i}{b + x_i} \tag{2.18}$$

is a non linear model because it cannot be expressed as a linear combination of a and b.

Multiple Regression

Suppose there are p parameters (variables) like absolute Blue magnitude (M_B), maximum rotation velocity (V_{max}), central density of the halo from the best-fit mass model (ρ_{00}), central surface brightness (μ_0), core radius of the halo (R_e), etc. for a number of spiral galaxies and we want to predict one of them on the basis of the others. Such a problem may be considered under multiple regression technique. Suppose there are p variables denoted by $x_1, x_2, \ldots, x_p$ where x_1 is dependent and of primary interest and we want to study how $x_2, x_3, \ldots, x_p$ jointly influence x_1. Here the idea is to build up a relationship between the dependent variable x_1 and the independent variables $x_2, \ldots, x_p$. Suppose then it is possible to establish an approximate relation of the form

$$x_1 = b_1 + b_2 x_2 + b_3 x_3 + \ldots + b_p x_p$$

Then by the least square method the estimates of the constants are given by

$$\begin{aligned} b_j &= -\frac{R_{1j}}{R_{11}} \frac{s_1}{s_j} \text{ for } j = 2, 3, \ldots p \\ \text{and } b_1 &= \bar{x_1} - \sum_{j=2}^{p} \frac{R_{1j}}{R_{11}} \frac{s_1}{s_j} \bar{x_j} \end{aligned}$$

where $\bar{x_j}$ and s_j are the mean and standard deviation of the jth variable x_j, R is the determinant of the symmetric correlation matrix

$$\begin{pmatrix} 1 & r_{12} & . & . & r_{1p} \\ r_{21} & 1 & . & . & r_{2p} \\ . & . & . & . & . \\ . & . & . & . & . \\ r_{p1} & r_{p2} & . & . & 1 \end{pmatrix}$$

and R_{ij} is the cofactor of r_{ij} in R.

Hence the multiple regression equation (also known as prediction equation) is given by

$$X_1 = \bar{x_1} - \frac{R_{12}}{r_{11}} \frac{s_1}{s_2} (x_2 - \bar{x_2}) \ldots - \frac{R_{1p}}{r_{11}} \frac{s_1}{s_p} (x_p - \bar{x_p})$$

The coefficient b_j is called the partial regression coefficient of x_1 on x_j for fixed $x_2, \ldots, x_{j-1}, x_{j+1}, \ldots, x_p$.

2.7 Multiple Correlation

In order to study the dependence of x_1 on a set of independent variables $x_2, \ldots, x_p$, we actually try to evaluate the influence of $x_2, \ldots, x_p$ on x_1. This may be computed by the simple correlation between x_1 and the predicted value of x_1 on the basis of $x_2 \ldots x_p$, denoted by $X_{1.23\ldots p}$, given by the multiple regression of x_1 on $x_2 \ldots x_p$. It is called the multiple correlation coefficient of x_1 on $x_2 \ldots x_p$ and is denoted by $r_{1.23\ldots p}$. By starting from the simple formula of Pearson's product moment correlation coefficient (2.7), this can be derived as

$$r_{1.23\ldots p} = \left(1 - \frac{R}{R_{11}}\right)^{1/2}$$

The only difference with the simple correlation coefficient is that $r_{1.2\ldots p}$ lies in between 0 and 1 instead of -1 to 1 as the covariance between x_1 and $X_{1.2\ldots p}$ is at the same time the variance of $X_{1.23\ldots p}$, which has to be non-negative. $r_{1.2\ldots p}$ may also be regarded as a measure of the efficiency of the multiple regression equation.

2.8 Random Variable

A random variable may be crudely defined as the value of a measurement associated with an experiment. For example, the number of sun spots per year may be treated as a random variable.

Definition: A random variable over a sample space is a function that maps every sample point (i.e. outcome of an experiment) to a real number.

Like ordinary variables, random variables also may be of two types, discrete and continuous. The most general way to express the nature of distribution a random variable is to compute the cumulative distribution function (c.d.f) defined as

$$F(x) = P[X \leq x]$$

Probability mass function (p.m.f) in case of discrete variables and probability density functions (p.d.f) in case of continuous variables are also used to describe the distribution of a random variables. Such functions give the probabilities associated with the different values or range of values of the random variable.

For a discrete variable (X), the pmf and cdf are defined as below. The pmf is defined as

$$f_X(x) = P[X = x]$$

where $f(x) \geq 0$ and $\sum_x f(x) = 1$, the sum being taken over all values of x having positive probabilities, known as the mass points of X. The cdf is defined as

$$F_X(x) = \sum_{X \leq x} f(x)$$

For a continuous random variable, the pdf $f(x)$ is defined as

$$\int_a^b f(x)dx = P[a < X < b]$$

Hence the function f itself has to be continuous or at least piece-wise continuous and the probability for X to take any particular value x must be zero.

Here also $f(x) \geq 0$ and $\int_R f(x)dx = 1$ where R is the domain of the possible values of x.

2.8.1 Some Important Discrete Distribution

Some discrete distributions which may be used as possible models for modelling astronomical parameters are as follows.

1. Binomial Distribution

   Suppose there are n objects in the sky and we introduce a variable

$$\begin{aligned} u_i &= 1 \text{ if the object is a planet} \\ &= 0 \text{ otherwise} \end{aligned}$$

   Then the variable $X = \sum_{i=1}^{n} u_i$ = number of objects which are planets out of n objects.

   Technically, if we denote the event of an objects being a planet by "success" and if it is known that the probability that an object will be a planet in the sky is p, then X denotes the number of successes out of n experiments (known as trials) with success probability for each trial being p (which is constant). If we further assume that the trials are independent, i.e. "the event that a particular object is planet" is independent of the event that "another object is a planet", then the distribution of the random variable X is given by the pmf

$$f(x) = \binom{n}{x} p^x q^{n-x} \quad x = 0, 1, \ldots n \tag{2.19}$$

Such a distribution is known as a binomial distribution. Here n is a positive number, $0 < p < 1$ and $q = 1 - p$. This distribution is determined by two unknown constants n and p, which are known as the parameters of the distribution.

Clearly, $f(x) \geq 0$ for all x and $\sum_{x=0}^{n} f(x) = (q+p)^n = 1$.

The first four moments of this distribution are given by

$$\begin{aligned} \mu_1{}' &= \text{mean} = np \\ \mu_2 &= \text{variance} = npq \\ \mu_3 &= npq(q-p) \\ \mu_4 &= 3n^2p^2q^2 + npq(1-6pq) \end{aligned} \tag{2.20}$$

This distribution is symmetrical, positively skew or negatively skew according as $p = 1/2, p < 1/2$ and $p > 1/2$.

2. Poisson Distribution

Consider a steady celestial source of constant luminosity that produces on the average a finite number of counts in a detector every second. The photons do not arrive with equal intervals. The average rate of arrival is controlled by a fixed probability of an event occurring in some fixed interval of time. There is some chance of an event occurring in every millisecond, but no certainty that the event will indeed arrive in any specific millisecond. This randomness leads to a variation in the number of counts detected in successive intervals. If we denote by X the number of counts in a particular interval, then X is said to follow a Poison distribution with pmf

$$f(x) = \frac{e^{-\lambda}\lambda^x}{x!} \quad x = 0, 1, 2, \ldots, \infty \tag{2.21}$$

where $f(x) \geq 0$ and $\sum_{x=0}^{\infty} f(x) = 1$

This distribution is determined by the only parameter λ.

The first four moments of this distribution are

$$\begin{aligned} \mu_1{}' &= \text{mean} = \lambda \\ \mu_2 &= \text{variance} = \lambda \\ \mu_3 &= \lambda \\ \mu_4 &= 3\lambda^2 + \lambda \end{aligned} \tag{2.22}$$

So, this distribution has the special property that the mean is equal to the variance.

The distribution is positively skew and leptokurtic.

3. Negative Binomial Distribution

The galaxy count-in-cell distribution characterizes the location of galaxies in space. It includes statistical information on voids and other underdense regions, on clusters of all shapes and sizes, on filaments, on counts of galaxies in cells of arbitrary shapes and sizes randomly located, etc. It can be shown that the overall galaxy count-in-cell distribution agrees with the negative binomial distribution.

The set-up of negative binomial is same as that of a binomial, i.e. there are a number of repeated trials which have only two outcomes ("success" and "failure") with constant probability of success (or failure) for each trial denoted by p and the trials are independent. Suppose the experiment is continued until r successes occur where r is specified in advance. Let us denote by X the total number of trials required to produce r successes. Then the pmf of the distribution of X is given by

$$P[X = x] = \binom{x-1}{r-1} p^r (1-p)^{x-r} \quad x = r, r+1, \ldots \infty \tag{2.23}$$

The negative binomial distribution can also be defined in an alternative manner. Let us denote by Y the number of failures before the rth success. Then $Y = X - r$ and the alternative pmf is given by

$$f(y) = P[Y = y] = \binom{r+y-1}{y} p^r (1-p)^y \quad y = 0, 1, 2, \ldots \infty \tag{2.24}$$

We will use the form (2.24).

Here also, $f(y) \geq 0$ and $\sum_{x=0}^{\infty} f(y) = 1$

The first four moments are given by

$$\begin{aligned} \mu_1{}' &= \frac{r(1-p)}{p} \\ \mu_2 &= \frac{r(1-p)}{p^2} \\ \mu_3 &= \frac{r(p-1)(p-2)}{p^3} \end{aligned} \tag{2.25}$$

$$\mu_4 = \frac{r(1-p)(6-6p+p^2+3r-3pr)}{p^4}$$

This distribution is positively skew and leptokurtic.

4. Uniform Distribution

Uniform distribution will occur in practice if, under the given experimental conditions, the different values of the random variable become equally likely. A simple way to uniformly distribute points on sphere is called the "hypercube rejection method". To apply this to a unit cube at the origin, choose co-ordinates (x, y, z) each uniformly distributed on the interval $[-1, 1]$. If the length of this vector is greater than 1, then reject it, otherwise normalize it and use it as a sample. By choosing uniformly distributed polar co-ordinates $\theta(0 < \theta < 360)$ and $\phi(0 < \phi < \pi/2)$, if the poles lie along the z-axis then the position on a unit hemisphere is

$$\begin{aligned} x &= \cos(\sqrt{\phi})\cos(\theta) \\ y &= \cos(\sqrt{\phi})\sin(\theta) \\ z &= \sin(\sqrt{\phi}) \end{aligned} \tag{2.26}$$

A whole sphere is obtained by simply randomizing the sign of z.

The p.m.f of a uniformly distributed random variable X over $(a, a + (k-1)h)$ is given by

$f(x) = \frac{1}{k}$ where $x = a, a+h, \ldots, a+(k-1)h$

Clearly, $f(x) \geq 0$ for all x

and $\sum_x f(x) = k \times \frac{1}{k} = 1$

The first four moments are given by

$$\begin{aligned} {\mu_1}' &= a + \frac{h(k-1)}{2} \\ \mu_2 &= h^2\frac{(k^2-1)}{12} \\ \mu_3 &= 0 \\ \mu_4 &= \frac{h^4}{240}(k^2-1)(3k^2-7) \end{aligned} \tag{2.27}$$

The distribution is symmetrical and highly platykurtic.

2.8.2 Some Important Continuous Distributions

Some important continuous type theoretical distributions that can be used to model Astronomical parameters and phenomenon are as follows:

1. Pareto (Power law) distribution

When the probability of measuring a particular value of some quantity varies inversely as a power of that value, the quantity is said to follow a power law or Zipf's law or the Pareto distribution. Mass density of stars in a star cluster follows the power law.

If X is a random variable with a Pareto distribution, then the probability density function of X is given by

$$f(x) = \frac{\alpha {x_m}^{\alpha}}{x^{\alpha+1}} \quad x_m < x < \infty \tag{2.28}$$

where x_m (positive) is the minimum possible value of X and α is a positive constant. For this distribution

$$\begin{aligned}
\text{Mean} &= \frac{\alpha x_m}{\alpha - 1} \text{ for } \alpha > 1 \\
\text{Median} &= x_m \sqrt[\alpha]{2} \\
\text{Mode} &= x_m \\
\text{Variance} &= \frac{{x_m}^2 \alpha}{(\alpha-1)^2(\alpha-2)} \text{ for } \alpha > 2 \\
\text{Skewness} &= \frac{2(1+\alpha)}{\alpha - 3}\sqrt{\frac{\alpha-2}{\alpha}} \text{ for } \alpha > 3
\end{aligned} \tag{2.29}$$

Depending on the value of α, the mean, variance and higher moments may not exist in some situation.

2. Normal distribution

Normal probability distribution, also known as Gaussian distribution refers to a family of distributions that are bell shaped.

Gaussian function approximates the shapes of many observables in astronomy, such as the profiles of seeing disks, the width of spectral lines and the distribution of noise in radio receivers. In error analysis, the Gaussian distribution is often used to determine the significance of a measurement is the presence of noise.

The distribution is defined by the probability density function

$$f(x) = \frac{1}{\sqrt{2\pi\sigma}} e^{-\frac{1}{2\sigma^2}(x-\mu)^2}, -\infty < x < \infty \tag{2.30}$$

The first four moments are given by

$$\begin{aligned} \mu_1{}' &= \mu \\ \mu_2 &= \sigma^2 \\ \mu_3 &= 0 \\ \mu_4 &= 3\sigma^4 \end{aligned} \tag{2.31}$$

The Normal distribution can be characterized by the two parameters μ and σ^2. This distribution is symmetrical and mesokurtic.

If the random variable X follows a Normal distribution with mean μ and variance σ^2, then the standardized random variable

$$z = \frac{X - \mu}{\sigma}$$

follows a standard normal distribution with mean zero and variance 1. The pdf of the distribution of z is given by

$$\phi(z) = \frac{1}{2\pi} e^{-z^2/2}, -\infty < z < \infty \tag{2.32}$$

3. Lognormal distribution

The variable X is said to have a log-normal distribution if $\ln X$ (or $\log X$) is normally distributed. Here X varies from 0 to ∞. It is also known as Galton distribution. A variable might be modelled as a lognormal if it can be thought of as the multiplicative product of many independent random variables each of which is positive.

The pdf of this distribution is given by

$$f(x) = \frac{1}{x\sqrt{2\pi\sigma^2}} e^{-\frac{(\ln x - \mu)^2}{2\sigma^2}} \quad 0 < x < \infty \tag{2.33}$$

$$\begin{aligned} \text{Mean} &= e^{\mu+\sigma^2/2} \\ \text{Median} &= e^{\mu} \\ \text{Mode} &= e^{\mu-\sigma^2} \\ \text{Variance} &= (e^{\sigma^2} - 1)e^{2\mu+\sigma^2} \\ \text{Skewness} &= (e^{\sigma^2} + 2)\sqrt{e^{\sigma^2} - 1} \end{aligned} \tag{2.34}$$

The distribution is positively skew and leptokurtic.

Chapter - 3

Sources of Astronomical Data

3.1 Introduction

Astronomy in recent past has developed a lot with the launch of several missions like GALEX (Galaxy Evolution Explorer), Kepler Space Telescope, Hubble Space Telescope (HST), etc. through which terabytes of data are available for preservation. This increasing proportionality of huge data demands data access efficiency. The implication of the above statement is important in a sense that most of the astrophysical phenomena are being observed in terms of light intensity as a function of wavelength or frequency which are snapshots of experiments which cannot be repeated as such. So one can easily understand why every single observation needs to be preserved. Thus, on the one hand, it requires the advent of more sophisticated observing instruments and sophisticated data management systems are to be developed, on the other hand, to complement the above challenge. With the above point in view astronomers have developed several Virtual data archives like SDSS (Sloan digital sky survey), MAST (Multi mission archive at STSCI), EXOSAT (European X-ray observatory Satellite), CGRO (Compton Gamma Ray Observatory), GALEX, Vizier, EDD, LEDA, HETE-2 (High Energy Transient Explorer), Chandra, Swift, ROSAT (Rontgen Satellite), WMAP (Wilkinson Microwave Anisotropy Probe), NED (NASA Extragalactic Data base), PDSC (Planetary Data System) etc. and in the making of future missions like ALMA (Atacama Large Millimeter Array), SKA (Square Kilometer Array) in 2025, etc.

In the following sections we will describe in brief some of the features of few data archives along with how data collected from different heavenly bodies can be accessed using various sites.

3.2 Sloan Digital Sky Survey

The general information on SDSS is in the website `www.sdss.org`. This virtual data archive contains huge amount of data on all objects starting from stars, stellar populations to galaxies and quasars in multi wavelength bands

A.K. Chattopadhyay, T. Chattopadhyay, *Statistical Methods for Astronomical Data Analysis*, Springer Series in Astrostatistics 3,
DOI 10.1007/978-1-4939-1507-1_3

u, g, r, i, z through a dedicated 2.5 m telescope at Apache Point Observatory, New Mexico. This telescope covers almost a quarter of the sky. Data release 8 (DR8) contains measurements on 500 million stars and galaxies with spectra of 2 million objects. It contains more than 1,20,000 quasars. In this site, on the LHS there are number of windows written, e.g. **Go to sdss.org/dr7/**. Among these windows if one clicks on **sdss.org/dr7/** and then **Database (CAS)** under **Data** the entire **tabular scheme** of SDSS appears showing the heads like **Site, Search Tools, Advanced Tools, Links** and **Help and Tutorials** (hereafter called **page 1**). There are several search techniques found under "Tools". When large amount of data are required **SQL Search** is chosen. If under **Tools** at page 1 (**skyserver.sdss.org/dr7/en/tools/search/**), one clicks **SQL Search**, a dialog box appears and one has to write a special query language known as "Structured Query Language" (SQL) corresponding to the requirement. Below some examples of SQL are given.

Example 1 Find ra, dec, u, g, r, i, z magnitudes of ten galaxies having redshift within 1 and 1.5.

Solution.

```
SELECT TOP 10

p.objid, p.ra, p.dec, p.u, p.g, p.r, p.i, p.z,

S.z, S.specClass

FROM photoObj AS p

JOIN SpecObj AS s ON s.bestobjid = p.objid

WHERE

S.z > 1

AND

S.z < 1.5

AND
```

S.specClass = 2

In the above example the index "2" is chosen for galaxy search. Similarly there are other indices which can be found clicking the following path at **page 1**:

Schema → Schema Browser → Views → SpecPhoto → specClass

There are other paths also.

The various indices used for different objects are

0 → Unknown object

1 → Star

2 → Galaxy

3 → Quasars

4 → High redshift quasars

5 → Blank sky

6 → Stars dominated by molecular bands

7 → Emission line galaxies

For acquaintance with various sample SQL queries one may click to **Sample SQL Queries** under **Help and Tutorials** menu at **page 1**.

It is possible to merge several SQL queries for getting photometric as well as spectral data of the same object.

Example 2 Find u, g, r, i, z magnitudes, ra, dec, IDS, along with equivalent widths of $H_\alpha, H_\beta, H_\delta$ absorption lines and corresponding errors of ten galaxies. Here u, g, r, i, z are photometric measurements and widths of $H_\alpha, H_\beta, H_\delta$ are spectral features.

Solution.

```
SELECT TOP 10

p.objid, p.ra, p.dec, p.u, p.g, p.r, p.i, p.z, S.z, S.specClass,

'Ha_6565', L.ew, L.ewErr, L.continuum,

'Hb_4863', L2.ew, L2.ewErr, L2.continuum,

'Hd_4103', L_Hd.ew, L_Hd.ewErr, L_Hd.continuum

FROM PhotoObj AS p

JOIN SpecObj AS S ON S.bestobjid = p.objid

JOIN SpecLine AS L ON S.SpecObjID = L.SpecObjID
JOIN SpecLine AS L2 ON S.SpecObjID = L2.SpecObjID

JOIN SpecLine AS L_Hd ON S.SpecObjID = L_Hd.SpecObjID

WHERE

L.Lineid = 6565

and L2.Lineid = 4863

and L_Hd.Lineid = 4103

and S.specClass = 2
```

For getting the spectrum of a galaxy one has to use **Get images** under **Search Tools**. Then click on any **plate** (the big white circular area on RHS contains numerous open dots and each dot corresponds to a plate containing spectra of 640 objects). Then among 640 objects corresponding to a particular plate, stars, galaxies, quasars are classified. So clicking any object of choice the corresponding spectra can be found.

If the co-ordinates (ra, dec) of any astronomical object is known, then its spectra and data on spectral properties can be obtained as follows.

1. First one has to follow the sequence starting from **DR7 Tools** on LHS at **page 1**

   Visual Tools → Navigate.

   A dialog box on LHS appears and one has to type the (ra, dec) of the object one requires.

   Then the image of the object will be found.

2. For having the spectra and other properties the following sequence is to be followed starting from **page 1**.

   Search Tools → Get images → Visual Tools → Explore.

   Then on LHS click on **ra,dec** gutter under **Search by**. A dialog box appears where one has to type ra and dec values of the object under query and then to press **OK** button.

   For spectral line indices one has to click **SpecLineIndex** under **SpecObj**, which is under **Summary**.

3. For having photometric properties one has to click the window under **PhotoObj** under **Summary** on the LHS of the dialog box.

The list of various line indices in SDSS observes can be found as follows from **page 1**:

Schema → Schema Browser → Views → SpecLine → line ID

Then a dialog box appears where line indices along with corresponding wavelengths are listed.

Moreover SDSS has its own tutorial pages for initial training. This can be found following the sequence under **Help** in the initial tabular scheme from **page 1**.

Help → SQL Tutorial and then clicking on **NEXT** subsequently.

SDSS has several projects, through which one can study various astrophysical problems. For this one has to click **projects** in the initial tabular schemes, at the top of the web page. The features about SDSS described above is just a snapshot of the numerous features which can be found exploring the various tools and links in initial tabular scheme (**cas.sdss.org/astrodr7/en/**). The transformation laws among SDSS and Johnson magnitudes can be found in `www.sdss.org/dr5/algorithms/sdssUBVRITransform.html`.

3.3 Vizier Service

This is a data archive where the observations of several catalogs are enlisted. The corresponding website is **vizier.u-strasbg.fr/viz-bin/VizieR**.

In the above site a dialog box appears and the query catalog is typed there, e.g. if one wants to find all observations on globular clusters so far enlisted in Vizier, one has to type "globular clusters". Then the list of all available tables of observations appears along with the corresponding literature. Then the viewer chooses the appropriate table and collects the data in a table.

3.4 Data on Eclipsing Binary Stars

It has been discussed in Chap. 1 that the intensity of light coming from two gravitationally bound stars, varies due to eclipse. Hence analysing these light curves properties of the stars can be studied in detail. For collecting data on observed light curves of these binary systems the following procedure is to be carried out.

1. The site `http://exoplanetarchive.ipac.caltech.edu/applications/ETSS/Kepler_index.html` contains **Kepler Light Curves** link. So clicking this link a dialog box appears.
2. If the Kepler ID of a binary star is known (which is an integer, e.g. 2141697) it is typed in the Kepler ID box and the **view** button is clicked.
3. Then a table appears where the second column (**StarID**) gives the list of all light curves corresponding to that binary star. Any one is clicked.
4. Then **Compute Periodogram** button is clicked. Then one source on RHS is selected and one item under "X axis" is chosen and it gives a table comprising of a number of periods computed for different values of the power spectrum along with p-values. A suitable period is chosen (generally with moderately low p-value) and the corresponding **Phased curve** is clicked.
5. Clicking the **Phased Curve download** button the light curve data, i.e. flux versus period table is found which can be stored for analysis.
6. The list of eclipsing binary stars (prsa catalog, Prsa et. al. 2011, AJ, 141, 83) is found in **http://keplerebs.villanova.edu/** or **archive.stsci.edu/kepler/ eclipsing_binaries.html**.

The latest light curve data can also be found from "Vizier" site, following the procedure discussed above.

3.5 Extra Galactic Distance Data Base (EDD) (edd.ifa.hawaii.edu/index.html)

This site gives the distance determination to galaxies within about 10,000 $\mathrm{km\,s^{-1}}$. In this site user can cross match the parameters collected from (1) the Virgo/Fornax SBF catalog of Blakslee et al., (2) the 2 MASS Large Galaxy Atlas, (3) the 2 MASS Redshift survey (2MRS). Arriving at this site the **NEXT** button should be pressed. Then a site appears where data on galaxies, dwarf galaxies, stars, Supernova Ia are stored under various catalogs. To cross match data from different catalogs, one has to select parameters from any one by clicking the **white empty box** on LHS and pressing the **Ctrl** on keyboard. Then other parameters are selected from other catalogs in the same way and finally the **submit** button is pressed below the last catalog selected.

3.6 Data on Pulsars

Australia Telescope National Facility (ATNF) collects radio observations of almost 2008 radio and millisecond pulsars and the table of parameters of these pulsars can be retrieved by clicking the white boxes adjoint to each parameter and pressing the **TABLE** button at the beginning/end of the site. Also a window **Pulsar Tutorial** is helpful for the users in this concern. The website is `http://www.atnf.csiro.au/people/pulsar/psrcat/`.

3.7 Data on Gamma Ray Bursts

Data on Gamma Ray Bursts are stored in BATSE, SWIFT and HETE catalogs. SWIFT is part of NASA's medium size explorer (MIDEX) program and has been launched on November, 2004. The official site is **swift.gsfc.nasa.gov/docs/swift/results** and then following the sequence, data on GRBs can be retrieved.

Swift Gamma-Ray Burst Table 4 which columns would you like to view → view.

The advantage of these GRBs is that their redshift values are also listed unlike BATSE.

The website of current BATSE catalog is

`www.batse.msfc.nasa.gov/batse/grb/catalog/current/`

The website of High Energy Transient Explorer (HETE-2) is

space.mit.edu/HETE/ and then press the **Bursts** button on the left and finally **Table of HETE Fregate Archival Data for Localized Bursts**.

3.8 Astronomical and Statistical Softwares

Aladin: It is a free interactive software and with this user can visualize images, superimpose entries from various catalogs. The user manual is freely available on its site **aladin.u-strasbg.fr**.

TOPCAT: TOPCAT is a free interactive software which can perform several types of graphics both in Astronomy and in Statistics. Having Java it can be downloaded.

Binary Maker 3.0: It is a priced software for modelling the light curves of eclipsing binaries.

STATA: This is a widely used commercial package especially used for research in Statistics, Economics, Sociology, Political science, Biomedicine, Epidemiology, etc. It has both the facilities of command-line interface, which facilitates replicable analyses and graphical user interface which uses menus and dialog boxes to give access to nearly all built-in commands. STATA stores the data set in random-access or virtual memory, which limits its use with extremely large data sets. Like R, STATA is regularly updated by incorporating newly developed statistical techniques.

S-PLUS: It is a priced statistical software for doing several statistical analysis including advanced multivariate statistical techniques like Cluster Analysis, Factor Analysis, Principal Component Analysis and several testings including ANOVA (Analysis of Variance), MANOVA (Multivariate Analysis of Variance), KS (Kolmogorov–Smirnov Test) test. It has also graphical advantage. It has both "script" and "menu driven" modes for computations.

MINITAB: It is also a priced statistical software performing all the above facilities.

SPSS: It is a priced statistical software with the advantage of a large memory space for handling with large data sets (e.g. 3–4 Lacs of objects at a time).

R: R is a free command based software for statistical analysis and also for graphics. It has variety of platforms like Windows and Linux. Since it is an open interface, everyday its content is increasing and as a result users can use recent statistical methods for their advanced research. There is also "User Manual" for the users once it is downloaded. Use of softwares will be discussed in detail in Chap. 11.

Virtual Observatory

The international Virtual observatory alliance (IVOA) was formed in June, 2002 in order to develop tools, systems and organizational structures for the proper utilization of the astronomical archives. About 20 countries are involved in this alliance and India is one of them. The VO-India (**vo.iucaa.ernet.in/∼voi/**) has several products like VO plot, VO Megaplot, VO Stat, VO Cat, VO Platform, VO Convert, etc. Among them the first three are particularly useful for different types of diagrammatic presentation of data and their statistical analysis. There are both stand alone and Web based versions which may be freely accessed. The input data files may be either in ASCII formed or in VOTABLE format.

Exercises

1. Download 10,000 galaxies having u, g, r magnitudes within redshift 1 and 2, along with $H_\alpha, H_\beta, H_\delta$ equivalent width absorption lines.
2. Draw H–R diagram of the globular cluster NGC2419, given its ra = 07 h 38 ' 08" .47 dec = +38 deg 52 ' 56".8. From the diagram compute its approximate age and the ratio between main sequence stars and white dwarfs.
3. Apply K-means cluster analysis with respect to the colours (u–g, g–r, r–u) and the equivalent widths of $H_\alpha, H_\beta, H_\delta$. Hence find the optimum group.
4. Collect the spectra of 10,000 stars from SDSS and assign indices to the several absorption lines. Now do a Hierarchical Cluster Analysis using Pearson's correlation instead of Euclidean distance. Draw a dendrogram to find the optimum grouping and compare it with the classification of stars in six groups O, B, A, F, G, K, M.
5. Using the simple modelling of binary star light curve (discussed in Chap. 1) find the ratio of luminosities of 100 binary star light curves downloaded from Prsa Catalog.

Chapter - 4

Statistical Inference

4.1 Population and Sample

Population: In any Statistical Analysis we are interested in some numerical characteristic of an aggregate of individuals rather than in the characteristics of the individuals themselves. Such an aggregate is called a "population" or "universe".

Sample: We know that in most of the situations it is not at all possible to study the entire population (just as the actual universe). In some cases it may be the infinite hypothetical population. So we have to remain content with the information gathered from a part of the population only. Such a part of the population by which we want to represent the entire population is called a sample.

Example: Classification of Spiral Galaxies on the basis of rotation curves.

Population: The class of all Spiral galaxies.

Sample: A small set of spiral galaxies selected by some method which is supposed to represent the population of Spiral galaxies.

We select some of the parameters (which are called variables in statistical language) to study the variation among spiral galaxies with respect to their rotation curves like:

D = distance from the observer; D_{25} = diameter at 25th$B$ mag arc s^{-2} isophote; h = disc scale length; Vmax = maximum rotational velocity outside r_{in}; M_B = absolute B magnitude; Bar = presence of a bar(1/0).

Electronic supplementary material The online version of this chapter (doi:10.1007/978-1-4939-1507-1_4) contains supplementary material, which is available to authorized users.

A.K. Chattopadhyay, T. Chattopadhyay, *Statistical Methods for Astronomical Data Analysis*, Springer Series in Astrostatistics 3, DOI 10.1007/978-1-4939-1507-1_4

4.2 Parametric Inference

The central problem of Statistics is to devise means of inferring the nature of the population from the known nature/distribution of the sample. This is similar to the classical problem of inductive inference, i.e. going from the particular to the general.

For astronomical observations, it is always difficult to accept them as samples from the corresponding population. Since the universe is unknown, it is always probable that the observations are not proper representative of the corresponding population. As a result, the inference derived from the observations may be far away from the reality. Although, for the known universe also the problem of proper sample selection is always a burning problem, in case of astronomical objects it is much more complicated. With the advancement of technology, at present more reliable observations are coming and the need for proper statistical techniques for drawing inferences is increasing day by day.

The problem of statistical inference includes two types of situations. In the first case the feature in which we are interested is totally unknown and we may want to make a guess about this feature completely on the basis of a random sample from the population. This type of problem is known as the problem of **estimation**.

In the second case some information of a tentative nature regarding the feature of the population may be available and we may want to see whether the information is acceptable on the basis of the random sample taken from the population. This type of problem is known as the problem of **testing of hypothesis**.

For first type (i.e. estimation) of problems, under the parametric set-up, the investigator is interested in the value of some parameters (these are different from physical parameters and are some unknown constants depending on which the base, scale and shape of the underlying population changes) which are completely unknown and he/she depends solely on the sample data to make a guess about the value of the unknown parameter. Again two procedures under this category may be distinguished. Under the first procedure, a single value may be used as the estimated value of the unknown parameter. This is called the method of **point estimation**. Under the second type of procedure, the investigator may compute two values, again on the basis of the sample data, and expect that the true value of the parameter lies within these two values (lower and upper confidence limits, respectively) with a high probability. This is known as the method of **interval estimation**.

4.2.1 Point Estimation

In point estimation we use the value of some statistic (function of the sample values), say T, as the estimator of the unknown parameter (function of the population values) under consideration, say θ where θ may be scalar or vector valued.

How one can choose a statistic as a good estimator for the corresponding parameter? Usually its quality is judged by the distribution of estimates which it yields, i.e. by the properties of its sampling distribution.

4.2.1.1 Unbiasedness

The first criterion is that the estimator should be *unbiased*. This means that it has no tendency to be regularly above or below the parameter so that the estimate is distributed in an unbiased manner about the true value of the parameter. Formally this means that the expectation of the proposed estimate should be equal to the parameter.

Unbiasedness: The Expectation of the proposed estimator T should be equal to the value of the unknown parameter θ, whatever the true value may be, i.e.

$$\textbf{Expectation}(T) = \theta \textbf{ for every possible value of } \theta$$

The second criterion is that for a good estimator the spread of its sampling distribution be as small as possible. So for every possible value of θ, the variance of T should be smaller than the variance of any other estimator satisfying the first criterion. Such an estimator is unique and is defined as the *Minimum variance Unbiased Estimator* (MVUE).

Minimum Variance Unbiased Estimator

An unbiased estimator T of the parameter θ is said to be a MVUE of θ if for **any other** unbiased estimator of θ, say, T^*

$$\textbf{Variance } (T) <= \textbf{Variance } (T^*) \textbf{ for every possible value of } \theta$$

Estimating Variance

Suppose $X_1, \ldots, X_n$ are independent and identically distributed random variables with expectation μ and variance σ^2. Let

$$\overline{X} = (X_1 + \ldots + X_n)/n$$

be the "sample average", and let

$$S^2 = \frac{1}{n}\sum_{i=1}^{n}(X_i - \overline{X})^2$$

be a "sample variance". Then S^2 is a "biased estimator" of σ^2 because

$$E(S^2) = \frac{n-1}{n}\sigma^2 \neq \sigma^2.$$

Note that when a transformation is applied to an unbiased estimator, the result is not necessarily itself an unbiased estimate of its corresponding population statistic. That is, for a non linear function f and an unbiased estimator U of a parameter p, $f(U)$ is usually not an unbiased estimator of $f(p)$. For example, the square root of the unbiased estimator of the population variance is not an unbiased estimator of the population standard deviation.

Bias is not the only consideration when choosing a statistic, however. Bias refers to the central tendency of the sampling distribution of a statistic, but the variance of the sampling distribution can also be an important consideration. Specially, statistics with smaller sampling variances will yield greater statistical power. For example, while S^2 above is more biased than the traditional sample variance

$$S^2_{sample} = \frac{1}{n-1}\sum_{i=1}^{n}(X_i - \overline{X})^2,$$

S^2 has a lower estimation variability than S^2_{sample} because the denominator dividing the sum of squares is larger in the calculation of S^2, resulting in a smaller scale of final values, and therefore lower estimation variability, than that of S^2_{sample}. Practically, this demonstrates that for some applications (where the amount of bias can be equated between gropups/conditions) it is possible that a biased estimator can prove to be a more powerful, and therefore useful, statistic.

4.2.1.2 Efficiency

As the sample size increases, the scatter of possible values of the sample mean from the actual (population) mean decreases so that the probability that a given value of sample mean differs by more than a fixed amount from the population mean decreases. We can therefore say that the accuracy of the estimator increases as the sample size increases. In other words, the variance of the sampling distribution of the estimator is inversely proportional to sample size.

This property of increasing accuracy with sample size is obviously desirable and an estimator which has that property is said to be consistent. Mathematically, this means that the probability that the estimate differs from parameters by less than an arbitrary small error approaches unity as the sample size increases to infinity. This property is analogous to convergence in mathematical sense. For example, the sample mean $\bar{x}$ is a consistent estimator of the population mean μ since

$$\sigma\bar{x} = \sigma/\sqrt{n} \text{ (which is the standard error of } \bar{x})$$

becomes smaller as n increases.

The property of consistency is concerned with the behaviour of an estimator as the sample size increases to infinity. We should note that a consistent estimator is not necessarily unbiased. Similarly as unbiased estimator is not necessarily consistent. However, many estimators are both unbiased and consistent. Whatever be the probability distribution of the observation, the sample mean is always an unbiased estimate of the population mean. However the variance (with divisor n) of a set of observations is a biased estimate of the population variance as shown earlier. But it is always consistent since its standard error is given by $\sqrt{(2/n)}\sigma^2$ which becomes smaller as n increases.

The sample mean is the most important property of a sample. If the sample is large, its mean has an approximate normal distribution even if the parent population is not normal. It is the reason for what so much attention is paid to the Normal Distribution.

4.2.1.3 Maximum Likelihood Estimator (MLE)

Let $f(x_1, x_2, \ldots, x_n|\theta)$ be the joint probability density function or probability mass function of the sample observations. When $x_1, x_2, \ldots, x_n$ are given it may be looked upon as a function of θ and denoted by $L(\theta)$. Under this method we take that value as the estimate of θ for which $L(\theta)$ is maximum.

$$\theta_{est} = \underset{\theta}{Max} L(\theta)$$

But the maximum has to be the global maximum and if the derivative does not exit at $\theta = \theta_{est}$ we will not get the estimator. This ML method is very popular and it also has the invariance property.

4.2.2 Interval Estimation

Under point estimation a single value is used to estimate the unknown population parameter. An alternative procedure is to give an interval within which it may be supposed to lie. This is called *interval estimation.*

Instead of assigning a single number to each sample and reporting the size of a typical error, the present method assigns an interval to each sample and reports the confidence level that the interval contains the parameter. *Confidence* is a technical term related to probability. Just as the standard error (SE) of an estimator measures the long-run average size of the error in repeated sampling, but the error for any particular sample could be smaller or larger than the SE, the confidence level is the long-run fraction of intervals that contain the parameter in repeated sampling, but the interval for any particular sample might or might not contain the parameter.

The statement "the interval [92, 94] contains the vale of the population parameter, at confidence level 90 %" does *not* mean that the probability that the population value is between 92 and 94 is 0.90. (The event that the interval [92, 94] contains the population value is not random: either the population value is between 92 and 94, or it is not.) Rather, the statement means that if we were to take samples of size n repeatedly and compute a 90 % confidence level for the population parameter's value from each sample of size n, the long-run fraction of intervals that contain the population value would converge to 90 %.

The length of the confidence interval and the confidence level measure how accurately we are able to estimate the parameter from a sample. If a short interval has high confidence, the data allow us to estimate the parameter accurately. Higher confidence generally requires a longer interval, and, shorter intervals generally have lower confidence levels. Conventional values for the confidence level of confidence intervals include 68, 90, 95, and 99 %, but sometimes other values are used.

The interpretation of confidence level for a particular interval is analogous to the interpretation of Standard Error (SE) for a particular value of the estimate. The SE is the square-root of the long-run average squared error of the estimator in repeated sampling, but for any particular sample, the error could be larger or smaller than the SE—and we will not know which unless we know the true value of the parameter. The confidence level measures the long-run fraction of intervals that contain the parameter in repeated sampling, but for any particular sample, the confidence interval either will or will not contain the parameter—and we will not know which unless we know the true value of the parameter.

4.3 Testing of Hypothesis

Some information regarding the underlying population may be available and we may want to see whether the information is tenable in the light of the sample taken from the population.

The information that we know previously and want to verify (nullify) is called the null hypothesis (H_0). The complementary part of H_0 is called the alternative hypothesis (H_1).

We perform the test on the basis of the given sample and possible errors are:

Rejecting H_0 when it is true (Type I error)

Accepting H_0 when it is false (Type II error).

	H_0 True	H_0 False
Accept H_0	Correct	Type II Error
Reject H_0	Type II Error	Correct

We try to fix the probability of type I error at a particular level and then try to minimize the probability of type II error or maximize the power (1−P(type II error)). The test is performed on the basis of a critical (rejection) region. If the value of the test static falls in the critical region, we reject H_0 otherwise we accept it.

Hence level of significance (α) may be defined as the maximum value of P(type I error). A test is of size α if under H_0 for at least one value of the parameter θ the level is attained.

4.3.1 *p*-Value

Each statistical test has an associated null hypothesis, the *p*-value is the probability that your sample could have been drawn from the population(s) being tested (or that a more improbable sample could be drawn) given the assumption that the null hypothesis is true. A *p*-value of 0.05, for example, indicates that you would have only a 5 % chance of drawing the sample being tested if the null hypothesis was actually true. *p*-Value corresponds to the level attained by a test.

Null Hypotheses are typically statements of no difference or effect. A *p*-value close to zero signals that your null hypothesis is false, and typically that a difference is very likely to exist. Large *p*-values closer to 1 imply that there is no detectable difference for the sample size used. A *p*-value of 0.05 is a typical threshold. *p*-Value may be defined as the minimum value of P(type I error).

Example:
Suppose the random variable X follows Normal (μ, σ^2) distribution.
To test $H_0 : \mu = \mu_0$ against $H_1 : \mu \neq \mu_0$,
where we assume that σ is known.
Let $X_1, X_2, .., X_n$ be a sample of size n and M be the Sample mean.

Critical region: $M \geq c$ (where c is the critical point)

Then level of significance
$\alpha = P[M \geq c|\mu = \mu_0]$
$= P[(M - \mu_0)\sqrt{n})/\sigma \geq (c - \mu_0)\sqrt{n}/\sigma]$
$= 1 - \Phi[(c - \mu_0)\sqrt{n}/\sigma] > 1 - \Phi[(M - \mu_0)\sqrt{n}/\sigma]$
$= p$ value (where Φ is the CDF of standard normal distribution).

Hence p value is the smallest level of significance.

4.3.2 One Sample and Two Sample Tests

One Sample (Test for One Mean Value)

Let $X_1, X_2, \ldots, X_n$ be a random sample drawn a Normal population with mean μ and sd σ. Students' t test is used to compare the unknown mean of the population (μ) to a known number (μ_0). So here the Null hypothesis is $H_0 : \mu = \mu_0$ against the alternative $H_1 : \mu \neq \mu_0$.

Test statistic (population standard deviation σ is known):

The formula for the Z-test is

$$Z = \sqrt{n}(\text{Sample mean} - \mu_0)/\sigma$$

Z has a Normal distribution with mean 0 and variance 1.

Test statistic (population standard σ deviation is unknown):

The formula to t test is $t = \sqrt{n}(\text{Sample mean} - \mu_0)/s$,

where s is the sample standard deviation.

The statistic t follows t distribution with $n - 1$ degrees of freedom, where n is the number of observations.

Decision of the z or t-Test: If the p-value associated with the z or t-test is small (usually set up at $p < 0.05$), there is evidence to reject the null hypothesis in favour of the alternative. In other words, there is evidence that the mean is significantly different than the hypothesized value i.e. **the test is significant**. If the p-value associated with the z or t-test is not small ($p > 0.05$), there is not enough evidence to reject the null hypothesis, and it may be concluded that there is evidence that the mean is not different from the hypothesized value i.e. **the test is not significant**.

Two Sample (Test for Equality of Two Means)

Suppose we have two independent samples The unpaired t method tests the null hypothesis that the population mean related to two independent, random samples from two independent approximately normal distributions are equal against the alternative that they are unequal (as in the one sample case).

Assuming equal variances, the test statistic is calculated as:

$$t = \frac{\overline{x_1} - \overline{x_2}}{\sqrt{s^2\left(\frac{1}{n_1} + \frac{1}{n_2}\right)}}$$

$$s^2 = \frac{\sum_{j=1}^{n_1}(x_j - \overline{x_1})^2 + \sum_{i=1}^{n_2}(x_i - \overline{x_2})^2}{n_1 + n_2 - 2}$$

where $\overline{x_1}$ and $\overline{x_2}$ are the sample means, s^2 is the pooled sample variance, n_1 and n_2 are the sample sizes and t follows Student t distribution with $n_1 + n_2 - 2$ degree of freedom.

Paired Sample (from Bivariate Normal Distribution)

The paired t test provides a hypothesis test of the difference between population means for a pair of n random samples whose differences are approximately normally distributed.

The test statistic is calculated as

$$t = \frac{\bar{d}}{\sqrt{s^2/n}}$$

where $\bar{d}$ is the mean difference, s^2 is the sample variance, n is the sample size and t follows a paired t distribution with n-1 degrees of freedom.

The decision can be taken exactly in a similar way as in the one sample situation.

4.3.3 Common Distribution Test

Quantile–Quantile Plot

The quantile–quantile (q–q) plot is a graphical technique for determining if two data sets come from populations with a common distribution.
A q–q plot is a plot of the quantiles of the first data set against the quantiles of the second data set. By a quantile, we mean the fraction (or percent) of points below the given value. That is, the 0.3 (or 30 %) quantile is the point at which 30 % of the data fall below and 70 % fall above that value.

A 45° reference line is also plotted. If the two sets come from a population with the same distribution, the points should fall approximately along this reference line. The greater the departure from this reference line, the greater the evidence for the conclusion that the two data sets have come from populations with different distributions.

Normality Tests

Probability Plot

Normal Test Plots (also called Normal Probability Plots or Normal Quartile Plots) are used to investigate whether process data exhibit the standard normal "bell curve" or Gaussian distribution (Fig. 4.1).

First, the x-axis is transformed so that a cumulative normal density function will plot in a straight line. Then, using the mean and standard deviation which are calculated from the data, the data is transformed to the standard normal values, i.e. where the mean is zero and the standard deviation is one. Then the data points are plotted along the fitted normal size.
The nice thing is it is not necessary to understand all the transformations. If the plotted points fit the normal line well, it can be safely assumed that the process data is normally distributed.

4.4 Empirical Distribution Function

In statistics, an empirical distribution function is a cumulative probability distribution function that concentrates probability $1/n$ at each of the n numbers in a sample.

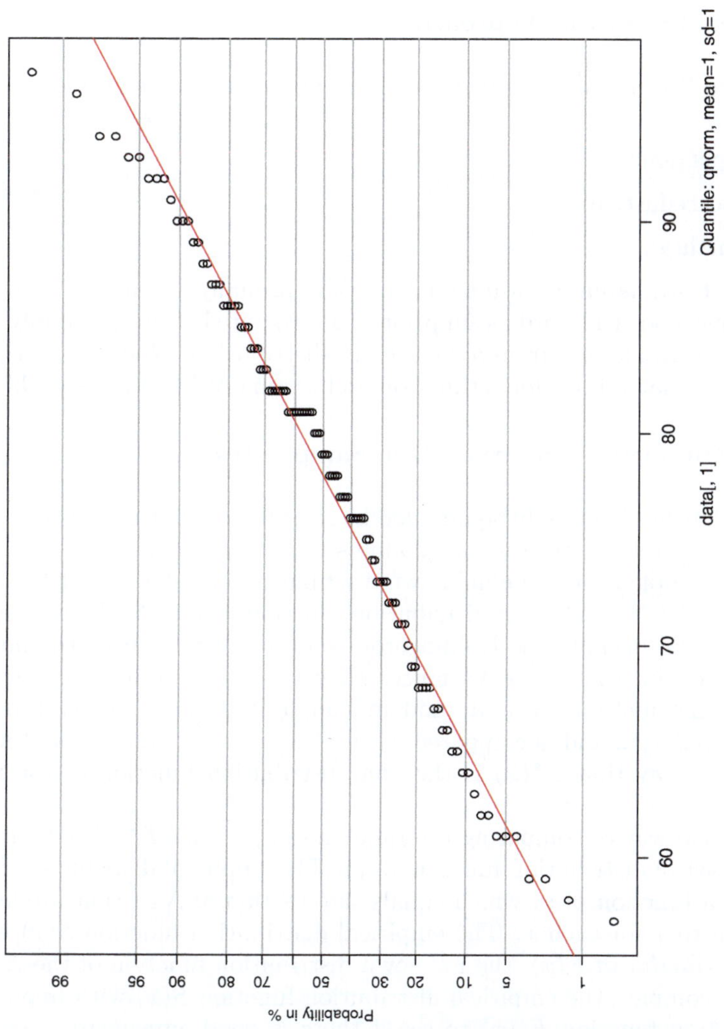

Figure 4.1 Normal probability plot

Let $x_1, x_2, \ldots, x_n$ be random variables with realizations. The empirical distribution function $F_n(x)$ based on sample $x_1, x_2, \ldots, x_n$ is a step function defined by

$$F_n(x) = \frac{\text{number of elements in the sample} \leq x}{n} = \frac{1}{n} \sum_{i=1}^{n} I(x_i \leq x),$$

where $I(A)$ is an indicator function.

4.5 Nonparametric Approaches

Standard Statistical techniques are optimal under the assumptions of:

1. Independence
2. Homoscedasticity
3. Normality

But the third assumption may not be valid in many situations. In nonparametric methods the third assumption is not required. Here we simply assume that the variables are from a continuous distribution. Nonparametric methods are available both for estimation and testing of hypothesis problems.

4.5.1 Kolmogorov–Smirnov One Sample Test

This is alternative to Chi-square goodness of fit test but can also be applied to small sample. The test for goodness of fit usually involves examining a random sample from some unknown distribution in order to test the null hypothesis that the unknown distribution function is in fact a known, specified function. We usually use Kolmogorov–Smirnov test to check the normality assumption in Analysis of Variance. However it can be used for other continuous distributions also. A random sample $X_1, X_2, \ldots, X_n$ is drawn from some population and is compared with $F^*(x)$ in some way to see if it is reasonable to say that $F^*(x)$ is the true distribution function of the random sample.

One logical way of comparing the random sample with $F^*(x)$ is by means of the empirical distribution function $S(x)$. The empirical distribution function $S(x)$ is a function of x, which equals the fraction of Xi's that are less than or equal to x for each x. The empirical distribution function $S(x)$ is useful as an estimator of $F(x)$, the unknown distribution function of the Xis.

We can compare the empirical distribution function $S(x)$ with hypothesized distribution function $F^*(x)$ to see if there is good agreement. One of the simplest measures is the largest distance between the two functions $S(x)$ and $F^*(x)$, measured in a vertical direction. This is the statistic suggested by Kolmogorov.

Let the test statistic T be the greatest (denoted by "sup" for supremum) vertical distance between $S(x)$ and $F(x)$. In symbols we say

$$T = \sup_x |F^*(x) - S(x)|$$

$$\begin{aligned} \text{For testing } H_0 : F(x) &= F^*(x) \text{ for all } x \\ H_1 : F(x) &\neq F^*(x) \text{ for at least one value of } x \end{aligned}$$

If T exceeds the 1-$á$ quantile as given in Biometrika table, Volume-2, then we reject H_0 at the level of significance $á$. The approximate p-value can be found by interpolation.

Example:
A random sample of size 10 is obtained: $X_1 = 0.621, X_2 = 0.503, X_3 = 0.203, X_4 = 0.477, X_5 = 0.710, X_6 = 0.581, X_7 = 0.329, X_8 = 0.480, X_9 = 0.554, X_{10} = 0.382$. The null hypothesis is that the distribution function is the uniform distribution function. The mathematical expression for the hypothesized distribution function is

$$\begin{aligned} F^*(x) &= 0, \text{ if } x < 0 \\ &\quad x, \text{ if } 0 \le x < 1 \\ &\quad 1, \text{ if } 1 \le x \end{aligned}$$

Formally, the hypotheses are given by
$H_0 : F(x) = F^*(x)$ for all x from $-\infty$ to ∞
$H_1 : F(x) = F^*(x)$ for at least one value of x
where $F(x)$ is the unknown distribution function common to the Xis and $F^*(x)$ is given by above equation.

The Kolmogorov test for goodness of fit is used. The critical region of size $\alpha = 0.05$ corresponds to values of T greater than the 0.95 quantile 0.409, obtained from Biometrika table for $n = 10$. The value of T is obtained by graphing the empirical distribution function $S(x)$ on the top of the hypothesized distribution function $F^*(x)$. The largest vertical distance is 0.290, which occurs at $x = 0.710$ because $S(0.710) = 1.000$ and $F^*(0.710) = 0.710$. In other words,

$$\begin{aligned} T &= \sup x|F^*(x) - S(x)| \\ &= |F^*(0.710) - S(0.710)| = 0.290 \end{aligned}$$

Since $T = 0.290$ is less than 0.490, the null hypothesis is accepted. In other words, the unknown distribution $F(x)$ can be considered to be of the form $F^*(X)$ on the basis of the given sample. The p value is seen, to be larger than 0.20.

4.5.2 Kolmogorov–Smirnov Two Sample Test

Perform a Kolmogorov–Smirnov two sample test that two data samples come from the same distribution. Note that we are not specifying what that common distribution is.

The two sample K–S test is a variation of one sample test. However, instead of comparing an empirical distribution function to a theoretical distribution function, we compare the two empirical distribution functions. That is,

$$D = \sup_x |S_1(x) - S_2(x)|$$

where S_1 and S_2 are the empirical distribution functions for the two samples. Note that we compute S_1 and S_2 at each point in both samples (that is both S_1 and S_2 are computed at each point in each sample).

The hypothesis regarding the distributional form is rejected if the test statistic, D, is greater than the critical value obtained from Biometrika table. There are several variations of these tables in the literature that use somewhat different scaling for the K–S test statistic and critical regions. These alternative formulations should be equivalent, but it is necessary to ensure that the test statistic is calculated in a way that is consistent with how the critical values were tabulated. For the R code, one may look at Chap. 11.

In order to increase the power of the K–S test near tails one may use the **Anderson Darling test.**

4.5.3 Shapiro–Wilk Test

Shapiro–Wilk is a standard test for normality. The test statistic W may be thought of as the correlation between given data and their corresponding normal scores, with $W = 1$ when the given data are perfectly normal in distribution. When W is significantly smaller than 1, the assumption of normality is not met. That is, a significant W statistic causes the researcher to reject the assumption that the distribution is normal. Shapiro–Wilk W is recommended for small and medium samples up to $n = 2{,}000$. For larger samples, the Kolmogorov–Smirnov test is recommended.

The Wilk–Shapiro test statistic is defined as:

$$W = \frac{(\sum_{i=1}^{n} w_i X_{(i)})^2}{\sum_{i=1}^{n}(X_i - \bar{X})^2}$$

where the summation is from 1 to n and n is the number of observations. The array X contains the original data, $X_{()}$s are the ordered data, X is the sample mean of the data, and $w' = (w_1, w_2, \ldots, w_n)$ or

$$w' = MV^{-1}[(M'V^{-1})(V^{-1}M)]^{-1/2}$$

M denotes the expected values of standard normal order statistics for a sample of size n and V is the corresponding covariance matrix.

W may be thought of as the squared correlation coefficient between the ordered sample values ($X_{()}$) and the w_i. The w_i are approximately proportional to the normal scores M_i. W is a measure of the straightness of the normal probability plot, and small values indicate departures from normality. For R code, we again refer Chap. 11.

4.5.4 Wilcoxon Rank-Sum Test

The Wilcoxon Rank Sum test can be used to test the null hypothesis that two populations X and Y have the same continuous distribution. We assume that we have independent random samples $x_1, x_2, \ldots, x_m$ and $y_1, y_2, \ldots, y_n$, of sizes m and n, respectively, from each population. We then merge the data and rank of each measurement from lowest to highest. All sequences of ties are assigned an average rank.

The Wilcoxon test statistic W is the sum of the ranks from population X. Assuming that the two populations have the same continuous distribution (and no ties occur), then W has a mean and standard deviation given by

$$\mu = m(m+n+1)/2$$

and

$$s = \sqrt{[mn(N+1)/12]},$$

where $N = m + n$.

We test the null hypothesis H_0: No difference in distributions. A one-sided alternative is H_a: first population yields lower measurements. We use this alternative if we expect or see that W is unusually lower than its expected value μ. In this case, the p-value is given by a normal approximation. We let $N \sim N(\mu, s)$ and compute the left-tail $P(N <= W)$ (using continuity correction if W is an integer).

If we expect or see that W is much higher than its expected value, then we should use the alternative H_a: first population yields higher measurements. In this case, the p-value is given by the right-tail $P(N >= W)$, again using continuity correction if needed. If the two sums of ranks from each population are close, then we could use a two-sided alternative H_a: there is a difference in distributions. In this case, the p-value is given by twice the smallest tail value ($2P(N <= W)$ if $W < \mu$, or $2P(N >= W)$ if $W > \mu$).

We note that if there are ties, then the validity of this test is questionable. For R code, see Chap. 11.

4.5.5 Kruskal–Wallis Two Sample Test

The Kruskal–Wallis test is a nonparametric test used to compare three or more samples. It is used to test the null hypothesis that all populations have identical distribution functions against the alternative hypothesis that at least two of the samples differ only with respect to location (median), if at all.
It is the analogue to the F-test used in analysis of variance. While analysis of variance tests depend on the assumption that all populations under comparison are normally distributed, the Kruskal–Wallis test places no such restriction on the comparison.

The Kruskal–Wallis test statistic for k samples, each of size n_i is:

$$T = \frac{1}{s^2}\left(\sum_{i=1}^{k} \frac{R_i}{n_i} - N\frac{(N+1)^2}{4}\right)$$

where N is the total number (all n_i) and R_i is the sum of the ranks (from all samples pooled) for the i-th sample and:

$$S^2 = \frac{1}{N-1}\left(\sum_{all} {R_{ij}}^2 - N\frac{(N+1)^2}{4}\right)$$

The null hypothesis of the test is that all k distribution functions are equal. The alternative hypothesis is that at least one of the populations tends to yield larger values than at least one of the other populations.

Either k population distribution functions are identical, or else some of the populations tend to yield larger values than other populations

The test statistic for the Kruskal–Wallis test is T. This value is compared to a table of critical values based on the sample size of each group. If T exceeds the critical value at some significance level (usually 0.05) it means that there is evidence to reject the null hypothesis in favour of the alternative hypothesis. For R code, see Chap. 11.

Example: The Anderson–Darling test determines whether a sample comes from a specified distribution. Given a set of observation $X_1, X_2, \ldots, X_n$ and their ordered values $X_{(1)} \leq X_{(2)} \ldots \leq X_{(n)}$, the Anderson–Darling (AD) statistic is given by

$$A^2 = -n - S^2$$

$$\text{where } S^2 = \sum_{k=1}^{n} \frac{2k-1}{n}\left[\ln(F(X_{(k)})) + \ln(1 - F(X_{(n+1-k)}))\right]$$

In Chattopadhyay and Chattopadhyay (2006, 2007) extinctions in colours and errors in ages of the globular clusters were tested for normally using Anderson–Darling statistic. This is an example of one sample nonparametric test under testing of hypothesis,

$$H_0 : \text{ The population is normal}$$

$$H_1 : \text{ The population is not normal}$$

The corresponding result is given in Table 4.1.

Name	Age	Errors in age	Errors in extinctions
Milky	Mean	1.59	1.95
Way	SD	0.44	0.62
Globular	AD statistic	0.45	0.11
Clusters	Remark	Good fit	Very good fit

Table 4.1 Example of one sample nonparametric test

From the above table it can be inferred that since both the error distributions are Normal, i.e. symmetric, the errors are supposed to be averaged out in final analysis and the results are thus not influenced by them.

Reference

Chattopadhyay, T., and A.K. Chattopadhyay. 2006. *Astronomical Journal* 131:2452.

Chattopadhyay, T., and A.K. Chattopadhyay. 2007. *Astronomy and Astrophysics* 472:131.

Chapter - 5

Advanced Regression and Its Applications with Measurement Error

5.1 Introduction

Regression analysis is used to study the relationship among variables. The target is to establish a causal effect of one variable upon another in case of simple regression in which only two variables are considered. In case of multiple regression the target is to study the effect of a number of variables on a single variable. Regression technique has been widely used in areas like econometrics, financial statistics, biostatistics, etc. In Astronomy also several authors have used this technique for prediction purpose. The problem of how to characterize spiral galaxies having extended rotation curve and how many parameters are necessary for this characterization has been studied by Chattopadhyay and Chattopadhyay (2006). Tully–Fisher relation is a relation for spiral galaxies between their luminosity and how fast they are rotating. The idea is that the bigger the galaxy is, faster it is rotating, i.e. if one knows the rotation velocity of the spiral galaxy, by using the Tully–Fisher relation it is easy to predict its intrinsic brightness. Again by comparing the intrinsic brightness with apparent magnitude, one can calculate its distance. By starting from virial theorem, Tully and Fisher suggested the following two forms

$$L \propto W^{\alpha}$$

$$\text{and} \quad L \propto R^{\beta}$$

where L is the intrinsic luminosity, W is a characterization of the motion of the body and R is a measure of linear size. α and β are proper constants. Kodaira and Kashikawa (2000) found a much tighter correlation among the three parameters given by

$$L \propto VR^2$$

Electronic supplementary material The online version of this chapter (doi:10.1007/978-1-4939-1507-1_5) contains supplementary material, which is available to authorized users.

A.K. Chattopadhyay, T. Chattopadhyay, *Statistical Methods for Astronomical Data Analysis*, Springer Series in Astrostatistics 3,
DOI 10.1007/978-1-4939-1507-1_5

where V is the rotational velocity of the galaxy. Such type of relations are used to derive the fundamental plane of galaxies. Regression techniques have been extensively used to establish fundamental planes with a less amount of scatter.

5.2 Simple Regression

Suppose we have data only on two variables, viz. intrinsic luminosity (L) and rotational velocity (V) and we want to predict L in terms of V by using the relation

$$L = a + bV + e \tag{5.1}$$

where a (intercept) and b (slope) are constants and e is the noise term reflecting other factors that influence luminosity.

The parameter (variable) L is termed the "dependant" or "effect" or "response" or "endogenous" variable. V is termed the "independent" or "cause" or "predictor" or "exogenous" or "explanatory" variable. In case of Astronomical data it is usually difficult to identify variables as "dependent"or "independent" due to lack of information. For such cases a symmetric relation is very much necessary.

At the outset of any regression analysis, one formulates some hypothesis about the relationship between the variables like (5.1). This may be linear or non linear. In particular if the dependent variable is categorical or binary one may use Poisson or logistic regression. Here the data set contains observations for L and V. The noise component e is unobservable and the constants a and b are unknown. In statistical term, these unknown constants a and b are known as parameters. Least square method is used to estimate these unknown constants where we minimize the sum of squares due to error (noise).

5.3 Multiple Regression

In case of simple regression we assume that the "effect" is the outcome of a single cause. But in practice the "effect" is usually the outcome of several "causes". For example, a galaxy with a higher luminosity has a larger central velocity dispersion (σ) or a galaxy with a larger size (viz. effective radius r_e) has fainter effective surface brightness ($< \mu_e >$). The above two point correlations are rather tight but scatter is still reduced using a three variable relation of the form

$$\log r_e = k + a \log \sigma + b < \mu_e > +e \tag{5.2}$$

where k, a, and b are unknown constants (parameters) and e is the noise term. The above relation is known as multiple regression.

In case of two variables (simple regression) it is easy to study the physical relationship (linear or non linear) by a two-dimensional diagram conventionally termed "scatter" diagram. Each point in the diagram represents an individual in sample. But in case of multiple regression it is not so easy. The task of estimating the parameters k, a and b is conceptually identical to the earlier task of estimating only a and b. The difference is that we can no longer think of regression as choosing a line in a two-dimensional diagram. With two explanatory variables we need three dimensions, and instead of estimating a line we are estimating a plane. Multiple regression analysis will select a plane so that the sum of squared errors is at a minimum. In this case the error is the vertical distance between the actual value of $\log r_e$ and the estimated plane. The intercept of that plane with the $\log r_e$ axis (where $\log \sigma$ and $< \mu_e >$ are at zero level) implies the constant term k, its slope in the $\log \sigma$ dimension implies the coefficient a and its slope in the $< \mu_e >$ dimension implies the coefficient b.

In model (5.2), k represents the log of the effective radius of a galaxy with unit central velocity dispersion and no effective surface brightness. Sometimes such situation may be imaginary and in such cases the value of k must be zero (i.e. no intercept). a captures the per unit effect on effective radius of central velocity dispersion and b captures per unit effect on effective radius of effective surface brightness.

With p explanatory variables, multiple regression analysis will estimate the equation of a hyperplane in space such that the sum of squared errors is minimized.

5.3.1 Estimation of Parameters in Multiple Regression

Consider a multiple regression equation with $(p-1)$ explanatory variables given by

$$X_1 = a + b_2X_2 + b_3X_3 + \ldots + b_pX_p + e_i \tag{5.3}$$

Here data will consist of p values corresponding to the p variables, for each of n individuals. We have to estimate the constants (parameters) $a, b_2, b_3 \ldots b_p$ by least square method. These parameters will be estimated by minimizing the error sum of squares given by

$$s^2 = \sum_{i=1}^{n} (X_{1i} - a - b_2X_{2i} - \ldots - b_pX_{pi})^2$$

Here p normal equations can be obtained by minimizing s^2 w.r.t $a, b_2 \ldots b_p$.

From the first normal equation we get

$$\overline{X}_1 = a + b_2\overline{X}_2 + \ldots + b_p\overline{X}_p \tag{5.4}$$

where $\overline{X}_i$ is the mean values of x_i $i = 1 \ldots p$.

If we denote by s_i and r_{ij}, the standard deviation of X_i and correlation coefficient between X_i and $X_j (i, j = 1 \ldots p)$ by solving the other $(p-1)$ normal equations we get the estimate of b_j as

$$\hat{b_j} = (-1)^{j-2} \frac{s_1}{s_j} \frac{\begin{vmatrix} r_{21} & r_{22} & \cdots & r_{2(j-1)} & r_{2(j+1)} & \cdots & r_{2p} \\ r_{31} & r_{32} & \cdots & r_{3(j-1)} & r_{3(j+1)} & \cdots & r_{3p} \\ r_{p1} & r_{p2} & \cdots & r_{p(j-1)} & r_{p(j+1)} & \cdots & r_{pp} \end{vmatrix}}{\begin{vmatrix} r_{22} & r_{23} & \cdots & \cdots & \cdots & \cdots & r_{2p} \\ r_{32} & r_{33} & \cdots & \cdots & \cdots & \cdots & r_{3p} \\ r_{p2} & r_{p3} & \cdots & \cdots & \cdots & \cdots & r_{pp} \end{vmatrix}}$$

If $R = ((r_{ij}))i, j = 1 \ldots p$ denotes the $p \times p$ correlation matrix of $x_1, x_2 \ldots x_p$, $|R|$ the corresponding determinant and R_{ij} the cofactor of r_{ij} in R, then

$$\begin{aligned} \hat{b_j} &= (-1)^{2j-1} \frac{s_1}{s_j} \frac{R_{1j}}{R_{11}} \\ &= -\frac{R_{1j}}{R_{11}} \frac{s_1}{s_j} \text{ for } j = 2, 3 \ldots p \end{aligned}$$

Hence from (5.4) we have

$$\hat{a} = \overline{x}_1 + \sum_{j=2}^{p} \frac{R_{1j}}{R_{11}} \frac{s_1}{s_j} \overline{x}_j \tag{5.5}$$

Thus the linear multiple regression equation (prediction equation) of X_1 on $X_2, X_3 \ldots X_p$ is given by

$$X_1 = \overline{x}_1 - \frac{R_{12}}{R_{11}} \frac{s_1}{s_2} (X_2 - \overline{x}_2) \ldots \frac{R_{1p}}{R_{11}} \frac{s_1}{s_p} (X_p - \overline{x}_p) \tag{5.6}$$

The coefficient b_j is called the partial regression coefficient of X_1 on X_j for fixed $X_2, \ldots X_{j-1}, X_{j+1}, \ldots X_p$.

Using the estimated regression coefficients we write the fitted regression equation as

$$\hat{X_1} = \hat{a} + \hat{b_2} X_2 + \ldots + \hat{b_p} X_p$$

For each observation in our data we can compute

$$\hat{X_{1i}} = \hat{a} + \hat{b_2} X_{2i} + \ldots + \hat{b_p} X_{pi}, i = 1 \ldots n \tag{5.7}$$

These are called fitted values.

The residuals are given by $e_i = X_{1i} - \hat{X_{1i}}, i = 1 \ldots n$. Assuming that e_is are independently and identically distributed as $N(0, \sigma^2)$, an unbiased estimate of error variance σ^2 is given by

$$\hat{\sigma^2} = \frac{SSE}{n-p} \tag{5.8}$$

where $SSE = \sum_{i=1}^{n}(X_{1i} - \hat{X_{1i}})^2 = \sum_{i=1}^{n} {e_i}^2$ is the residual sum of squares. Here the denomination $(n-p)$ is called degrees of freedom. It is equal to the number of observations minus the number of estimated regression coefficients.

5.3.2 Goodness of Fit

After fitting the linear model to the given data set, an assessment is necessary for the goodness of fit. The strength of linear relationship between X_1 and the set of predictors $X_2, X_3 \ldots X_p$ can be examined through the examination of the correlation coefficient of X_1 versus $\hat{X_1}$ given by

$$r(X_1, \hat{X_1}) = \frac{\sum(X_{1i} - \overline{X}_1)(\hat{X_{1i}} - \overline{\hat{X}}_1)}{\sqrt{\sum(X_{1i} - \overline{X}_1)^2}\sqrt{\sum(\hat{X_{1i}} - \overline{X}_1)^2}} \tag{5.9}$$

where $\overline{X}_1$ is the mean of the response variable X_1 and $\overline{\hat{X}}_1$ is the mean of the fitted values. The coefficient of determination $R^2 = [r(X_1, \hat{X_1})]^2$ is also given by

$$R^2 = 1 - \frac{SSE}{SST} = 1 - \frac{\sum(X_{1i} - \hat{X_{1i}})^2}{\sum(X_{1i} - \overline{X}_1)^2} \tag{5.10}$$

Thus, R^2 may be interpreted as the proportion of the total variability in the response variable X_1 that can be accounted for by the set of predictor variables $X_2, X_3 \ldots X_p$. Here R is called the multiple correlation coefficient. It measures the relationship between one variable X_1 and a set of variables $X_2, X_3 \ldots X_p$.

When the model fits the data well, it is clear that the value of R^2 is close to unity. In the absence of any linear relationship between X_1 and $X_2, X_3 \ldots X_p$, R^2 will be near to zero. But in some situations a large value of R^2 may not mean that the model fits the model well. For those situations a more detailed analysis is necessary.

Another quantity, known as adjusted R^2, denoted by R^2_α is also used to judge goodness of fit. It is defined as

$$R^2_\alpha = 1 - \frac{(SSE)/n-p}{SST/n-1}$$

$$= 1 - \frac{n-1}{n-p}(1 - R^2) \tag{5.11}$$

5.3.3 Regression Line Through the Origin

Let us consider two regression lines of the forms

$$X_1 = a + b_2X_2 \tag{5.12}$$

$$\text{and } X_1 = b_2X_2 \tag{5.13}$$

The first one is a regression line with an intercept and the second one is a regression line passing through the origin. This is called a no-intercept model. Sometimes the line may be forced to go through the origin because of the subject matter or other external considerations. One has to make the choice between the two models with care. Here goodness of fit of the two models should be judged in terms of the residual sum of squares rather than the R^2 value as the R^2 values obtained from the two models are not comparable. In the first case the R^2 value is based on the deviations from the sample mean whereas in the second case R^2 is based on the deviations measured about zero.

5.4 Effectiveness of the Fitted Model

After fitting an appropriate regression model it is necessary to perform some investigation to check the effectiveness of the fitted model. So graphical presentation may be used to check linearity and normality assumptions. For this one may use normal probability plot where ordered standardized residuals are plotted against the ordered normal scores. Under normality assumption this plot should be a straight line with a zero intercept and slope one.

One should also check whether the fit is not overly determined by one or few influential observations. To find influential observations, leverage values have important role. Consider the fitted model

$$\hat{X}_{1i} = \hat{a} + \hat{b}_2X_{2i} + \ldots + \hat{b}_pX_{pi}(i = 1 \ldots n)$$

and the corresponding ordinary least square residuals

$$e_i = X_{1i} - \hat{X}_{1i}$$

The fitted values can also be written in an alternative form

$$\hat{X}_{1i} = p_{i1}X_{11} + p_{i2}X_{12} + \ldots + p_{in}X_{1n}(i = 1 \ldots n) \tag{5.14}$$

where p_{ij}'s are quantities which depend only on the values of independent variables. In simple regression of the form $y_i = \hat{a} + \hat{b}x_i(i = 1 \ldots n)$

p_{ij} is given by

$$p_{ij} = \frac{1}{n} + \frac{(x_i - \overline{x})(x_j - \overline{x})}{\sum (x_i - \overline{x})^2} \tag{5.15}$$

In multiple regression p_{ij}'s are the elements of that "hat" or projection matrix. Here the p_{ii} value is known as the leverage value for the i-th observation. There are several distance measures to identify influential observations. One of them is Cook's distance defined as

$$C_i = \left(\frac{r_i^2}{p}\right)\left(\frac{p_{ii}}{1 - p_{ii}}\right), i = 1 \ldots n \tag{5.16}$$

$$\text{where } r_i = \frac{e_i}{\hat{\sigma}\sqrt{1 - p_{ii}}}$$

A large value of C_i indicates that the point is influential. Points with C_i value greater than the 50 % point of the F distribution with p and $(n - p)$ degrees of freedom be classified as influential point. As a practical rule, one may take that point as influential whose Cook's distance measure is greater than 1. After identifying the influential observations, these should be critically examined. One may refit the model by excluding the influential observations to see the effect of these points.

Another measure was proposed by Hadi (1992) on the basis of the fact that the influential observations are outliers either in the response variable or in the predictors or both. The proposed measure is

$$H_i = \frac{p_{ii}}{1 - p_{ii}} + \frac{p}{1 - p_{ii}} \frac{d_i^2}{1 - d_i^2} i = 1 \ldots n \tag{5.17}$$

where $d_i = \frac{e_i}{\sqrt{SSE}}$ is the normalized residual. Observations with large values of H_i are usually treated as influential.

5.5 Best Subset Selection

As the least square estimates often have low bias but large variance, prediction accuracy can sometimes be improved by shrinking or setting some coefficients to zero. Further with a large number of predictors sometimes we like to determine a smaller subset that exhibits the strongest effect. There are a number of approaches to variable subset selection with linear regression. Best subset regression finds for each m, the subset of size m that gives smallest residual sum of squares, $m = 1, 2 \ldots p$, where p is the total number of predictors. Problem of choosing m involves the trade off between bias and variance. But for large values of p, this method becomes infeasible. There are many other methods also. Some of them are discussed below.

5.5.1 Forward and Backward Stepwise Regression

Forward subset selection starts with the intercept and then subsequently adds into the model gradually those predictors which improve the fit most rapidly. Although it involves a large amount of computation for large values of p, there are several reasons for preferring it. Even if p is larger than n (total number of observations), we can always compute the forward stepwise sequence. Further forward stepwise is a constrained search and has lower variance.

Backward stepwise selection starts with the full model and subsequently deletes the predictor that has the least impact on the fit. Backward solution can be used only when $n > p$ but forward stepwise method can always be used.

There are stepwise selection strategies that consider both forward and backward waves at each step and select the best of the two. For the R code we refer Chap. 11. The corresponding data file UCD1.txt is given in the Appendix.

Like forward stepwise method, there is another method known as forward stepwise regression. At each step the algorithm identifies the variable which has the largest correlation with the current residual. It then computes the simple linear regression coefficient of the residual on this chosen variable and then adds it to the current coefficient for that variable. This is continued till none of the variables have correlation with this residuals. This method can take many more than p steps to reach the least square fit.

5.5.2 Ridge Regression

Ridge regression shrinks the regression coefficients by imposing a penalty on their size. The ridge coefficients minimize a penalized residual sum of squares. The ridge regression coefficients can be obtained by minimizing

$$\sum_{i=1}^{n}(X_{1i} - a - b_2X_{2i} - \ldots - b_pX_{pi})^2 \text{ subject to } \sum_{j=2}^{p} b_j^2 \leq t \tag{5.18}$$

where t is a known constant known as size constraint. When there are many correlated variables in a linear regression model their coefficients are likely to be poorly determined and show high variance. By imposing size constraint on the coefficients this problem can be minimized. This method is applicable to the situation where p is very large compared to n. For R code we refer Chap. 11.

5.5.3 Least Absolute Shrinkage and Selection Operator (LASSO)

Like ridge regression method, LASSO is also a shrinkage method where the L_2 penalty function is replaced by L_1 penalty function. The LASSO estimate of the regression coefficients can be obtained by minimizing

$$\sum_{i=1}^{n}(X_{1i} - a - b_2X_{2i} - \ldots - b_pX_{pi})^2 \text{ subject to } \sum_{j=2}^{p}|b_j| \leq t \qquad (5.19)$$

This L_1 norm constraint makes the solution non linear in the X_{1i} and there is no closed form expression. The problem is of the form of a quadratic programming problem. Here the solution largely depends on the choice of t. If t is chosen larger than $t^* = \sum_{2}^{p}|\hat{b}_j|$, where $\hat{b}_j, j = 2\ldots p$ are the least square estimates, then the LASSO estimates will be same as the least square estimates. Alternatively, for $t = t/2$, the least square estimates are shrunk by about 50 % on average.

5.5.4 Least Angle Regression (LAR)

Efron et al. (2004) introduced this concept which can be looked upon as an improvement over the forward stepwise regression. At first step LAR identifies the variable having the largest correlation with the response variable. Then moves the coefficient of this variable continuously towards its least square value so that the correlation with the residual decreases in absolute value. Another variable then enters the active set as soon as its correlation with the residual becomes as much as that of the first variable. Then the coefficients of both the variables are moved together in a way that keeps their correlations decreasing. This process is continued until all the variables are in the model and ends at the full least square fit. The LAR algorithm can be described as follows.

(a) Standardize the predictors to have mean zero and unit norm. Start with the residual $e = X_1 - \overline{X}_1, b_2 = b_3 = \ldots = b_p = 0$.

(b) Find the predictor $X_j (j = 2\ldots p)$ having the largest correlation with e.

(c) Move b_j from 0 towards its least square coefficient until some other competitor variable X_k has as much correlation with the current residual as does X_j.

(d) Move b_j and b_k in the direction defined by their joint least square coefficients of the current residual on (X_j, X_k) until some other variable X_l has as much correlation with the current residual.

(e) Continue this way until all $p - 1$ predictors have been entered.

5.6 Multicollinearity

In multiple regression, the underlying assumption is that the predictor variables are not strongly interrelated. Regression coefficient measures the change in the response variable when the corresponding predictor variable is changed by one unit and all other predictor variables are held constant. But if the independent variables are strongly related then this interpretation may not remain valid. When there is no linear relationship among the predictor variables they are said to be orthogonal. The lack of orthogonality is not a serious problem. But in some cases the causes are so strongly interrelated that the regression result becomes ambiguous.

The condition of nonorthogonality is known as the problem of multicollinearity. Usually it is very difficult to identify multicollinearity in a data set. Multicollinearity increases the standard errors of the coefficients. As a result some of the coefficients of independent variables may be found insignificant whereas without multicollinearity and with lower standard errors, the same coefficients might have been found to be significant. Thus we may say that multicollinearity makes some variables statistically insignificant while they should be otherwise significant. Variance inflation factors (VIF) measure the amount by which the variance of estimated coefficients are increased over the cases of no correlation among the predictor variables. If no two predictor variables are correlated, then all the VIFs will be 1. If VIF for one of the variables is greater than 5, we say that there is multicollinearity associated with that variable. As a practical rule, if there are two or more variables having VIF around or greater than 5, one of these variables must be removed from the regression model.

The VIF can be computed in the following manner. Consider the linear model with p independent variables $Y = b_0 + b_1X_1 + b_2X_2 + \dots b_pX_p + e$. Then the estimated variance of the estimate of b_j is given by

$$\hat{Var}(\hat{b}_j) = \frac{s^2}{(n-1)\hat{Var}(X_j)} \frac{1}{1-R_j^2}$$

where R_j^2 is the value of multiple R^2 for the regression of X_j on other predictors, s^2 is the mean square error and n is the sample size.

$\frac{1}{1-R_j^2}$ is known as the VIF of $\hat{b}_j$. We calculate p different VIFs, one for each $X_i (i = 1 \dots p)$ by first running an ordinary least square regression that has X_i as a function of all other predictors. Then compute the VIF for $\hat{b}_i (i = 1 \dots p)$ by using the formula

$$VIF = \frac{1}{1-R_i^2}$$

A common rule of thumb is that if $VIF(\hat{b}_i) > 5$ then multicollinearity is high.

One can tackle multicollinearity by increasing the sample size as when sample size is increased standard error decreases. Another way is to remove the most inter correlated variables from the analysis. But this method may contradict the basic assumption that these variables were there due to the theory of the model.

5.7 Regression Problem in Astronomical Research (Mondal et al. 2010)

Regression analysis is a widely used method in astronomical research. It is used for two purposes: (1) to develop a quantitative relationship among astronomically observed properties of a set of objects and (2) to predict the values of a particular property in terms of other properties of that set, e.g. relations between X-ray temperatures and velocity dispersions for galaxy clusters, the colour–luminosity relations for field galaxies, period luminosity relation for variable stars, Tully–Fisher relation (maximum rotation velocity vs. luminosity relation) for galaxies and other Fundamental Plane (FP) (later discussed in detail) relations when considering more than two variables. In case of two variables X and Y, one is treated as independent and the other dependent and ordinary least squares method gives a single linear regression of the dependent variable Y against the independent variable X, denoted by OLS($Y|X$). OLS($Y|X$) is one which minimizes the sum of squares of the Y residuals and predicts Y in terms of X. For astronomical purpose several problems are faced with the above choice. If the choice of the independent variable is not clear, then there is alternative option of OLS($X|Y$) and the distinction between these two approaches is often not clear (Bandiera and Hunt 1989) though a third robust process is discussed by Branham (1982) and Lutz (1983) which is not least squares procedure at all. So one single relationship treating X and Y symmetrically is required. In the above discussion measuremental errors have not been considered. Measuremental errors are the errors which arise in the measurement process of the instrument, e.g. signal to noise ratio, repeated measurements of some property etc. Heteroscedastic measuremental errors are mentioned in Isobe et al. (1990) as well as in Feigelson and Babu (1992).

In the paper of Isobe et al. (1990), the authors deal with data having no measuremental errors. Feigelson and Babu (1992) performed regression between two variables including measuremental errors. Non linear regression using ORDPACK (Boggs et al. 1990; Press et al. 1986) has been performed also. Akritas and Bershady (1996) have developed regression regarding two variables including known measuremental errors, (a) allowing measuremental errors of both the variables, (b) allowing measuremental errors of the variables to be dependent on each other, (c) measuremental errors depending on measurement and (d) finding other symmetric lines, e.g. bisector and orthogonal regression, etc. Then these techniques are applied to various astrophysical regression situations like colour–luminosity relation for field galaxies, Tully–Fisher relation and Tolman test as mentioned above.

In the work by Mondal et al. (2010), the regression line has been extended to regression plane of three variable y_1, y_2 and y_3, on the one hand, including measuremental errors of all the variables and, on the other hand, finding any symmetric plane in which any of the three variables can be considered as dependent and other two as independent variables. The extension from regression line to regression plane is an important aspect of studying Fundamental Plane of galaxies. There are various characteristics of galaxies which are correlated, e.g. a galaxy with a higher luminosity has a larger central velocity dispersion (σ) (Faber and Jackson 1976) or a galaxy with a larger size (viz. effective radius r_e) has fainter effective surface brightness ($< \mu_v >$) (Kormendy 1977). The usefulness of these correlations is when a characteristic that can be determined without prior knowledge of galaxy's distance, e.g. central velocity dispersion and it is correlated with a characteristic, such as luminosity or in turn effective radius, that can be determined only for galaxies with known distances, then, with this correlation, one can determine the distances to distant galaxies which is a difficult task in astronomy.

The above two point correlations are rather tight but the scatter is still reduced using a three variable relation of the form $\log r_e = a \log \sigma + b < \mu_v > +c$ (Dressler et al. 1987; Djorgovski and Davis 1987). This relation is known as fundamental plane (FP). It is applicable for giant early type galaxies and extends to faint and low-mass galaxies (Nieto et al. 1990). Following the above argument if σ, $< \mu_v >$ and angular r_e (in seconds of arc) can be measured for distant galaxies, then the distances (D in kpc) of these galaxies can be measured from the above FP relation as r_e in the above relation is linear (in kpc) and r_e (linear) $\sim Dr_e$ (angular). In the above relation $\log r_e$ is considered as a dependent and σ and $< \mu_v >$ are considered as independent, as in ordinary regression for three variables case. But actually these three parameters are intrinsically independent of each other, so the concept of dependent and independent variables is not applicable to the above situation. Hence we are in need of a symmetric relation (symmetric fundamental plane) where the concept of dependent or independent variables does not come into the picture and any variable can be expressed in terms of the remaining ones. This symmetric plane then plays the role of the so-called FP used generally in Astrophysics. In their work, Feigelson and Babu (1992) have found such a unique symmetric line while considering two variables (X, Y) and the symmetric line is the angular bisector of the two regression line $(Y|X)$ and $(X|Y)$ with minimum variance. But they have not considered the case of three variables. In the work of Mondal et al. (2010) the concept of symmetric line has been extended to symmetric plane involving three variables although the choice is not unique.

5.7.1 Regression Planes and Symmetric Regression Plane

Following Akritas and Bershady (1996) the regression planes are derived extending from two to three variables in the following manner that incorporates measuremental errors. Let the observations and their variance–covariance matrices corresponding to variables of interest be denoted by

$$(Y_{1i}, Y_{2i}, Y_{3i}, V_i), i = 1, \ldots, n \tag{5.20}$$

where for each i, V_i is a symmetric 3×3 matrix with elements

$$V_i = \begin{pmatrix} V_{11,i} & V_{12,i} & V_{13,i} \\ & V_{22,i} & V_{23,i} \\ & & V_{33,i} \end{pmatrix}$$

The observed data are related to unobserved intrinsic values of the variables (X_{1i}, X_{2i}, X_{3i})

by the relation

$$\begin{aligned} Y_{1i} &= X_{1i} + \in_{1i} \\ Y_{2i} &= X_{2i} + \in_{2i} \\ Y_{3i} &= X_{3i} + \in_{3i}, i = 1, 2, \ldots, n \end{aligned} \tag{5.21}$$

where the errors $(\in_{1i}, \in_{2i}, \in_{3i})$ are the measuremental errors corresponding to unobserved true values X_{1i}, X_{2i} and X_{3i}. Y_{1i}, Y_{2i} and Y_{3i} are the observed values of the variables (also known as surrogates). They have a joint trivariate distribution with assumed mean vector $0^{1\times 3}$ and dispersion matrix V_i. In this model, we allow V_i to depend on (Y_{1i}, Y_{2i}, Y_{3i}) and thus implicity on (X_{1i}, X_{2i}, X_{3i}). However, we assume that V_i is the only aspect of the distribution of $(\in_{1i}, \in_{2i}, \in_{3i})$ that depends on (Y_{1i}, Y_{2i}, Y_{3i}). This implies that, on the basis of the above assumption, given $V_i, (\in_{1i}, \in_{2i}, \in_{3i})$ are independent of (X_{1i}, X_{2i}, X_{3i}). The intuitive meaning of the above technical assumption is that $\in_{1i}$, $\in_{2i}$, $\in_{3i}$ are equally likely to be positive or negative for any values of X_{1i}, X_{2i}, X_{3i} and the sizes of their absolute values are governed (in addition to the type of the measuremental error distribution) by the magnitudes of $V_{11,i}, V_{22,i}, V_{33,i}$, which are given. In most cases the measuremental errors for (X_{1i}, X_{2i}, X_{3i}) are assumed independent implying that the off diagonal elements of the matrix V_i are zeros and observed data are of the form $(Y_{1i}, Y_{2i}, Y_{3i}, V_{11,i}, V_{22,i}, V_{33,i})$ where $V_{kk,i}$ denote the variances of $\in_{ki}, k = 1, 2, 3$. It is further assumed that the true values of the variables follow the multiple regression model corresponding to the i-th observation,

$$X_{1i} = a_1 + b_1 X_{2i} + c_1 X_{3i} + e_{1i} \tag{5.22}$$

where e_{1i} (i.e., model error) is assumed to have zero mean and finite variance. The variance or standard deviation of e_{1i} is known as the "intrinsic scatter".

When it is not known that which one should be the dependent variable, then the other two possible models are

$$X_{2i} = a_2 + b_2 X_{1i} + c_2 X_{3i} + e_{2i}$$

$$X_{3i} = a_3 + b_3 X_{1i} + c_3 X_{2i} + e_{3i}$$

For three variables (where dependent variable is unknown) we have to consider the regression equations $(X_1|X_2, X_3), (X_2|X_1, X_3)$ and $(X_3|X_1, X_2)$.

If we use model (5.22), without measuremental errors the estimates of the unknown constants are (using (5.21)):

$$\begin{aligned}
\hat{b}_1 &= \frac{1}{d}\Bigg[\sum_{i=1}^{n}(Y_{1i}-\overline{Y}_1)(Y_{2i}-\overline{Y}_2)\sum_{i=1}^{n}(Y_{3i}-\overline{Y}_3)^2 - \\
&\quad \sum_{i=1}^{n}(Y_{1i}-\overline{Y}_1)(Y_{3i}-\overline{Y}_3)\sum_{i=1}^{n}(Y_{2i}-\overline{Y}_2)(Y_{3i}-\overline{Y}_3)\Bigg] \\
\hat{c}_1 &= \frac{1}{d}\Bigg[\sum_{i=1}^{n}(Y_{1i}-\overline{Y}_1)(Y_{3i}-\overline{Y}_3)\sum_{i=1}^{n}(Y_{2i}-\overline{Y}_2)^2 - \\
&\quad \sum_{i=1}^{n}(Y_{1i}-\overline{Y}_1)(Y_{2i}-\overline{Y}_2)\sum_{i=1}^{n}(Y_{2i}-\overline{Y}_2)(Y_{3i}-\overline{Y}_3)\Bigg] \\
\hat{a}_1 &= \overline{Y}_1 - \hat{b}_1\overline{Y}_2 - \hat{c}_1\overline{Y}_3
\end{aligned}$$

where

$$d = \sum_{i=1}^{n}(Y_{2i}-\overline{Y}_2)^2\sum_{i=1}^{n}(Y_{3i}-\overline{Y}_3)^2 - \left[\sum_{i=1}^{n}(Y_{2i}-\overline{Y}_2)(Y_{3i}-\overline{Y}_3)\right]^2$$

These are the simple usual regression estimates (OLS).

But when measuremental errors are present in all variables, then the covariance matrix changes with subtractions, i.e., Modified covariance matrix = covariance matrix of the data – covariance matrix of the error.

Some formulae:

$$
\begin{aligned}
var(Y_{1i}) &= var(X_{1i}) + E(V_{11,i}) \\
var(Y_{2i}) &= var(X_{2i}) + E(V_{22,i}) \\
var(Y_{3i}) &= var(X_{3i}) + E(V_{33,i})
\end{aligned}
\tag{5.23}
$$

If there is no repeated observation, $E(V_{11,i}), E(V_{22,i}), E(V_{33,i})$ are replaced by $V_{11,i}, V_{22,i}, V_{33,i}$, respectively.

$$
\begin{aligned}
cov(Y_{1i}, Y_{2i}) &= cov(X_{1i}, X_{2i}) + E(V_{12,i}) \\
cov(Y_{2i}, Y_{3i}) &= cov(X_{2i}, X_{3i}) + E(V_{23,i}) \\
cov(Y_{1i}, Y_{3i}) &= cov(X_{1i}, X_{3i}) + E(V_{13,i})
\end{aligned}
\tag{5.24}
$$

If there is no repeated observation, $E(V_{12,i}), E(V_{23,i}), E(V_{13,i})$ are replaced by $V_{12,i}, V_{23,i}, V_{13,i}$, respectively.

Now the modified estimates are (using (5.21), (5.23), (5.24)):

$$
\begin{aligned}
\hat{b}_1 &= \frac{1}{d}\Bigg[(cov(Y_1, Y_2) - \sum_{i=1}^{n} V_{12,i})(var(Y_3) - \sum_{i=1}^{n} V_{33,i}) - \\
&\quad (cov(Y_1, Y_3) - \sum_{i=1}^{n} V_{13,i})(cov(Y_2, Y_3) - \sum_{i=1}^{n} V_{23,i})\Bigg] \\
\hat{c}_1 &= \frac{1}{d}\Bigg[(cov(Y_1, Y_3) - \sum_{i=1}^{n} V_{13,i})(var(Y_2) - \sum_{i=1}^{n} V_{22,i}) - \\
&\quad (cov(Y_1, Y_2) - \sum_{i=1}^{n} V_{12,i})(cov(Y_2, Y_3) - \sum_{i=1}^{n} V_{23,i})\Bigg] \\
\hat{a}_1 &= \overline{Y}_1 - \hat{b}_1\overline{Y}_2 - \hat{c}_1\overline{Y}_3
\end{aligned}
$$

where

$$d = \left[\left\{var(Y_2) - \sum_{i=1}^{n} V_{22,i}\right\}\left\{var(Y_3) - \sum_{i=1}^{n} V_{33,i}\right\} - \left\{cov(Y_2, Y_3) - \sum_{i=1}^{n} V_{23,i}\right\}^2\right]$$

Here we make assumption on

$$cov(X_{ki}, e_{li}) = 0$$

where

$$k = 1, 2, 3 \ ; \ l = 1, 2, 3$$

Similarly the least squares estimates can also be obtained for the models $(X_2|X_1, X_3)$ and $(X_3|X_1, X_2)$, respectively.

In the standard OLS model without measuremental errors, we assume that model errors are independently and identically distributed as Gaussian with mean 0 and variance σ^2. Also X_i's become identical with Y_i's as $\in_{1i}$'s are zero (viz. Eq. (5.21)).

5.7.2 The Symmetric Regression Plane with Intercept

The three regression planes without model error can be written as

$$y_1 = a_1 + b_1 Y_2 + c_1 Y_3 \tag{5.25}$$

$$y_2 = a_2 + b_2 Y_1 + c_2 Y_3 \tag{5.26}$$

$$y_3 = a_3 + b_3 Y_1 + c_3 Y_2 \tag{5.27}$$

where y_1, y_2 and y_3 are the predicted values of Y_1, Y_2 and Y_3, respectively. Let these planes intersect at (a, b, c). Then the line of intersection of any two planes among the three and passing through $\mathrm{O}(a, b, c)$ is OA (say) is given by (Fig. 5.1)

$$\frac{x-a}{\alpha'} = \frac{y-b}{\beta'} = \frac{z-c}{\gamma'}$$

where α', β' and γ' are direction cosines of OA. Now, OA is perpendicular to the normals of the intersecting planes $(Y_1|Y_2, Y_3)$, $(Y_2\ Y_1, Y_3)$ (say) so,

$$\alpha' - b_1\beta' - c_1\gamma' = 0$$

$$-b_2\alpha' + \beta' - c_2\gamma' = 0$$

From the above equations ratio of α', β' and γ' can be found. Let A be any point on

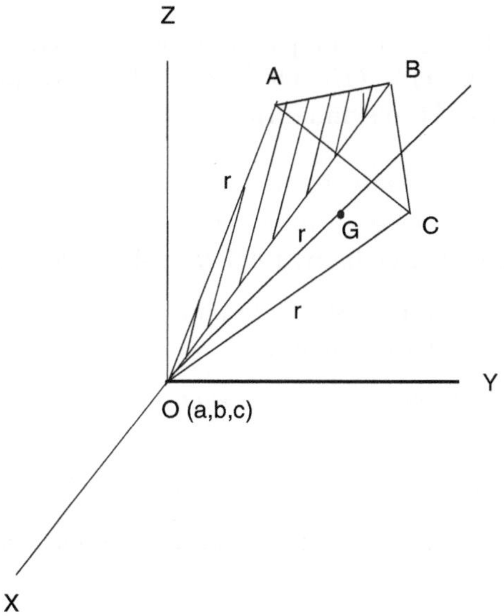

Figure 5.1 Regression planes OABO $(Y_1|Y_2, Y_3)$, OACO $(Y_2|Y_1, Y_3)$ and OBCO $(Y_3|Y_1, Y_2)$

OA at a distance r (say) (choice of r is arbitrary). Then A has the co-ordinates $A(\alpha' r + a, \beta' r + b, \gamma' r + c)$. Similarly we can choose other two points B and C at a distance r from O, lying on the lines of intersection of other regression planes such that $B(\alpha'' r+a, \beta'' r+b, \gamma'' r+c)$, $C(\alpha''' r+a, \beta''' r+b, \gamma''' r+c)$. Then the co-ordinates of the centroid G of the tetrahedron OABC has the co-ordinates, $G([\alpha' + \alpha'' + \alpha''' r + 4a]/4, [\beta' + \beta'' + \beta''' r + 4b]/4, [\gamma' + \gamma'' + \gamma''' r + 4c]/4)$. So the equation of OG is

$$\frac{x-a}{[(\alpha' + \alpha'' + \alpha''')r]/4} = \frac{y-b}{[(\beta' + \beta'' + \beta''')r]/4} = \frac{z-c}{[(\gamma' + \gamma'' + \gamma''')r/4}$$

Hence the required symmetric plane perpendicular to OG and passing through O is

$$(\alpha' + \alpha'' + \alpha''')(x-a) + (\beta' + \beta'' + \beta''')(y-b) + (\gamma' + \gamma'' + \gamma''')(z-c) = 0$$

i.e.

$$Ax + By + Cz + D = 0, \tag{5.28}$$

where A, B, C, D are constants.

The choice of the plane is not unique. Here we have considered tetrahedron with three sides (viz. OA, OB, OC) equal, but asymmetry may lead to different such choices of the symmetric planes.

References

Akritas, M.G., and M.A. Bershady. 1996. *The Astrophysical Journal* 470:706.

Bandiera, R., and L. Hunt. 1989. *Data analysis in astronomy III*, ed. V. Di Gesu et al., 47. New York: Plenum.

Boggs, P.T., et al. 1990. *ACM Transactions on Mathematical Software* 15:348.

Branham, R.L., Jr. 1982. *Astronomical Journal* 87:928.

Chattopadhyay, T., and A.K. Chattopadhyay. 2006. *Astronomical Journal* 131:2452.

Djorgovski, S., and M. Davis. 1987. *The Astrophysical Journal* 313:59.

Dressler, A., et al. 1987. *The Astrophysical Journal* 313:42.

Efron, B., T. Hastie, I. Johnstone, and R. Tibshirani. 2004. *Annals of Statistics* 32(2):407.

Faber, S.M., and R.E. Jackson. 1976. *The Astrophysical Journal* 204:668.

Feigelson, D., and G.J. Babu. 1992. *The Astrophysical Journal* 397:55.

Hadi, A.S. 1992. *Journal of the Royal Statistical Society* B 54:761.

Isobe, T., et al. 1990. *The Astrophysical Journal* 364:104.

Kodaira, K., and N. Kashikawa. 2000. *Astrophysical Journal* 531:665.

Kormendy, J. 1977. *The Astrophysical Journal* 204:668.

Lutz, T.E. 1983. *Statistical methods in astronomy*, ed. C. Jaschek et al., 179. Noordwijk: ESA Sci & Tech. Pub.

Mondal, S., B. Warule, and T. Chattopadhyay. 2010. *Calcutta Statistical Association Bulletin* 62(247–248):277.

Nieto, J.-L., et al. 1990. *Astronomy & Astrophysics* 230:L17.

Press, W.H., et al. 1986. *Numerical recipes: the art of scientific computing.* Cambridge: Cambridge University Press.

Chapter - 6

Missing Observations and Imputation

6.1 Introduction

Statistical analysis with missing data is an important problem as the problem of missing observation is very common in many situations. During the last two decades different methods have been developed to tackle the situation. One possible way to handle missing values is to remove either all features or all objects that contain missing values. Another possibility is imputation where we fill in the missing values by inferring new values for them. The imputation method may not be applicable to some astronomical data sets as the missing value may arise from physical process and imputing missing values is misleading and can skew subsequent analysis of data. For example, the Lyman break technique (Giavalisco 2002) can identify high-redshift galaxies based on the absence of detectable emissions in bands corresponding to the FUV rest frame of the objects.

6.2 Missing Data Mechanism

Under the regression set-up with predictor X and response Y missing value problems often arise. To decide how to handle missing value problems, primarily we need to know why these values are missing. We may define four general missingness mechanisms.

6.2.1 Missingness Completely at Random (MCAR)

A variable value is missing completely at random if the probability of missingness is the same for all units. Under the regression set-up if the missing values are independent of both X and Y then these are called missing completely at random.

A.K. Chattopadhyay, T. Chattopadhyay, *Statistical Methods for Astronomical Data Analysis*, Springer Series in Astrostatistics 3,
DOI 10.1007/978-1-4939-1507-1_6

6.2.2 Missingness at Random (MAR)

Most missingness is not completely at random. A more general assumption, missing at random, is that the probability of a variable value is missing depends only on variable information. Under the regression set up, if the missing value depends on X but not on Y then these are called missing at random.

If X is age and Y is income of a group of persons, then if the probability of recording an income value is the same for all individuals irrespective of their age and income then the missing values are called MCAR. But if the probability of recording an income value varies according to the age of the respondent but does not vary according to the income of the respondent within an age group then the missing data are missing at random but not observed at random (MAR).

6.2.3 Missingness that Depends on Unobserved Predictors and the Missing Value Itself

Missingness is no longer at random if it depends on information that has not been recorded and this information also predicts the missing values. A particularly difficult situation arises when the probability of missingness depends on the variable itself. Under the regression set-up this type of situation arises when probability of response depends on both X and Y.

For statistical inference with missing information, we usually assume that the missingness pattern is MCAR or MAR.

Generally we try to include as many predictors in the model as possible so that the "missing at random" assumption is reasonable.

6.3 Analysis of Data with Missing Values

6.3.1 Complete Case Analysis

A direct approach is to exclude the missing observations from the analysis. Here we initially delete all units for which the outcomes or any of the inputs are missing. But this approach introduces significant bias in the final result. If many variables are included in a model, there may be very few complete cases so that most of the data would be excluded from the analysis. Further if the units with missing values differ systematically from the completely observed units, the final results are very much likely to be biased.

6.3.2 Imputation Methods

Under this approach we fill in the missing values by some suitable method. Different imputation techniques have been developed by several authors. This filled in data increases the precision of the estimates but may introduce a different kind of bias. Under single imputation, each missing values in a data set is replaced with one randomly imputed value whereas under multiple imputation each missing value is replaced by several imputed values in order to reflect uncertainty about the imputation model. Different single imputation methods are discussed below.

6.3.2.1 Mean Imputation

Let Y_{ij} be the ith observation corresponding to the jth variable and some of Y_{ij} values are missing for the jth variable Y_j. Under mean imputation Y_{ij} is estimated by $\overline{Y}_j^{(j)}$, the mean of the recorded values of the variable Y_j. Let us denote by n_j, the number of recorded observations for the jth variable Y_j and by $s_{jj}^{(j)}$, the estimated variance from the recorded values. Under MCAR $s_{jj}^{(j)}$ is a consistent estimator of the true variance.

The variance by considering both observed and imputed values is given by

$$\frac{(n_j - 1)s_{jj}^{(j)} + 0}{n - 1}$$

So, the sample variance obtained by using the filled in data under estimates the variance by a factor $(n_j - 1)/(n - 1)$.

Similarly it can be proved that the covariance values are also underestimated.

Mean imputation may also severely distort the distribution of the variable.

6.3.2.2 Hot Deck Imputation (Andridge and Little 2010)

Hot deck imputation involves replacing missing values of one or more variables for a non-respondent with observed values from a respondent that is similar to the non-respondent with respect to characteristics observed by both cases. The similar respondent may be randomly selected from a set of potential similar respondents. This is known as random hot deck method. Otherwise a single similar respondent is identified and values are imputed from that respondent. This particular similar respondent (known as nearest neighbour) is selected on the basis of some metric. This is known as deterministic hot deck method. Hot deck method does not use model fitting for the variable to be imputed and thus is less sensitive to model misspecification than an imputation method based on a parametric model. For this the hot deck method is very popular. But hot deck method heavily depends on the choice of metric to find similar respondent(s) for the respondents with missing values. One can choose such a metric on the basis of the available covariates in the following manner. Let $x_i = (x_{i1}, x_{i2}, \ldots, x_{im})$ be the values of m covariates for the observation i to create adjustment cells and $c(x_i)$ denote the cell in the cross-classification in which subject i falls. Then matching the observation i with missing values by the observation j may be based on the metric

$$d(i,j) = \begin{cases} 0 & \text{if} \quad j \in c(x_i) \\ 1 & \text{if} \quad j \notin c(x_i) \end{cases}$$

Other possible matric choices may be

1. Maximum absolute distance

$$d(i,j) = \max_k |x_{ik} - x_{jk}|$$

where the x_k values have been suitably scaled to make differences comparable.

2. Mahalanobis distance

$$d(i,j) = (x_i - x_j)'[\hat{var}(x_i)]^{-1}(x_i - x_j)$$

where $\hat{var}(x)$ is an estimate of the covariance matrix of x_i.

Once a metric is chosen there are several ways to define the set of similar observations for each observation with missing values. One possible method is to define the set of similar observations j with $d(i,j) < \delta$ for a pre specified maximum distance δ corresponding to the missing observation i. One similar observation j is then selected by a random draw from the set. Alternatively if the closest observation to i (denoted by j) is selected, the method is called a deterministic or nearest neighbour hot deck.

6.3.2.3 Cold Deck Imputation (Shao 2000)

A cold deck method imputes a non-respondent of Y variable by reported values from anything other than Y values. It may take values from a covariate and/or from a previous survey. Cold deck imputation is opposite to hot deck imputation in which a non-respondent is imputed by a respondent from the same variable in the current survey. Suppose we have a sample s selected from a finite population P consisting of some units represented by $i = 1 \ldots N$. The observed values are given by $\{Y_i, i \in r\}, r \subset s$. Suppose also that we have auxiliary data x_i's observed for all $i \in s$ and $x_i > 0$. Then the simplest cold deck imputes a non-respondent $Y_i, i \in s - r$ by x_i and the resulting Horvitz–Rhompson estimator of Y is

$$\hat{Y} = \sum_{i \in r} w_i Y_i + \sum_{i \in s-r} w_i x_i \tag{6.1}$$

where w_i is the survey weight associated with the ith sampled unit.

6.3.2.4 Warm Deck Imputation

This method is also known as ratio imputation. Under this imputation method, k imputation cells P_k are created such that $P_1 U P_2 \ldots U P_k = P$, according to a categorical auxiliary variable (as in hot deck) which is observed for every $i \in s$ and is different from X such that for every k the following model is valid.

$$Y_i = \beta_k x_i + x_i^{1/2} e_i \;\; i \in P_k \tag{6.2}$$

where β_k is an unknown parameter, e_i is independent of x_i with $E(e_i) = 0$ and $v(e_i) = \sigma_k^2 > 0$ which is unknown. Then, within imputation cell k, a non respondent Y_i is imputed by $\hat{\beta}_k x_i$, where

$$\hat{\beta}_k = \sum_{i \in r_k} w_i Y_i / \sum_{i \in r_k} w_i x_i \tag{6.3}$$

r_k is the set of respondent in the kth imputation cell and w_i is the survey weight associated with the ith sampled unit.

For dealing with data involving missing values, so far we have discussed methods like complete case analysis, hot deck, cold deck methods for imputation, etc. One may also use methods like maximum likelihood, Bayesian, multiple imputation, etc.

6.4 Likelihood Based Estimation: EM Algorithm

In general, there is no difference between ML estimation for incomplete data and complete data. But for incomplete data asymptotic standard error values are not reliable as the large sample normality is not readily applicable. Let us introduce a new notation to denote the observed and missing part of the complete data Y as $Y = (Y_{obs}, Y_{mis})$. Then the joint density function is given by

$$f_\theta(Y) = f_\theta(Y_{obs}, Y_{mis})$$

Hence the marginal density of observed values is given by

$$f_\theta(Y_{obs}) = \int f_\theta(Y_{obs}, Y_{mis}) dY_{mis}$$

Then the likelihood of θ based on the data Y_{obs} (ignoring the missing data) is given by

$$L_\theta(Y_{obs}) \propto f_\theta(Y_{obs}) \tag{6.4}$$

Inference about θ can be based on the likelihood $L_\theta(Y_{obs})$ under the assumption that incomplete data can be ignored. For each component of Y we define a missing data indicator.

If $Y = (Y_{ij})^{n \times k}$ be the matrix of n observations measured for k variables. We define a response indicator

$$\begin{aligned} R_{ij} &= 1 \text{ if } Y_{ij} \text{ is observed} \\ &= 0 \text{ if } Y_{ij} \text{ is missing} \end{aligned}$$

Here $R = (R_{ij})^{n \times k}$ is considered as a matrix of random variables and the joint distribution of R and Y is given by

$$f_{\theta,\psi}(Y, R) = f_\theta(Y) f_\psi(R|Y)$$

The conditional distribution of R given Y is indexed by an unknown parameter ψ which specified the missing data mechanism.

$$f_{\theta,\psi}(Y_{obs}, R) = \int f_\theta(Y_{obs}, Y_{mis}) f_\psi(R|Y_{obs}, Y_{mis}) dY_{mis} \tag{6.5}$$

Hence the likelihood of θ and ψ is given by

$$L_{\theta,\psi}(Y_{obs}, R) \propto f_{\theta,\psi}(Y_{obs}, R) \tag{6.6}$$

Now inference for θ may depend either by $L_\theta(Y_{obs})$ given by (6.4) or by $L_{\theta,\psi}(Y_{obs}, R)$ given by (6.6).

If $f_\psi(R|Y_{obs}, Y_{mis}) = f_\psi(R|Y_{obs}) \ldots (6.6.1)$ i.e. distribution of missing data mechanism does not depend on the missing values, then inference for θ from $L_\theta(Y_{obs})$ will be same as that from $L_{\theta,\psi}(Y_{obs}, R)$.

Equation (6.6.1) implies that probability of a particular component of Y is missing does not depend on the value of that component, i.e. missing values are missing at random. So the likelihood based inferences that ignore the missing data, missing value mechanism is required to be MAR and not necessarily MCAR.

If the likelihood is differentiable and unimodel, ML estimates can be found by solving the likelihood equation

$$\frac{\delta l_n L_\theta(Y_{obs})}{\delta\theta} = 0 \tag{6.7}$$

When a closed form of (6.7) cannot be found iterative methods can be applied. Let $\theta^{(0)}$ be the initial estimate, e.g. based on completely observed observations. Let $\theta^{(t)}$ be the estimate at the tth iteration. Then by Newton–Raphson method

$$\theta^{(t+1)} = \theta^{(t)} + I^{-1}(\theta^{(t)}|Y_{obs})\frac{\delta l_n L_\theta(Y_{obs})}{\delta\theta} \tag{6.8}$$

where $I(\theta|Y_{obs}) = -\frac{\delta^2 l_n L_\theta(Y_{obs})}{\delta\theta^2}$.

If the log likelihood function is concave and unimodel, then the sequence $\theta^{(t)}$ converges to the ML estimate of θ. It will converge in one step if $l_n L_\theta(Y_{obs})$ is a quadratic function of θ. For expectation Maximization (EM) algorithm (Dempster and Laird 1977) we do not require to calculate the second derivative. Basically there are only two steps, viz. E-step and M-step. Under E-step we calculate the conditional expectation of the missing data given the observed data and the current estimated values of the parameters. Then we substitute these expected values for missing values. Under M-step we calculate the ML estimate of θ assuming that there is no missing value.

The steps of E-M algorithm are as follows:

1. Replace missing values by estimated values.
2. Estimate parameters of the distribution.
3. Reestimate the missing values assuming that the new parameter estimates are correct values.
4. Reestimate the parameters and so on until convergence.

6.5 Multiple Imputation

Under multiple imputation, instead of filling in a single value for each missing value, Rubin (1987) proposed that each missing value should be replaced by a set of plausible values that represent the uncertainty about the right value to impute. Multiple imputation inference involves these distinct steps.

1. The missing data are filled in k times to generate k complete data sets.
2. The k complete data sets are analysed by using standard procedures.
3. The results from the k complete data sets are combined for the inference.

References

Andridge, R.R., and J.A.R. Little. 2010. *International Statistical Review* 78(1):40.

Dempster, A.P., N.M. Laird, and D.B. Rubin. 1977. *Journal of the Royal Statistical Society Series B* 39(1):1.

Giavalisco, M. 2002. *Annual Reviews of Astronomy and Astrophysics* 40:579.

Rubin, D.B. 1987. *Multiple imputation for nonresponse in surveys.* New York: Wiley.

Shao, J. 2000. *Survey Methodology* 26:79.

Chapter - 7

Dimension Reduction and Clustering

7.1 Introduction

For multivariate analysis with p variables the problem that often arises is the ambiguous nature of the correlation or covariance matrix. When p is moderately or very large it is generally difficult to identify the true nature of relationship among the variables as well as observations from the covariance or correlation matrix. Under such situations a very common way to simplify the matter is to reduce the dimension by considering only those variables (actual or derived) which are truly responsible for the overall variation. Important and useful dimension reduction techniques are Principal Component Analysis (PCA), Factor Analysis, Multidimensional Scaling, Independent Component Analysis (ICA), etc. Among them PCA is the most popular one. One may look at this method in three different ways. It may be considered as a method of transforming correlated variables into uncorrelated one or a method of finding linear combinations with relatively small or large variability or a tool for data reduction. The third criterion is more data oriented. In PCA primarily it is not necessary to make any assumption regarding the underlying multivariate distribution but if we are interested in some inference problems related to PCA then assumption of multivariate normality is necessary. The eigen values and eigen vectors of the covariance or correlation matrix are the main contributors of a PCA. The eigen vectors determine the directions of maximum variability whereas the eigen values specify the variances. In practice, decisions regarding the quality of the principal component approximation should be made on the basis of eigen value–eigen vector pairs. In order to study the sampling distribution of their estimates the multivariate normality assumptions became necessary as otherwise it is too difficult. Principal components are a sequence of projections of the data. The components are constructed in such a way that they are uncorrelated and ordered in variance. The components of a p-dimensional data set provide a sequence of best linear approximations. As only a few of

Electronic supplementary material The online version of this chapter (doi:10.1007/978-1-4939-1507-1_7) contains supplementary material, which is available to authorized users.

A.K. Chattopadhyay, T. Chattopadhyay, *Statistical Methods for Astronomical Data Analysis*, Springer Series in Astrostatistics 3,
DOI 10.1007/978-1-4939-1507-1_7

such linear combinations may explain a larger percentage of variation in the data, one can take only those components instead of p variables for further analysis.

More recently, ICA has emerged as a strong competitor to PCA and factor analysis. ICA finds a set of source data that are mutually independent, PCA finds a set of data that are mutually uncorrelated. ICA was primarily developed for non-Gaussian data in order to find independent components responsible for a larger part of the variation. ICA separates statistically independent original source data from an observed set of data mixtures.

Factor analysis is used to describe the covariance relationship among many variables in terms of a few underlying, but unobservable, random quantities called factors. Factor analysis can be used in situations where the variables can be grouped according to correlations so that all variables within a particular group are highly correlated among themselves but have relatively small correlation with variables in a different group. Thus each group of variables represents a single underlying factor. Factor analysis can be considered as an extension of PCA.

7.2 Principal Component Analysis

A PCA is concerned with explaining the variance–covariance structure through a few linear combinations of the original variables. Its general objectives are

1. data reduction
2. interpretation

Reduce the number of variables from p to k $(p > k)$. Let the random vector $X' = (X_1 \ldots X_p)$ have the covariance matrix Σ with eigen values $\lambda_1 \geq \lambda_2 \ldots \geq \lambda_p \geq 0$.

Consider the linear combinations

$$Y_1 = l_1{}'X = l_{11}X_1 + l_{21}X_2 + \ldots + l_{p1}X_p$$

$$Y_2 = l_2{}'X = l_{12}X_1 + l_{22}X_2 + \ldots + l_{p2}X_p$$

$$Y_p = l_p{}'X = l_{1p}X_1 + l_{2p}X_2 + \ldots + l_{pp}X_p$$

$$\begin{aligned} \text{Then } var(Y_i) &= l_i{}'\Sigma l_i \quad i = 1, 2, \ldots, p \\ cov(Y_i Y_k) &= l_i{}'\Sigma l_k \quad i, k = 1, 2, \ldots, p \end{aligned}$$

The first principal component is the linear combination with the maximum variance, i.e. it maximizes

$$V(Y_1) = {l_1}'\Sigma l_1$$

It is clear that $V(Y_1) = {l_1}'\Sigma l_1$ can be increased by multiplying any l_1 by some constant.

It is better to restrict attention to coefficient vectors of unit length.

First principal component is the linear combination ${l_1}'X$ that maximizes $var({l_1}'X)$ subject to ${l_1}'l_1 = 1$.

Second principal component is the linear combination ${l_2}'X$ that maximizes $var({l_2}'X)$ subject to ${l_2}'l_2 = 1$ and $cov({l_1}'X, {l_2}'X) = 0$ and so on.

Theorem 7.1 Let Σ be the covariance matrix associated with the random vector $X' = (X_1 \ldots X_p)$. Let Σ have the eigen value–eigen vector pairs $(\lambda_1, e_1)(\lambda_2, e_2)\ldots(\lambda_p, e_p)$ where $\lambda_1 \geq \lambda_2 \geq \ldots \geq \lambda_p \geq 0$. Then ith principal component is given by

$$Y_i = {e_i}'X = e_{1i}X_1 + e_{2i}X_2 + \ldots + e_{pi}X_p \;\; i = 1 \ldots p.$$

With these choices

$$\begin{aligned} V(Y_i) &= {e_i}'\Sigma e_i = \lambda_i \;\; i = 1 \ldots p \\ cov(Y_i Y_k) &= {e_i}'\Sigma e_k = 0 \;\; i \neq k \end{aligned}$$

If some λ_is are equal the choice of the corresponding coefficient vectors is same e_is and hence Y_is are not unique.

Proof: We know that $\max\limits_{l \neq 0} \frac{l'\Sigma l}{l'l} = \lambda_1$ is attained where $l = e_1$

But ${e_1}'e_1 = 1$ since the eigen vectors are normalized. Thus

$$\max_{l \neq 0} \frac{l'\Sigma l}{l'l} = \lambda_1 = \frac{{e_1}'\Sigma e_1}{{e_1}'e_1} = {e_1}'\Sigma e_1 = var(Y_1)$$

Similarly

$$\max_{l \perp e_1 \ldots e_k} \frac{l'\Sigma l}{l'l} = \lambda_{k+1} \;\; k = 1, \ldots, p-1.$$

For the choice $l = e_{k+1}$ with ${e_{k+1}}'e_i = 0 \;\; i = 1 \ldots k$ and $k = 1 \ldots p-1$,

$${e_{k+1}}'\Sigma e_{k+1} / {e_{k+1}}'e_{k+1} = {e_{k+1}}'\Sigma e_{k+1} = var(Y_{k+1})$$

But ${e_{k+1}}'(\Sigma e_{k+1}) = \lambda_{k+1}{e_{k+1}}'e_{k+1} = \lambda_{k+1}$

$$\begin{aligned} \text{So, } var(Y_{k+1}) &= \lambda_{k+1} \\ cov(Y_i Y_k) &= {e_i}'\Sigma e_k = {e_i}'\lambda_k e_k = \lambda_k {e_i}'e_k = 0 \;\; i \neq k \end{aligned}$$

Theorem 7.2 Let $X' = (X_1 \ldots X_p)$ have covariance matrix Σ with eigen value–eigen vector pair $(\lambda, e_1) \ldots (\lambda_p, e_p)$ where $\lambda_1 \geq \lambda_2 \ldots \lambda_p \geq 0$.

Let $Y_1 = e_1{}'X \;\; Y_2 = e_2{}'X \ldots Y_p = e_p{}'X$ be the principal components. Then

$$\begin{aligned} \sigma_{11} + \sigma_{22} \ldots + \sigma_{pp} &= \sum_1^p var(X_i) \\ &= \lambda_1 + \ldots + \lambda_p = \sum_1^p var(Y_i) \end{aligned}$$

Proof: We can write

$$\Sigma = P \triangle P' \quad \triangle = Diag(\lambda_1 \ldots \lambda_p)$$

$$P = (e_1 \ldots e_p) \quad PP' = P'P = I \quad Tr(\Sigma) = \sigma_{11} + \ldots + \sigma_{pp}$$

$$\begin{aligned} Tr(\Sigma) &= Tr(P \triangle P') = Tr(\triangle PP') \\ &= Tr(\triangle) = \lambda_1 + \ldots + \lambda_p \end{aligned}$$

$$\text{Thus } \sigma_{11} + \ldots + \sigma_{pp} = V(X_1) + \ldots + V(X_p)$$

$$\text{or, } \sum_1^p V(X_i) = Tr(\Sigma) = Tr(\triangle) = \sum_1^p var(Y_i)$$

or, total population variance

$$= \sigma_{11} + \ldots + \sigma_{pp} = \lambda_1 + \ldots + \lambda_p$$

Hence proportion of total variance due to the kth principal component is

$$= \frac{\lambda_k}{\lambda_1 + \ldots + \lambda_p}, \quad k = 1 \ldots p$$

If most (80–90 %) of the total population variance for large p can be attributed to the first one, two or three components, then there components can replace the original p variables.

The magnitude of e_{ki} measures the importance of the kth variable to the ith principal component irrespective of other variables.

Result: If $Y_1 = e_1{}'X \;\; Y_2 = e_2{}'X \ldots Y_p = e_p{}'X$ are principal components obtained from the covariance matrix Σ, then

$$\rho_{Y_i, X_k} = \frac{e_{ki}\sqrt{\lambda_i}}{\sqrt{\sigma_{kk}}} \quad i, k = 1 \ldots p$$

where (λ_i, e_i) are eigen value–eigen vector pairs of Σ.

Proof: Let $l_k{}' = [0, 0 \ldots 0, 1, 0, \ldots 0]$

So that $X_k = l_k{}'X$

and $cov(X_k Y_i)$

$= cov(l_k{}'X, e_i{}'X)$

$= l_k{}'\Sigma e_i$

$= l_k{}'\lambda_i e_i = \lambda_i l_k{}' e_i = \lambda_i e_{ki}$

$$\Rightarrow \rho_{Y_i X_i} = \frac{cov(Y_i X_k)}{\sqrt{V(Y_i)}\sqrt{V(X_k)}} = \frac{\lambda_i l_{ki}}{\sqrt{\lambda_i}\sqrt{\sigma_{kk}}} = \frac{e_{ki}\sqrt{\lambda_i}}{\sqrt{\lambda_{kk}}} \quad i, k = 1 \ldots p$$

7.2.1 An Example Related to Application of PCA (Babu et al. 2009)

Here analysis is based on four sets of GCs in Milky Way which have appropriate photometric and structural parameter values.

This consists of 50 GCs taken from Recio-Blanco et al. (2006). All the photometric data come from HST/WFPC2 observations in the F439W and F555W bands, the WFPC2 equivalents of the B and V filters, which are suited for a generic survey and constitute the best choice to identify new anomalous HBs. The parameters used for study are:

$\log T_{effHB}$: Maximum effective temperature along the HB. The effective temperature of an astronomical object is the temperature it would have if it acted like a black body, absorbing all the incoming radiation received at its surface and reradiating it all back to space.

M_V: Absolute magnitude measured using V band filter. Magnitude is the scale of brightness. Absolute magnitude is the apparent magnitude if the object is placed at a distance of 10 parsec.

c: Central concentration $= \log(r_t/r_c)$, where r_t is the tidal radius and r_c is the core radius.

R_{gc}: Distance from galactic centre in kpc (1 kpc = 1,000 parsec).

t_{rh}: Logarithm of core relaxation time at half-light radius. The relaxation time measures the time for the velocity of an object to be changed by gravitational perturbations from other objects. When the objects are in relaxed state, equipartition of kinetic energy occurs. The sizes of galaxies are difficult to measure since they don't possess clearly defined boundaries. Most galaxies simply get fainter and fainter in their outer regions, and the apparent size of the galaxy depends almost entirely on the sensitivity of the telescope used and the length of time for which the object is observed. To overcome this ambiguity, astronomers define the "half-light" or "effective" radius (re) as the radius within which half of the galaxy's luminosity is contained.

r_c: Core radius.

μ_v: Central surface brightness per square arc seconds.

[Fe/H]: Cluster metallicity. The metallicity of an object is the proportion of its matter made up of chemical elements other than hydrogen and helium.

Γ_{col}: Collisional parameter. The collisional parameter is defined as the probability of collisions, per unit time, for one star in the cluster and it was derived via the formula (King 2002):

$\Gamma_{col} = \frac{\log[5\times10^{-15}\sqrt{\sigma^3 r_c}]}{N_{star}}$, where $\sigma = 10^{[-0.4\times(\mu_v-26.41)]}$, N_{star} = Total number of stars in the cluster.

ρ_0: central luminosity density in Solar luminosities per cubic parsec.

r_h: half-light radius in parsec.

t_{rc}: core relaxation time in year.

He: initial helium abundance. It was taken from Salaris et al. (2004). They estimated the initial He content in about 30 % of the Galactic globular clusters (GGCs).

Hertzsprung–Russell diagram or the colour–magnitude diagram (CMD) is that in which the absolute magnitudes (intrinsic luminosity) of stars are plotted against their surface temperatures or colours. The scatter plot shows a strong correlation between luminosity and surface temperature among the average-size stars known as main sequence stars, with hot, blue stars having the highest luminosities and relatively cool, red stars having the lowest ones. The horizontal branch (HB) (already discussed in previous section) is a unique astrophysical tool for understanding and the interpretation of the CMDs of globular clusters. Its wider colour (temperature) distribution is usually called the HB morphology.

Fusi Pecci et al. (1993) defined some HB morphology parameters, viz. HBR, HB_{RE} , L_t, $(B-V)_{peak}$, BT, DT which we have used in our work. They are defined as follows:

HBR: (B-R)/(B+V+R)

HB_{RE}: The HB Red Extreme, defined as the intrinsic $(B-V)_0$ colour of the point.

L_t: The total length of the HB measured from HB_{RE} down to the blue end of the HB.

$(B-V)_{peak}$: The dereddened colour of the peak of the HB star distribution, measured by dividing the whole length into bins starting from HB_{RE} and counting the stars populating each bin, perpendicular to the adopted ridge line.

BT: The length of the Blue Tail, measured along the ridge line of the HB starting from $(B-V)_{peak}$ down to the adopted blue HB extreme.

A powerful tool in the cluster classification was introduced by Dickens (1972). He defined seven HB types [Dickens type (DT)] from type 1, corresponding to blue HB, up to type 7, with completely red HB.

7.2.1.1 The Correlation Vector Diagram (Biplot)

A matrix of rank 2 can be displayed as a biplot which consists of a vector for each row and a vector for each column, chosen so that each element of the matrix is exactly the inner product of the vectors corresponding to its row and its column (Gabriel 1971). If the rank of a matrix is higher, it can be displayed by a biplot of a matrix of rank 2 that approximates the original matrix. In PCA, a biplot can be used to show inter unit distances and indicate the clustering of units, at the same time to display the variances and correlations of the parameters.

Any matrix of observations y of dimension $m \times n$ (in the present work n is the number of galaxies and m is the number of parameters) can be written by singular value decomposition as

$$y = \sum_{i=1}^{r} \lambda_i p_i q_i', (\lambda_1 \geq \lambda_2 \geq \ldots \geq \lambda_r) \tag{7.1}$$

where r is the rank of the matrix y and λ_i, p_i & $q_i{}'$ are the singular value, singular column and singular row, respectively. Then by applying the method of least squares fitting a matrix of rank 2 approximation to y is

$$y = \sum_{i=1}^{2} \lambda_i p_i q_i{}', \tag{7.2}$$

and the corresponding measure of goodness of fit is given by

$$\rho(2) = \frac{{\lambda_1}^2 + {\lambda_2}^2}{\sum_{i=1}^{r} {\lambda_i}^2} \tag{7.3}$$

If $\rho(2)$ is close to 1, then the biplot will be a good approximation to y. If the variance–covariance matrix be denoted by $S^{m\times n} = \frac{1}{n} y' y = (s_{ij})$ and the correlation matrix by $R^{m\times n} = (r_{ij})$, then it can be shown that

$$y^{m\times n} \sim G'^{n\times 2} H'^{2\times m}, \tag{7.4}$$

where

$$G = ({p_1}^{n\times 1} {p_2}^{n\times 1})\sqrt{n} = ({g_1}^{n\times 1} {g_2}^{n\times 1}) \tag{7.5}$$

and

$$H = (1/\sqrt{n})(\lambda_1 q_1 \lambda_2 q_2) = ({h_1}^{m\times 1} {h_2}^{m\times 1}) \tag{7.6}$$

Further,

$$s_{ij} \sim h_i{}' h_j,$$
$$s_j{}^2 \sim ||h_j||^2$$
$$r_{ij} \sim \cos(h_i, h_j).$$

Using PCA with the present data set, we see that if we take HBR, HB_{RE}, DT, L_t, BT and $(B-V)_{peak}$ by excluding $\log T_{effHB}$ from the set then in the first component there are HBR, HB_{RE}, DT, $(B-V)_{peak}$ while in second component there are L_t and BT (Fig. 7.1). Here, Figs. 7.1, 7.2, 7.3, and 7.4 are the biplots (correlation vector diagrams) corresponding to PCA.

At the second step we have chosen two representative morphological parameters from the two components, namely HBR and L_t and studied their variations with respect to intrinsic parameters [Fe/H] and M_v. Initially, we have chosen these two intrinsic parameters. Later, we have included more independent parameters through stepwise multiple regression technique for a better prediction of the morphology parameter. Fist, we have taken HBR, L_t and [Fe/H] (Fig. 7.2). Here, PCA shows that [Fe/H] belongs to the same component with HBR. Then we have taken HBR, L_t and M_v. PCA shows that M_v is in the same component with L_t (Fig. 7.3). From this it may be inferred that the choice of HBR is not sensitive to variation in M_v while choice

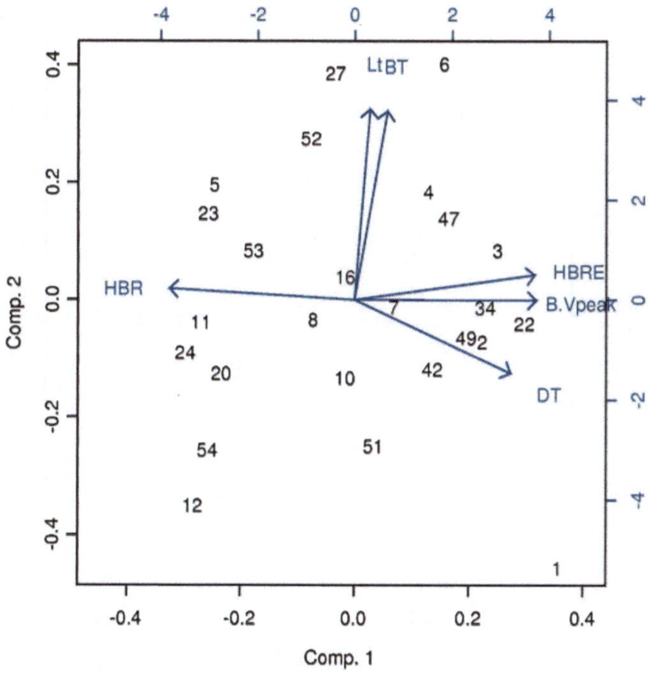

Figure 7.1 PCA for the HB morphology parameters HBR, HB_{RE}, DT L_t, BT and $(B-V)_{peak}$ of data set

of L_t is not sensitive to variation in [Fe/H] values. Finally, we have chosen L_t, HBR, [Fe/H], M_v and $\log T_{effHB}$ together and here from PCA it appears that $\log T_{effHB}$ has contribution to two different components of which in one component there is [Fe/H] and in the other there is M_v (Fig. 7.4). Thus, as a result $\log T_{effHB}$ seems to be sensitive to both of the independent parameters [Fe/H] and M_v. Hence, it may be concluded that $\log T_{effHB}$ may be selected as the proper HB morphology parameters for comparison.

The R code for PCA and Biplot is given below. The corresponding data file is given in the Appendix.

```
data ← read.table ("C:\\Users\\Tanuka \\Desktop\\
NGC5128new \\.txt",
header = TRUE)
cor (data)
eigen (cor (data))
prcomp (data, cor = TRUE)
summary (pc.cr ← prcomp (data, cor = TRUE)
```

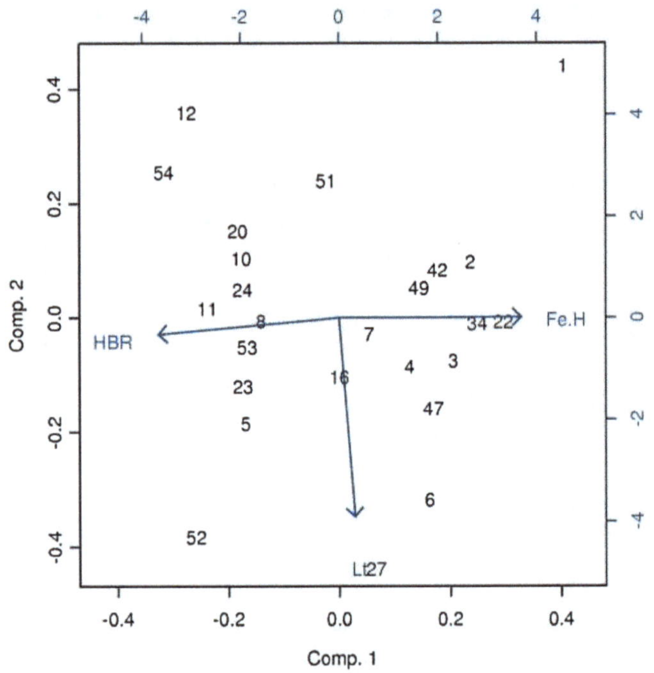

Figure 7.2 PCA for the HB morphology parameters HBR, [Fe/H] and L_t of data set

```
pc.cr ← prcomp (data, cor = TRUE, scores = TRUE)
x ← pc.cr$rotation
x
t ← pc.cr$scores
t
plot (t[,1], t[,2])
biplot (pc.cr)
```

7.3 Independent Component Analysis

ICA has emerged as a strong competitor to PCA and factor analysis. ICA was primarily developed for non-Gaussian data in order to find independent components (rather than uncorrelated as in PCA) responsible for a larger part of the variation. ICA separates statistically independent component data, which is the original source data, from an observed set of data mixtures. All information in the multivariate data sets are not equally important. We need to extract the most useful information. ICA extracts and reveals useful information form the whole data set. This technique has been applied in various fields like speech processing, brain imaging, stock predictions, etc.

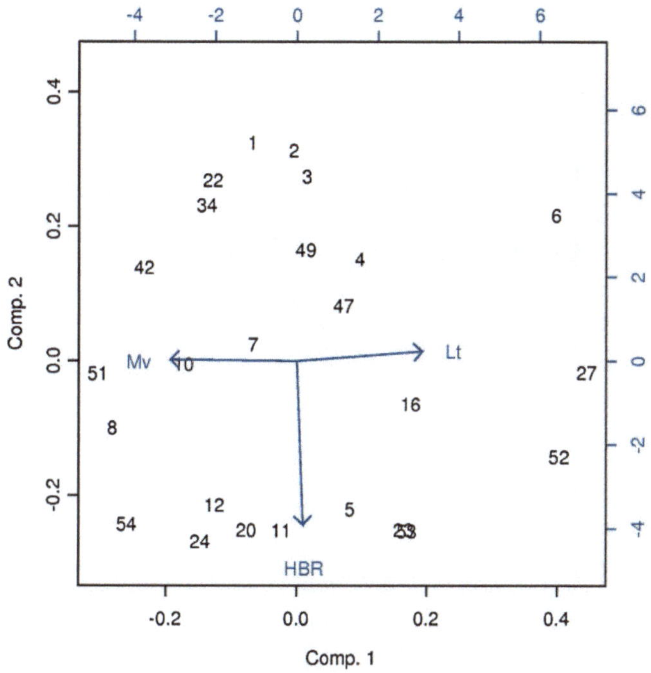

Figure 7.3 PCA for the HB morphology parameters HBR, M_v and L_t of data set

Suppose there are n observations on each of p correlated variables. Let us denote the data matrix by $X^{n \times p}$. By singular value decomposition one can write $X = UDV'$. Writing $S = \sqrt{n}U$ and $A' = DV'/\sqrt{n}$, we have $X = SA'$ and hence each of the columns of X is a linear combination of the columns of S. Now since U is orthogonal and assuming that the columns of X have mean zero, it is easy to show that the columns of S have zero mean, unit variance and they are uncorrelated. In terms of random variables we can interpret the PCA as an estimate of a latent variable model $X = AS$. But given any orthogonal $p \times p$ matrix R, we can write $X = AS = AR'RS = A_1S_1$, where $A_1 = AR', S_1 = RS$ and $cov(S_1) = Rcov(S)R' = 1$. Hence it is impossible to identify any particular latent variable as a unique underlying source.

ICA was most clearly stated by Comon (1994). Formally, the classical ICA model is of the same form:

$$X = AS \tag{7.7}$$

where A is the nonsingular mixing matrix. So A^{-1} is the unmixing matrix. The main goal of ICA is to estimate the unmixing matrix A^{-1} and thus to

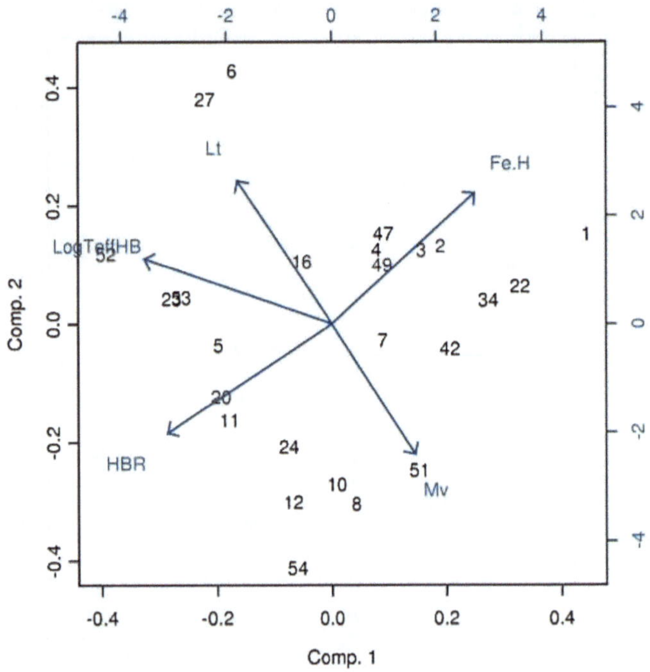

Figure 7.4 PCA for the HB morphology parameters L_t, HBR, [Fe/H], M_v and $\log T_{effHB}$ of data set

recover the hidden source using $S_k = A_k^{-1}X$, where A_k^{-1} is the kth row of A^{-1}. Lack of correlation determines the second order cross moments (covariance) of a multivariate distribution while, in general statistical independence determines all of the cross moments. These extra moment conditions allow us to identify the elements of the matrix A uniquely. Since the multivariate Gaussian distribution is determined by the second order moments alone, it is the exception, and any Gaussian independent component can be determined only up to a rotation. Hence the identifiability problem related to PCA or factor analysis can be avoided if we assume that S_i's are independent and non-Gaussian.

In the model, it is assumed that the data variables are lines or non linear mixtures of some latent variables, and the mixing system is also unknown. Equation (7.7) can be written as:

$$X_i = a_{i1}S_1 + a_{i2}S_2 + \ldots + a_{ip}S_p, \quad i = 1, 2, \ldots, p. \qquad (7.8)$$

The S_i's are statistically mutually independent, where a_{ij}'s are the entries of the nonsingular matrix A. All we observe are the random variables X_i, and we have to estimate both the mixing coefficients a_{ij} and the independent components S_i, using the X_i.

There are many computer algorithms for performing ICA. A first step in those algorithms is to whiten (sphere) the data. This means that any correlations in the data are removed, i.e. the data are forced to be uncorrelated. Mathematically speaking, we need a linear transformation V such that $Z = VX$, where $E(ZZ') = 1$. This can be easily accomplished by choosing $V = C^{-1/2}$, where $C = E(XX')$.

After sphering, the separated data can be found by an orthogonal transformation on the whitened data Z. ICA can be carried out in different ways like maximization of non-Gaussianity, minimization of mutual information, etc. Here we have concentrated on maximization of non-Gaussianity.

7.3.1 ICA by Maximization of Non-Gaussianity

$$X = AS, \quad VX = VAS$$
$$\Rightarrow X = (VA)S, \tag{7.9}$$

which implies that Z_i is closer to Gaussian that S_i. S_i is estimated by Z_i through maximization of non-Gaussianity. From Eq. (7.9) we can write

$$S = WZ, \tag{7.10}$$

where $W = (VA)^{-1}$.

We can measure non-Gaussianity by *Negentropy* (Hyvarinen et al. 2001). The entropy of a discrete source S with possible values $s_1, s_2, \ldots, s_n$ is defined as:

$$H(S) = -\sum_{i=1}^{n} p_s(s_i) \log p_s(s_i), \tag{7.11}$$

p_s is the mass function of S. On the other hand, for a continuous source, S, the entropy is called differential entropy which is given by:

$$H(S) = -\int p_s(\eta) \log p_s(\eta) d\eta, \tag{7.12}$$

$p_s(\eta)$ is the density function of S. Negentropy is the difference between the differential entropy of a source S from the differential entropy of a Gaussian source with the same covariance of S. It is denoted by $J(S)$ and defined as follows:

$$J(S) = H(S_{Gauss}) - H(S), \tag{7.13}$$

where S_{Gauss} is a Gaussian random variable with the same variance as S. It can be proved that among all random variables with equal variances, Gaussian variables have the maximum entropy (Hyvarinen et al. 2001). As such, when applied to a sample the expectations are replaced by data averages.

Negentropy is always non-negative, and it is zero if and only if S has a Gaussian distribution. Negentropy has an interesting property that it is invariant for invertible linear transformation. It is also a robust measure of non-Gaussianity. Here we estimate S by maximizing the distance of its entropy from Gaussian entropy as the noises are assumed to be Gaussian and if the signals are non-Gaussian only then they can be separated from the noise. If the signals are Gaussian, then ICA will not work.

7.3.2 Approximation of Negentropy

One drawback of negentropy is that it is very difficult to compute. That's why it needs to be approximated (Hyvarinen et al. 2001). The approximation is given by:

$$J(S) \propto (E[G(S)] - E[G(S_{Gauss})])^2, \tag{7.14}$$

where G is a non-quadratic function. In particular, G should be so chosen that it does not grow too fast. Two popular choices of G are:

$$G_1(S) = \frac{1}{a} \log \cosh(aS)$$

$$G_2(S) = -e^{-s^2/2} \tag{7.15}$$

where $1 \leq a \leq 2$ is some suitable constant, which is often taken equal to 1.

7.3.3 The FastICA Algorithm

There are many algorithms which do ICA like FastICA, ProDen (Hastie and Tibshirani 2003), KernelICA, etc. The fastICA algorithm is a commonly used one, including industrial applications. This algorithm was developed by Hyvarinen and Oja (2000). In this method the independent components are estimated one by one. This algorithm converges very fast and is very reliable. This algorithm is also very easy to use. Our objective is to maximize $J(S)$. Now this is equivalent to maximizing $E[G(WZ)]$ as given in Eq. (7.13) under the constraint $||W|| = 1$. For the sake of notational and computational complicity, we consider one particular component. We are interested in finding out the optima of $E[G(W_K^T Z)]$ under the constraint $||W|| = 1$, where $W_K^T Z$ is the kth component of WZ. This optimization problem can be solved by the Lagrange multiplier method. The objective function is

$$O(W_K) = E[G(W_K^T Z)] - \beta(W_K^T W_K - 1).$$

We take the derivative of $O(W_K)$ with respect to W_K, set it to zero and get

$$F(W_K) = \frac{\partial O(W_K)}{\partial W_K} = E[Zg(W_K^T Z)] - \beta W_K = 0,$$

where $g(\cdot)$ is the derivative of the function $G(\cdot)$. We solve this system of equations iteratively by the Newton–Raphson method:

$$W_{K+1} = W_k - J_F^{-1}(W_K)F(W_K),$$

where $J_F(W_K)$ is the Jacobian of function $F(W_K)$, which is given by

$$J_F(W_K) = \frac{\partial F}{\partial W_K} = E[ZZ^T g'(W_K^T Z)] - \beta I.$$

The first term on the right-hand side of above equation can be approximated as

$$E[ZZ'g'(W_K^T Z)] \approx E[ZZ']E[g'(W_K^T Z)] = E[g'(W_K^T Z)]I.$$

Thus the Jacobian becomes diagonal.

$$J_F(W_K) = [E\{g'(W_K^T Z)\} - \beta]I.$$

From expression ( ), the Newton–Raphson iteration becomes

$$W_{k+1} = W_K - \frac{1}{E[g'(W_K^T Z)] - \beta}[E\{Zg(W_K^T Z)\} - \beta W_K].$$

Multiplying both sides by the scaler $\beta - E[g'(W_K^T Z)]$ we get

$$\begin{aligned} W_{K+1}[\beta - E[g'(W_K^T Z)]] &= W_K[\beta - E[g'(W_K^T Z)]] \\ &+ E[Zg(W_k^T Z)] - \beta W_K \end{aligned}$$

implying,

$$\begin{aligned} W_{K+1}[\beta - E[g'(W_K^T Z)]] &= \beta W_K - E[g'(W_K^T Z)]W_K \\ &+ E[Zg(W_K^T Z)] - \beta W_K \end{aligned}$$

and finally,

$$W_{K+1} = E[Zg(W_K^T Z)] - E[g'(W_K^T Z)]W_K.$$

It is to be noted that we are using the representation W_{K+1} for the left-hand side, while its value is actually multiplied by a scaler. This is taken care of by renormalization, in the FastICA algorithm.

7.3.4 ICA Versus PCA

Both ICA and PCA are used for analysing large data sets. Whereas ICA finds a set of source data that are mutually independent, PCA finds a set of data that are mutually uncorrelated. ICA was originally developed for separating mixed audio signals into independent sources. In this paper we make the comparison by analysing GC data.

The purpose of PCA is to reduce the original data set of two or more sequentially observed variables by identifying a small number of meaningful components. To start, the data are organized in an $n \times p$ matrix $X' = [x_1, \ldots, x_p]$, where the x_i vector describes the n globular clusters (GCs) for the ith parameter. To be precise in this discussion and without loss of generality, we can assume that the data matrix has dimension $n \geq p$ with rank $r \leq p$. The data are then centred by subtracting the average from each value. The procedure consists in finding the eigen values and eigen vectors of the covariance matrix. Eigen values correspond to the variance of PCs and eigen vectors correspond to the loadings of the different parameters in a particular component.

PCA, based on the linear correlation between data points, shows a way to extract parameters or variables which are linearly uncorrelated. Although required to be linearly uncorrelated, because of the higher order correlations these parameters are not necessarily independent.

ICA (Hyvarinen et al. 2001; Stones 2004) is based on the basic assumption that the source components are statistically independent in the sense that the value of one gives no information about the values of the others. For non-Gaussian variables, the p.d.f.s need all moments to be specified, and higher order correlations must be taken into account to establish independence. It is expected that for a non-Gaussian situation ICA will perform better than PCA in terms of the homogeneity of data corresponding to the different groups formed with respect to the important components.

We must fix the number of independent components to be sought. In practice, before the ICA algorithm is applied, the observed data are often preprocessed to remove the correlation between the observed variables, which is called whitening. The FastICA algorithm uses PCA as the whitening method. At present there is no better method available to automatically determine the optimum number of ICs. In this paper, the number of ICs is determined by the number of PCs chosen (Albazzaz and Wang 2004). PCA is performed to determine this number and find a breakpoint in the eigen value spectrum. If there is no clear breakpoint in the spectrum, we can keep the number of leading components that carry some arbitrary percentage of the variance (Hyvarinen et al. 2001). In our work it is difficult to find a clear breakpoint. Under such a situation, generally, the number of PCs is fixed which have a cumulative variance greater than 50–80 %. Here, we fix the cumulative variance as 80 % and as a result we have decided to take four components for further analysis.

7.3.5 An Example (Chattopadhyay et al. 2013)

Analysis is based on the sample of globular clusters (GCs) of the early-type central giant elliptical galaxy in the Centaurus group, NGC 5128, whose structural parameters have been derived by McLaughlin et al. (2008). The distance is that adopted by McLaughlin et al. (2008), namely 3.8 Mpc. The sample consists of 130 GCs (three outliers have been excluded during cluster analysis) whose available structural and photometric parameters are tidal radius (R_{tid}, in pc), core radius (R_c, in pc), half light radius (r_h, in pc), central volume density ($\log \rho_0$, in $M_\odot$ pc^{-3}), $\sigma_{p,0}$ (predicted line of sight velocity dispersion at the cluster centre, in km s^{-1}), two-body relaxation time at the model projected half mass radius (t_{rh}, in year), galactocentric radius (R_{gc}, in kpc), the concentration ($c \sim \log(R_{tid}/R_c)$), the dimensionless central potential of the best fitting model (W_0), the extinction-corrected central surface brightness at F606W bandpass (μ_0 in mag arcsec^{-2}), V surface brightness averaged over $r_h((\mu_0))_h$ in mag arcsec^{-2}, the logarithm of integrated model mass ($\log M_{tot,}$ in $M_\odot$), Washington T_1 magnitude, extinction corrected colour $(C - T_1)_0$ and metallicity determined from colour ([Fe/H], index).

The radial velocities (V_r, in km s^{-1}) are available for 50 GCs (Woodley et al. 2007). The position angles (ψ, east of north) are available for all 127 GCs (Woodley et al. 2007). The ages and metallicities ([Z/H], in dex) using Lick indices of some of the GCs have been used from Chattopadhyay et al. (2009).

The entire data set of 130 GCs with all the parameters (used from the literature as well as derived by the authors) are listed in Chattopadhyay et al. (2009).

In order to test the normality of the distribution pattern of a variable, the Shapiro–Wilk test is used. This test was published in 1965 by Shapiro and Wilk (1965). The null hypothesis of this test is that the data are normally distributed. The test statistic is $W = \frac{(\sum_{i=1}^{n} a_i x_{(i)})^2}{\sum_{i=1}^{n} (x_i - \bar{x})^2}$, where n is the number of observations, $x_{(i)}$ are the ordered sample value ($x_{(1)}$ is the smallest) and a_i are constants generated from the means, variances and covariances of the order statistics of a sample of size n from a normal distribution. But in our case the data set used is multivariate. So, in this paper we have used the multivariate extension of the Shapiro–Wilk test. Here the null hypothesis was that the entire data set follows a multivariate normal distribution. Here the test statistic is based on the ordered sample vectors and the vector of constants generated from the mean vector and the dispersion matrix of the vector of order statistics of a sample of size n from a multivariate normal distribution. If the p-value is less than the chosen level of significance, the null hypothesis is rejected or in other words, the data are not multivariate normally distributed. We found that the p-value of the test was 4.231×10^{-14}, which is too small. Thus the null hypothesis has been rejected at the 5 % level of significance and we could conclude that the data set does not follow a multivariate normal distribution.

On the basis of PCA of the present data set the total percentage of variation by the first four components was found to be almost 87 % and if we add the fifth component, this amount becomes almost 94 %. As the increase is not significant we have stopped at four components. Then we have done cluster analysis (CA) by k-means clustering on the basis of principal components and independent components. Three groups (viz. G1, G2 and G3) have been found as a result of CA with respect to the four principal and independent components, respectively (found in the present analysis). In this work, we have done clustering on the basis of principal and independent components, respectively, whereas in our previous work, we have done clustering on the basis of three significant parameters, viz. V surface brightness averaged over $r_h((\mu_v)_h)$, half-light radius (r_h), and predicted line of sight velocity dispersion at the cluster centre ($\sigma_{p,0}$), but these three parameters are extracted through PCA and as such that classification may also be treated as the classification through PCA and it is also apparent from the corresponding leadings of the PCs. We have applied the same technique as in Chattopadhyay et al. (2009), viz. the work by Sugar and James (2003) to find the optimum number of clusters in both cases.

Under this method, we have first determined the structures of subpopulations (clusters) for varying number of clusters taking $K = 1, 2, 3, 4$, etc. For each such cluster formation, we have computed the values of a distance measure $d_k = (1/p) \min_x E[(x_k - c_k)'(x_k - c_k)]$ which is defined as the distance of the x_k vector (values of the parameters) from the centre c_k (which is estimated as the mean value), p is the order of the x_k vector. Then to find the optimum number of clusters the following steps are followed. Let us denote by d'_k the estimate of d_k at the Kth point. Then d'_k is the minimum achievable distortion associated with fitting K centres to the data. A natural way of choosing the number of clusters is to plot d'_k versus K and look for the resulting distortion curve. This curve is always monotonic decreasing. Initially, one would expect much smaller drops, i.e. a levelling off for K greater than the true number of clusters because past this point adding more centres simply partitions within groups rather than between groups. According to Sugar and James (2003) for a large number of items the distortion curve when transformed to an appropriate negative power (e.g. $p/2$) will exhibit a sharp "jump" (if we plot k versus transformed d'_k). Then the jumps have been calculated in the transformed distortion as $J_k = (d_k'^{-p/2} - d_{k-1}'^{-p/2})$. The optimum number of clusters is the value of k at which the distortion curve levels off as well as its value associated with the largest jump.

It has been found that the number of GCs in G1 are 7 and 5, in G2 are 85 and 95, and in G3 are 35 and 27 in clustering with respect to the principal and independent components, respectively. Here, the number of the GCs in group G1 is significantly lower compared to those in the other two groups G2 and G3 in both cases, unlike the comparable numbers of members in all three groups found by Chattopadhyay et al. (2009). This difference may be due to the use of a model based clustering technique (Qiu et al. 2007) on a non-Gaussian data set in Chattopadhyay et al. (2009).

Within cluster sum of squares (WSS) is a tool to check whether a clustering is good or poor. It is the sum of squares of the distances of the observations within cluster from the cluster centre, which is called the cluster centroid. Mathematically,

$$WSS = \sum_{x \in C_i} (x - r_i)^2.$$

where the summation is taken over the observations x within the cluster C_i and r_i is the cluster centroid. A good clustering yields clusters where they have a small within cluster sum of squares (and a high between cluster sum of squares). In other words, we choose the best clustering, in terms of the minimum within cluster sum of squares. From the analysis based on the within cluster sum of squares, we can say that on the basis of this metric, the ICA classification is much better in comparison with the PCA classification

Table

Comparison of within cluster sum of squares

	Within cluster sum of squares (IC)	Within cluster sum of squares (PC)	Within cluster sum of squares (model based clustering (Chattopadhyay et al. 2009))
G1	12.676	374.130	409.362
G2	210.673	2561.471	1745.402
G3	122.977	1351.861	437.855

and the previous model based clustering method. We may further try to investigate how much worse the PCA classification is on the basis of a confusion matrix assuming that the ICA classification is much closer to the actual situation. As the number of observations in G1 obtained corresponding to PCA and ICA are 7 and 5, respectively (which are very much insignificant compared to the total size) we have constructed the confusion matrix on the basis of G2 and G3, i.e. on the basis of $(127 - 12 = 115)$ GCs. A confusion matrix contains information about actual and confused classifications done by a classification.clustering system. The performance of such systems is commonly evaluated using the data in the matrix. The diagonal elements represent correctly classified observations while the cross-diagonal elements represent misclassified observations or confusions. The following is the confusion matrix for our analysis:

		PC		Total
		G2	G3	
IC	G2	84	10	94
	G3	0	21	21

From the above confusion matrix, we can say that out of 94 GCs, which are expected to be in G2, 84 are correctly classified by the PCA method and out of 21 GCs, which are expected to be in G3, 21 are correctly classified by the PCA method and hence the percentage of confusion is around 10 and 0, respectively. The R code for ICA is as below.

```
data2 ← read.table (“C:\\Users\\Tanuka \\Desktop\\
NGC5128new \\.txt”,
header = TRUE)
data2
library(fastICA)
fastICA(data2, 4, fun = c(“log cosh(S)”), alpha =1.0,
maxit = 200, tol = 1e-04)
```

The data file NGC5128new.tex is given in the Appendix.

7.4 Factor Analysis

Factor analysis is a statistical method used to study the dimensionality of a set of variables. In factor analysis, latent variables represent unobserved constructs and are referred to as factors or dimensions. Factor analysis attempts to identify underlying variables, or factors, that explain the pattern of correlations within a set of observed variables. Factor analysis is often used in data reduction to identify a small number of factors that explain most of the variance that is observed in a much larger number of manifest variables.

Suppose the observable random vector X with p components has mean vector μ and covariance matrix Σ. In the factor model we assume that X is linearly dependent upon a few(m$<$ p) unobservable random variables $F_1 F_2, \ldots,$ F_m called common factors and p additional sources of variation $\in_1, \in_2, \ldots, \in_p$ called the errors (or specific factors).

Then the factor model is

$$\underset{p\times 1}{X} = \underset{p\times 1}{\mu} + \underset{p\times m}{L}\ \underset{m\times 1}{F} + \underset{p\times 1}{\in} \tag{7.16}$$

$$\begin{aligned}
X_1 - \mu_1 &= l_{11}F_1 + l_{12}F_2 + \ldots + l_{1m}F_m + \in_1 \\
X_2 - \mu_1 &= l_{21}F_1 + l_{22}F_2 + \ldots + l_{2m}F_m + \in_2 \\
&\vdots \\
X_p - \mu_p &= l_{p1}F_1 + l_{p2}F_2 + \ldots + l_{pm}F_m + \in_p
\end{aligned}$$

The coefficients l_{ij}s are called the loading of the ith variable on the jth factor so the matrix L is the matrix of factor loadings. Here $\in_i$ is associated only with the ith response X_i. The p deviations $X_1 - \mu_1 \ldots X_p - \mu_p$ are expressed in terms of $p+m$ random variables $F_1, F_2, \ldots, F_m, \in_1, \ldots \in_p$ which are unobservable (but in multivariate regression independent variables can be observed).

With some additional assumption on the random vectors F and $\in$, the model implies certain covariance relationships which can be checked.

We assume that

$$E(F) = 0^{m\times 1} \quad cov(F) = E(FF') = I^{m\times m}$$

$$E(\in) = 0^{p\times 1} \quad cov(\in) = E(\in\in') = \psi = \begin{pmatrix} \psi_1 & 0 & \ldots & 0 \\ 0 & \psi_2 & \ldots & 0 \\ 0 & 0 & \ldots & \psi_p \end{pmatrix}$$

$$\text{and } cov(\in, F) = E(\in, F) = 0^{p\times m} \tag{7.17}$$

The model $X - \mu = LF + \in$ is linear in the common factors. If the p response of X are related to the underlying m factors in a non linear form, then the covariance structure $LL' + \psi$ may not be adequate. The assumption of linearity is inherent here.

These assumption and the relation (7.16) constitute the orthogonal factor model.

The orthogonal factor model implies a covariance structure for X.

$$\begin{aligned}
\text{Here } (X-\mu)(X-\mu)' &= (LF+\in)(LF+\in)' \\
&= (LF+\in)((LF)'+\in') \\
&= LF(LF)' + \in(LF)' + LF\in' + \in\in' \\
&= LFF'L' + \in F'L' + LF\in' + \in\in'
\end{aligned}$$

$$\begin{aligned}\Sigma &= \text{covariance matrix of } X \\ &= E(X-\mu)(X-\mu)' \\ &= LE(FF')L' + E(\in F)'L' + LE(F\in') + E(\in\in') \\ &= LIL' + \psi = LL' + \psi\end{aligned}$$

$$\text{Again } (X-\mu)F' = (LF+\in)F' = LFF' + \in F'$$

$$\text{or, } cov(X,F) = E(X-\mu)F' = E(LF+\in)F' = LE(FF') + E(\in F') = L$$

Now $\Sigma = LL' + \psi$ implies

$$\left.\begin{aligned} var(X_i) &= {l_{i1}}^2 + \ldots + {l_{im}}^2 + \psi_i \\ cov(X_i X_k) &= l_{i1}l_{k1} + \ldots + l_{im}l_{km}\end{aligned}\right\} \quad (7.18)$$

$$cov(XF) = L \Rightarrow cov(X_i F_j) = l_{ij}$$

$$\Rightarrow V(X_i) = \sigma_{ii} = {l_{i1}}^2 + \ldots + {l_{im}}^2 + \psi_i$$

Let ith communality $= {h_i}^2 = {l_{i1}}^2 + \ldots + {l_{im}}^2$

Then $\sigma_{ii} = {h_i}^2 + \psi_i \;\; (i = 1 \ldots p)$

${h_i}^2 =$ sum of squares of loadings of ith variable on the m common factors.

Given a random sample of observations $\overset{p\times 1}{x_1}, x_2 \ldots \overset{p\times 1}{x_n}$, the basic problem is to decide whether Σ can be expressed in the form (7.18) for reasonably small value of m, and to estimate the elements of L and ψ.

Here the estimation procedure is not so easy. Primarily we have from the sample data estimates of the $\frac{p(p+1)}{2}$ distinct elements of the upper triangle of Σ but on the RHS of (7.18) we have $pm + p$ parameters, pm for L and p for ψ. The solution will be indeterminate unless $\frac{p(p+1)}{2} - p(m+1) \geq 0$ or $p > 2m$. Even if this condition is satisfied L is not unique.

Proof: Let $\overset{m \times m}{T}$ be any $\perp$ matrix so that $TT' = T'T = I$

Then (7.16) can be written as

$$X - \mu = LF + \in = LTT'F + \in = L^*F^* + \in \tag{7.19}$$

$$\text{where } L^* = LT \text{ and } F^* = T'F$$

$$\text{Since } E(F^*) = T'E(F) = 0$$

$$\text{and } cov(F^*) = T'C\dot{o}v(F)T = T'T = I$$

It is impossible to distinguish between loadings L and L^* on the basis of the observations on X. So the vectors F and $F^* = T'F$ have the same statistical properties and even if the loadings L and L^* are different they both generate the same covariance matrix Σ, i.e.

$$\Sigma = LL' + \psi = LTT'L' + \psi = L^*L^{*'} + \psi \tag{7.20}$$

The above problem of uniqueness is generally resolved by choosing an orthogonal rotation T such that the final loading L satisfies the condition that $L'\psi^{-1}L$ is diagonal with positive diagonal elements. This restriction requires L to be of full rank m. With a valid ψ, viz. one with all positive diagonal elements it can be shown that the above restriction yields a unique L.

7.4.1 Method of Estimation

Given n observations vectors $x_1 \ldots x_n$ on p generally correlated variables, factor analysis seeks to verify whether the factor model (7.16) with a small number of factors adequately represent the data.

The sample covariance matrix is an estimator of the unknown covariance matrix Σ. If Σ appears to deviate significantly from a diagonal matrix, then a factor model can be used and the initial problem is one of estimating the factor loadings l_{ij} and the specific variances. ψ_i.

Principal Component Method

Let Σ has eigen value–eigen vector pairs (λ_i, e_i) with $\lambda_1 \geq \lambda_2 \geq \ldots \lambda_p \geq 0$. Then by spectral decomposition

$$\Sigma = \lambda_1 e_1 e_1{}' + \lambda_2 e_2 e_2{}' + \ldots + \lambda_p e_p e_p{}' \tag{7.21}$$

$$= (\sqrt{\lambda_1} e_1 \ldots \sqrt{\lambda_p} e_p) \begin{pmatrix} \sqrt{\lambda_1} e_1{}' \\ \vdots \\ \sqrt{\lambda_p} e_p{}' \end{pmatrix}$$

$$= \overset{p\times p}{L} \; \overset{p\times p}{L'} + 0^{p\times p}$$

[in (7.21) $m = p$ and jth column of $L = \sqrt{\lambda_j} e_j$]

Apart from the scale factor $\sqrt{\lambda_j}$, the factor loadings on the jth factor are the population jth principal component.

The approximate representation assumes that the specific factors $\in$ are of minor importance and can be ignored in factoring Σ. If specific factors are included in the model, their variances may be taken to be the diagonal elements of $\Sigma - LL'$.

Allowing for specific factors, the approximation becomes

$$\Sigma = LL' + \psi$$

$$= (\sqrt{\lambda_1} e_1 \sqrt{\lambda_2} e_2 \ldots \sqrt{\lambda_m} e_m) \begin{pmatrix} \sqrt{\lambda_1} e_1{}' \\ \sqrt{\lambda_2} e_2{}' \\ \vdots \\ \sqrt{\lambda_m} e_m{}' \end{pmatrix} + \begin{pmatrix} \psi_1 & 0 & \ldots & 0 \\ 0 & \psi_2 & \ldots & 0 \\ \vdots & \vdots & \vdots & \vdots \\ 0 & 0 & \ldots & \psi_p \end{pmatrix} \tag{7.22}$$

where $m \leq p$

(we assume that last $p - m$ eigen values are small)

and $\psi_{ii} = \sigma_{ii} - \sum_{j=1}^{m} l_{ij}{}^2$ for $i = 1 \ldots p$.

For the principal component solution, the estimated factor loadings for a given factor do not change as the number of factors is increased. If $m = 1$

$$L = \left(\sqrt{\lambda_1} \widehat{e_1}\right)$$

if $m = 2$

$$L = \left(\sqrt{\widehat{\lambda_1}} \widehat{e_1} \sqrt{\widehat{\lambda_2}} \widehat{e_2}\right)$$

where $(\widehat{\lambda_1, e_1})$ and $(\widehat{\lambda_2, e_2})$ are the first two eigen value–eigen vector pairs for S (or R).

By the definition of $\widehat{\psi}_i$ the diagonal elements of S are equal to the diagonal elements of $\widehat{L}\widehat{L}' + \psi$. How to determine m?

The choice of m can be based on the estimated eigen values.

Consider the residual matrix $S - (LL' + \psi)$

Here the diagonal elements are zero and if the off diagonal elements are also small we may take that particular value of m to be appropriate.

Analytically we chose that m for which

$$\text{Sum of squared entries of } (S - (LL' + \psi)) \leq \widehat{\lambda}^2_{m+1} + \ldots + \widehat{\lambda}^2_p \tag{7.23}$$

Ideally the contribution of the first few factors to the sample variance of the variables should be large. The contribution to the sample variance s_{ii} from the first common factor is ${l_{ii}}^2$. The contribution to the total sample variance $s_{11} + \ldots s_{pp} = Tr(S)$ from the first common factor is

$$\widehat{l}^2_{11} + \widehat{l}^2_{21} + \ldots + \widehat{l}^2_{p1} = (\sqrt{\lambda_1}\widehat{e}_1)'(\sqrt{\lambda_1}\widehat{e}_1) = \widehat{\lambda}_1$$

Since the eigen vectors $\widehat{e}_1$ has unit length.

In general

$$\begin{pmatrix} \text{Proportion of total} \\ \text{sample variance due} \\ \text{to the } j\text{th factor} \end{pmatrix} = \begin{cases} \frac{\widehat{\lambda}_j}{s_{11}+\ldots+s_{pp}} & \text{for a factor analysis of } S \\ \frac{\widehat{\lambda}_j}{p} & \text{for a factor analysis of } R \end{cases} \tag{7.24}$$

Criterion (7.24) is frequently used as a heuristic device for determining the appropriate number of common factors. The value of m is gradually increased until a suitable proportion of the total sample variance has been explained.

Other Rules Used in Package

No. of eigen value of R greater than one (when R is used)

No. of eigen value of S that are positive (when S is used)

7.4.2 Factor Rotation

If $\widehat{L}$ be the $p \times m$ matrix of estimated factor loadings obtained by any method, then

$$L^* = \widehat{L}T \text{ where } TT' = T'T = I$$

is a $p \times m$ matrix of **rotated loadings**.

Moreover the estimated covariance (or correlation) matrix remains unchanged since

$$\widehat{L}\widehat{L}' + \widehat{\psi} = \widehat{L}TT'\widehat{L}' + \widehat{\psi} = \widehat{L}^*\widehat{L}^{*'} + \widehat{\psi}$$

The above equation indicates that the residual matrix $S_n - \widehat{L}\widehat{L}' - \widehat{\psi} = S_n - \widehat{L}^*\widehat{L}^{*'} - \widehat{\psi}$ remains unchanged. Moreover the specific variances $\widehat{\psi}_i$ and hence the communication $\widehat{h}_i^2$ are unaltered. Hence mathematically it is immaterial whether $\widehat{L}$ or L^* is obtained.

Since the original loadings may not be readily interpretable, it is usual practice to rotate them until a "simple structure" is achieved.

Ideally we should like to see a pattern of loadings of each variable loads highly on a single factor and has small to moderate loading on the remaining factors.

The problem is to find an **orthogonal rotation** which compounds to a "simple structure".

These can be achieved if after rotation the orthogonality of the factor still exists. This is maintained if we perform orthogonal rotation. Among these (1) Varimax rotation, (2) Quartimax rotation, (3) Equamax rotation are important.

Oblique rotation does not ensure the orthogonality of factors after rotation. There are several algorithm like oblimax, Quartimin, etc.

1. Varimax Rotation

Orthogonal Transformation on L

$$L^* = LT \ \ TT' = I$$

L^* is the matrix of orthogonally rotated loadings and let $d_j = \sum_{i=1}^{p} {l^*}_{ij}^2$ $j = 1 \ldots m$

Then the following expression is maximized

$$\sum_{j=1}^{m}\left\{\sum_{i=1}^{p}\left({l^*}_{ij}^{4} - {d_j}^2/p\right)\right\}$$

Such a procedure tries to give either large (in absolute value) or zero values in the columns of $\mathbf{L}^*$. Hence the procedure tries to produce factors with either a strong association with the responses or no association at all.

The communality

${h_i}^2 = \sum_{j=1}^{m} {l^*}_{ij}^{2} = \sum_{j=1}^{m} {l_{ij}}^2$ remains constant under rotation.

2. Quatrimax Rotation

The factor pattern is simplified by forcing the variables to correlate highly with one main factor (the so called G-factor of 1Q studies) and very little with remaining factors. Here all variables are primarily associated with a single factor.

Interpretation of results obtained from factor analysis is usually difficult. Many variables show significant coefficient magnitudes on many of the retained factors (coefficient greater than $|.60|$ are often considered large and coefficients of $|0.35|$ are often considered moderate), especially on the first factor.

For good interpretation factor rotation is necessary. The objective of the rotation is to achieve the most "simple structure" through the manipulation of factor pattern matrix.

The most simple structure can be explained in terms of five principles of factor rotation.

1. Each variable should have at least one zero (small) loading.
2. Each factor should have a set of linearly independent variables whose factor loadings are zero (small).
3. For every pair of factors, there should be several variables where loadings are zero (small) for one factor but not the other.
4. For every pair of factors, a large proportion of variables should have zero (small) loading on both factors whenever more than about four factors are extracted.
5. For every pair of factors, there should only be a small number of variables with nonzero loadings on both.

In orthogonal rotation

1. Factors are perfectly uncorrelated with one another.
2. Less parameters are to be estimated.

3. Promax Rotation

Factors are allowed to be correlated with one another.

Step I. Rotate the factors orthogonally.

Step II. Get a target matrix by raising the factor coefficients to an exponent (3 or 4). The coefficients become smaller but absolute distance increases.

Step III. Rotate the original matrix to a best fit position with the target matrix.

Here many moderate coefficients quickly approaches zero than the large coefficients ($\geq .6$).

References

Albazzaz, H., and X.Z. Wang. 2004. *Industrial and Engineering Chemistry Research* 43(21):6731.

Babu, J., et al. 2009. *The Astrophysical Journal* 700:1768.

Chattopadhyay, A.K., T. Chattopadhyay, E. Davoust, S. Mondal, and M. Sharina. 2009. *The Astrophysical Journal* 705:1533.

Chattopadhyay A.K., S. Mondal, and T. Chattopadhyay. 2013. *Computational Statistics & Data Analysis* 57:17.

Comon, P. 1994. *Signal Processing* 36:287.

Dickens, R.J. 1972. *Monthly Notices of Royal Astronomical Society* 157:281

Fusi Pecci, F., et al. 1993. *Astronomical Journal* 105:1145.

Gabriel, K.R. 1971. *Biometrika* 5:453.

Hastie, T., and R. Tibshirani. 2003. In *Independent component analysis through product density estimation in advances in neural information processing system*, vol. 15, ed. Becker, S., and K. Obermayer, 649–656. Cambridge, MA: MIT Press.

Hyvarinen, A., and E. Oja. 2000. *Neural Networks* 13(4–5):411.

Hyvarinen, A., J. Karhunen, and E. Oja. 2001. *Independent component analysis*. New York: Wiley.

King, I.R. 2002. *Introduction to Classical Stellar Dynamics*. Moscow: URSS.

McLaughlin, D.E., et al. 2008. *Monthly Notices of the Royal Astronomical Society* 384:563.

Qiu, D., and A.C. Tamhane. 2007. *Journal of Statistical Planning and Inference* 137:3722

Recio-Blanco, A., et al. (2006). *Astronomy & Astrophysics* 452:875

Salaris, M., et al. 2004. *Astronomy & Astrophysics* 420:911.

Shapiro, S.S., and M.B. Wilk. 1965. *Biometrika* 52(3–4):591.

Stones, V. 2004. *Independent component analysis: a tutorial introduction*. Bradford Books. Cambridge: The MIT Press.

Sugar, A.S., and G.M. James. 2003. *Journal of the American Statistical Association* 98:750.

Woodley, K.A., et al. 2007. *Astronomical Journal* 134:494.

Chapter - 8

Clustering, Classification and Data Mining

8.1 Introduction

Cluster analysis is an art of finding groups in variables or observations. Over the last 50 years different algorithms and computer programs have been developed for cluster analysis. Generally clustering algorithms can be divided into two principal types, viz. partitioning and hierarchical methods.

Clustering is different from the classification methods. Classification is concerned with a known number of groups and the operational objective is to assign new observations to one of these groups. Cluster analysis is a more primitive technique where no assumption is made concerning the number of groups or group structure.

8.2 Hierarchical Cluster Technique

Cluster Analysis, also called data segmentation, has a variety of goals. All relate to grouping of segmenting a collection of objects (also called observations, individuals, cases, or data rows) into subsets or "clusters", such that those within each cluster are more closely related to one another than objects assigned to different clusters. Central to all of the goals of cluster analysis is the notion of degree of similarity (or dissimilarity) between the individual objects being clustered. There are two major methods of clustering, viz., hierarchical clustering and k-means clustering.

In hierarchial clustering the data are not partitioned into a particular cluster in a single step. Instead, a series of partitions takes place, which may run from a single cluster containing all objects to n cluster each containing a single object. Hierarchical clustering is subdivided into agglomerative methods, which proceed by series of fusions of the n objects into groups, and divisive methods, which separate n objects successively into finer groupings.

Electronic supplementary material The online version of this chapter (doi:10.1007/978-1-4939-1507-1_8) contains supplementary material, which is available to authorized users.

A.K. Chattopadhyay, T. Chattopadhyay, *Statistical Methods for Astronomical Data Analysis*, Springer Series in Astrostatistics 3,
DOI 10.1007/978-1-4939-1507-1_8

Agglomerative techniques are more commonly used. Hierarchical clustering may be represented by a two-dimensional diagram known as dendrogram which illustrates the fusions or divisions made at each successive stage of analysis.

8.2.1 Agglomerative Methods

An agglomerative hierarchical clustering procedure produces a series of partitions of the data, $C_n, C_{n-1}, \ldots, C_1$. The first C_n consists of n single object "clusters", the last C_1 consists of single group containing all n cases.

At each particular stage the method joins together the two clusters which are closest together (most similar). At the first stage, of course, this amounts to joining together the two objects that are closest together, since at the initial stage each cluster has one object.

Differences between methods arise because of the different ways of defining distance (or similarity) between clusters. Several agglomerative techniques will now be described in detail.

8.2.2 Distance Measures

A key step in a hierarchical clustering is to select a distance measure. A simple measure is Manhattan distance, equal to the sum of absolute distances for each pair of variables. The name comes from the fact that in a two-variable case, the variables can be plotted on a grid that can be compared to city streets, and the distance between two points is the number of blocks a person would walk.

A more common measure is Euclidean distance, computed by finding the square of the distance between each pair of variables, summing the squares, and finding the square root of that sum. In the two-variable case, the distance is analogous to finding the length of the hypotenuse in a triangle; that is, it is the distance as the crow flies. A review of cluster analysis in health psychology research found that the most common distance measure in published studies in that research area is the Euclidean distance or the squared Euclidean distance.

Manhattan or Euclidean distance measures are useful for continuous data. But for nominal, ordinal or in particular binary data these dissimilarity measures are not applicable. In case of binary data (with two possible values 0 and 1), for computing similarity or dissimilarity between two objects i and j with respect to n variables one may start with the following contingency table.

object i

		1	0	Total
object j	1	a	b	$a+b$
	0	c	d	$c+d$
	Total	$a+c$	$b+d$	$a+b+c+d=n$

Then a similarity measure is given by

$$s(i,j) = \frac{a+d}{a+b+c+d}$$

and a dissimilarity (distance) measure is given by

$$d(i,j) = \frac{b+c}{a+b+c+d}$$

To calculate distance between two clusters it is required to define two representative points from the two clusters. Different methods have been proposed for this purpose. Some of them are listed below.

8.2.3 Single Linkage Clustering

One of the simplest methods is single linkage, also known as the nearest neighbour technique. The defining feature of the method is that distance between clusters is defined as the distance between the closest pair of objects, where only pairs consisting of one object from each cluster are considered.

In the single linkage method, $D(r,s)$ is computed as

$$D(r,s) = \text{Min } d(i,j)$$

where object i is in cluster r and the object j is in cluster s.

Here the distance between every possible object pair $(i.j)$ is computed, where object i is in cluster r and object j is in cluster s. The minimum value of these distances is said to be the distance between clusters r and s. In other words, the distance between two clusters is given by the value of the shortest link between the clusters. At each stage of hierarchical clustering, the clusters r and s, for which $D(r,s)$ is minimum, are merged.

8.2.4 Complete Linkage Clustering

The complete linkage, also called farthest neighbour, clustering method is the opposite of single linkage. Distance between clusters is now defined as the distance between the most distant pair of objects, one from each cluster.

In the complete linkage method, $D(r, s)$ is computed as

$$D(r, s) = \text{ Max } d(i, j)$$

where object i is in cluster r and object j is cluster s.

Here the distance between every possible object pair (i, j) is computed, where object i is in cluster r and object j is in cluster s and the maximum value of these distances is said to be the distance between clusters r and s. In other words, the distance between two clusters is given by the value of the longest link between the clusters. At each stage of hierarchical clustering, the clusters r and s, for which $D(r, s)$ is minimum, are merged.

8.2.5 Average Linkage Clustering

Here the distance between two clusters is defined as the average of distances between all pairs of objects, where each pair is made up of one object from each group.

In the average linkage method, $D(r, s)$ is computed as

$$D(R, s) = T_{rs}/(N_r * N_s)$$

where T_{rs} is the sum of all pairwise distances between cluster r and cluster s. N_r and N_s are the sizes of the clusters r and s, respectively. At each stage of hierarchical clustering, the clusters r and s, for which $D(r, s)$ is the minimum, are merged.

8.3 Partitioning Clustering: k-Means Method

The k-means clustering algorithm assigns each point to the cluster whose centre (also called centroid) is nearest. The centre is the average of all the points in the cluster that is, its co-ordinates are the arithmetic mean for each dimension separately over all the points in the cluster. The algorithm is roughly (MacQueen 1967) as follows.

Choose the number of clusters, k. Randomly generate k clusters and determine the cluster centres, or directly generate k seed points as cluster centres. Assign each point to the nearest cluster centre on the basis of Euclidean distance. Recompute the new cluster centres. Repeat until some convergence criterion is met (usually that the assignment hasn't changed). The main advantages of this algorithm are its simplicity and speed which allow it to run on large data sets. Its disadvantage is that it does not yield the same result with each run, since the resulting clusters depend on the initial random assignments. It maximizes inter-cluster (or minimizes intra-cluster) variance, but does not ensure that the result has a global minimum of variance.
In order to minimize this problem one may use group average method proposed by Milligan (1980), in order to choose the initial seeds.

The advantages of partitioning method are as follows:

(a) A partitioning method tries to select best clustering with k groups which is not the goal of hierarchical method.

(b) A hierarchical method can never repair what was done in previous steps.

(c) Partitioning methods are designed to group items rather than variables into a collection of k clusters.

(d) Since a matrix of distances (similarities) does not have to be determined and the basic data do not have to be stored during the computer run partitioning methods can be applied to much larger data sets.

For k-means algorithm (Hartigan 1975) the optimum value of k can be obtained in different ways.

By using k-means algorithm first determine the structures of sub populations (clusters) for varying number of clusters taking $k = 2, 3, 4$, etc. For each such cluster formation compute the values of a distance measure $d_k = (1/p) \min_x E[(x_k - c_k)'(x_k - c_k)]$ which is defined as the distance of the x_k vector (values of the variables) from the centre c_k (which is estimated as mean value). p is the order of the x_k vector. Then the algorithm for determining the optimum number of clusters is as follows (Sugar and James 2003). Let us denote by d'_k the estimate of d_k at the kth point. Then d'_k is the minimum achievable distortion associated with fitting k centres to the data. A natural way of choosing the number of clusters is to plot d'_k versus k and look for the resulting distortion curve. This curve is always monotonic decreasing. Initially one would expect much smaller drops, i.e. a levelling off for k greater than the true number of clusters because past this point adding more centres simply partitions within groups rather than between groups. According to Sugar and James (2003) for a large number of items the distortion curve when transformed to an appropriate negative power ($p/2$ in our case), will exhibit a sharp "jump" (if we plot k versus transformed d'_k). Then calculate the jumps in the transformed distortion as $J_k = \left(d_k'^{-(p/2)} - d_{k-1}'^{-(p/2)}\right)$.

The optimum number of clusters is the value of k at which the distortion curve levels off as well as its value associated with the largest jump. The distortion curve and jump curve for GRB data (Sect. 8.5) are shown in Figs. 8.1 and 8.2.

8.4 Classification

Discriminant analysis and classification are multivariate techniques concerned with separating distinct sets of objects and with allocating new objects to

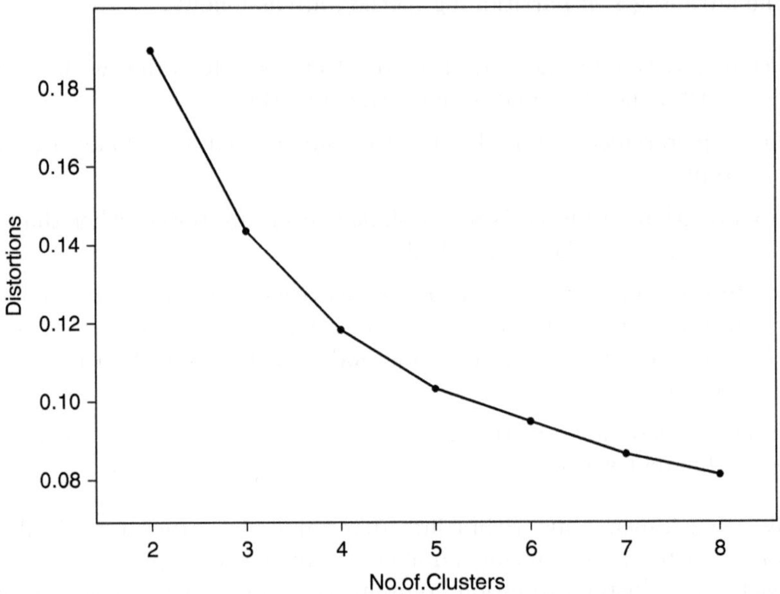

Figure 8.1 xy diagram for the number of clusters (k) and distortions (d'_k) for the sample

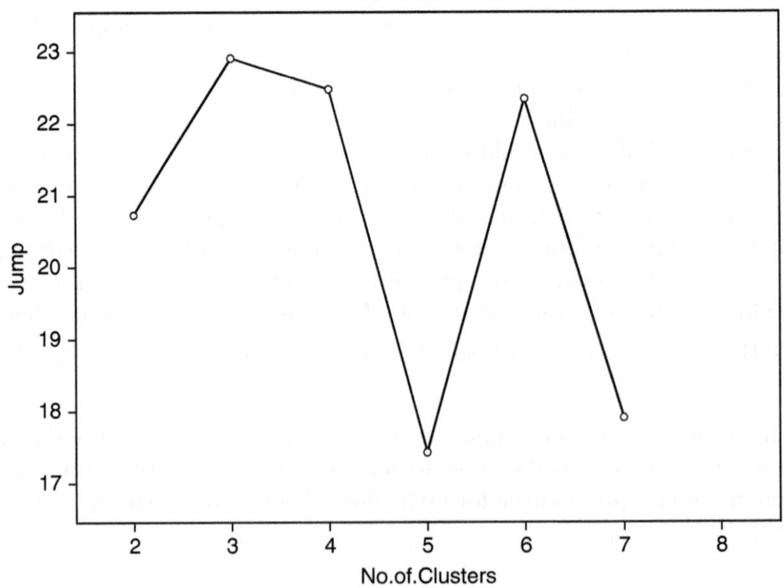

Figure 8.2 xy diagram for the number of clusters (k) and jumps for the sample

previously defined groups. Once the optimum classification (clustering) is obtained by applying the method discussed under previous section one can verify the acceptability of the classification by computing classification/ misclassification probabilities for the different observations. Although the k-means clustering method is purely a data analytic method, for classification it may be necessary to assume that the underlying distribution is Multivariate Normal. The method can be illustrated as follows for two populations (clusters). The method can be easily generalized for more than two underlying populations. Let $f_1(x)$ and $f_2(x)$ be the probability density functions associated with the $p \times 1$ random vector X for the populations π_1 and π_2, respectively. An object must be assigned to either π_1 and π_2. Let Ω be the sample space. Let us denote by x the observed value of X. Let R_1 be that set of x values for which we classify objects as π_1 and $R_2 = \Omega - R_1$ be the remaining x values for which we classify objects as π_2. Since every object must be assigned to one and only one of the two groups, the sets R_1 and R_2 are disjoint and exhaustive. The conditional probability of classifying an object as π_2 when in fact it is from π_1 (error probability) is

$$p(2 \mid 1) = P[X \in R_2 \mid \pi_1] = \int_{R_2} f_1(x)dx$$

Similarly the other error probability can be defined. Let p_1 and p_2 be the error probabilities of π_1 and π_2, respectively ($p_1 + p_2 = 1$). Then the overall probabilities of correctly and incorrectly classifying objects can be derived as

$$\begin{aligned} &P\text{ (correctly classified as } \pi_1)\\ &\quad = P[\text{observation actually comes from } \pi_1 \text{ and is correctly classified as } \pi_1]\\ &\quad = P[X \epsilon R_1|\pi_1]p(\pi_1) = p[X \epsilon R_1|\pi_1)p_1 \end{aligned}$$

$$\begin{aligned} P\text{ (misclassified as } \pi_1) &= P[X \epsilon R_1|\pi_2]p(\pi_2)\\ &= P[X \epsilon R_1|\pi_2)p_2 \end{aligned}$$

The associated cost of misclassification can be defied by a cost matrix

		Classified π_i	as π_2
True population	π_1	0	$C(2 \mid 1)$
	π_2	$C(1 \mid 2)$	0

For any rule, the average or expected cost of misclassification (ECM) is given by

$$ECM = C(2 \mid 1)p(2 \mid 1)p_1 + C(1 \mid 2)p(1 \mid 2)p_2$$

A reasonable classification rule should have ECM as small as possible.

Rule: The regions R_1 and R_2 that minimize the ECM are defined by the value of x for which the following inequalities hold.

$$R_1 : \frac{f_1(x)}{f_2(x)} \geq \frac{C(1 \mid 2)p_2}{C(2 \mid 1)p_1}$$

$$R_2 : \frac{f_1(x)}{f_2(x)} < \frac{C(1 \mid 2)p_2}{C(2 \mid 1)p_1}$$

If we assume $f_1(x)$ and $f_2(x)$ are multivariate normal with mean vectors μ_1 and μ_2 and covariance matrices $\sum_1$ and $\sum_2$, respectively, then a particular object with observation vector x_0 may be classified according to the following rule (under the assumption $\sum_1 = \sum_2$):

Allocate x_0 to π_1 if

$$(\mu_1 - \mu_2)' \sum^{-1} x_0 - \frac{1}{2}(\mu_1 - \mu_2)' \sum^{-1} (\mu_1 + \mu_2) \geq In[\frac{C(1 \mid 2)p_2}{C(2 \mid 1)p_1}]$$

allocate x_0 to π_2 otherwise. If we choose $C(1 \mid 2) = C(2 \mid 1)$ and $p_1 = p_2$, then the estimated minimum ECM rule for two Normal populations will be as follows: Allocate x_0 to π_1 if

$$(\overline{x_1} - \overline{x_2})' S_{pooled}^{-1} x_0 - \frac{1}{2}(\overline{x_1} - \overline{x_2})' S_{pooled}^{-1} (\overline{x_1} + \overline{x_2}) \geq 0$$

[where ($\overline{x_1}$ and $\overline{x_2}$) are sample mean vectors of the two populations and S_{pooled} is pooled (combined) sample covariance matrix]. Allocate x_o to π_2 otherwise.

8.5 An Example (Chattopadhyay et al. 2007)

We have taken data on 1,594 GRBs from the BATSE current GRB catalogue, given in the website (Vizier) for public use. There are thirteen variables of astrophysical interest, Galactic longitudes (l_i) and latitudes (b_i), two measures of burst durations, the times within which 50 % (T_{50}) and 90 % (T_{90}) of the flux arrive, three peak fluxes, $P_{64}, P_{256}, P_{1024}$ measured in 64, 256 and 1,024 ms bins, respectively, four time integrated fluences $F_1 - F_4$, in the 20–50, 50–100, 100–300 keV and 300+ keV spectral channels, respectively, and peak counts of photons over the time $T_1(C_p)$ and limiting count of photons that triggers detection (Clim), of which first 11 variables have nonzero values for all these 1,594 bursts. Out of these 13 variables some composite variables have been constructed which are widely used for statistical analysis. They are total fluence $F_T = F_1 + F_2 + F_3 + F_4$, spectral hardness $H_{32} = F_3/F_2$ and $H_{321} = F_3/(F_1 + F_2)$, isotropy parameters, dipole moments $\cos\theta_i = \cos l_i, \cos b_i$ (using four parts formula), quadruple moments $\sin^2 b_i - 1/3$ and parameter for testing homogeneity $V/V_{max} = (C_p/C_{lim})^{-3/2}$

(Schmidt et al. 1988). For the present study we have used six variables $\log T_{50}, \log T_{90}, \log P_{256}, \log F_T, \log H_{32}, \log H_{321}$ for the classification purpose and calculated the mean values of homogeneity and isotropy parameters for the classified groups to test the nature of spatial distribution of these groups. We have taken initially these six variables which have been used by most of the authors (Mukherjee et al. 1998; Hakkila et al. 2000) for comparative study. For classification of GRBs we have used partitioning algorithm (k-means clustering).

We have taken 14 GRBs of HETE 2 catalog and 36 GRBs of Swift Satellite data having known redshifts for classification with respect to two parameters, viz., duration (T_{90}) and total fluence (F_T). The fluence of HETE 2 data is in between 30 and 400 keV and those of Swift data is between 15 and 150 keV. The fluence of BATSE data is from 30 keV and above 400 keV. This may introduce an error or at most of 10 %. The other four parameters are not available at present for HETE 2 and Swift GRBs.

8.5.1 Cluster Analysis of BATSE Sample and Discriminant Analysis

For analysis of the data using the methods discussed under above sections we have used statistical packages like MINITAB, R and C-program codes. For k-means clustering we have used Euclidean distance assigning a GRB to the cluster where centroid (mean) is nearest. For discrimination we have used linear discriminant analysis. A GRB is classified into a cluster if the Mahalanobis distance of the observations to the cluster mean is the minimum. An assumption is made that the covariance matrices are equal for all clusters (Table 8.1).

In Fig. 8.2 the maximum "jump" is seen to be at $k = 3$, so that the three class classification is the optimum classification in the present scheme. In Table 8.2 the group means and standard errors of the parameters for all the classes are shown, the corresponding values of the isotropy and homogeneity parameters are also shown for each group for the available values. The groups are arranged according to increasing values of duration ($< T_{90} >$) as clusters I, II and III, respectively. It is seen from Table 8.2 that all the expected values of the isotropy parameters except quadruple moments for class II and class III lie within 1σ level. So it is concluded that all the three classes are isotropically disturbed. For calculating $< V/V_{\max} >$ parameter the available values are used only which is obviously less than the sample size. It is seen that $< V/V_{\max} >$ value for cluster III is very small implying that cluster III is the most inhomogeneous and extremely deep population class having very high redshifts, whereas class I and class II are more or less uniform. The most uniform class is the intermediate class I having shortest duration.

Table 8.1 Correlation matrix for the composite parameters for the present sample

Parameters	$\log T_{50}$	$\log_{90}$	$\log H_{32}$	$\log P_{256}$	$\log F_T >$	$\log H_{321}$
$\log T_{50}$	1					
$\log T_{90}$	0.97	1				
$\log P_{256}$	−0.02	0.03	1			
$\log F_T$	0.64	0.67	0.58	1		
$\log H_{32}$	−0.40	−0.40	0.13	−0.07	1	
$\log H_{321}$	−0.41	−0.41	0.15	−0.08	0.96	1

Table 8.2 The group and standard errors for the various parameters in three class classification

Cluster	Cluster 1		Cluster 2		Cluster 3	
No. of members	423		622		549	
Parameters	Mean	Std error	Mean	Std error	Mean	Std error
$< \log T_{50} >$	−0.724	0.020	0.727	0.016	1.356	0.019
$< \log T_{90} >$	−0.296	0.022	1.196	0.015	1.806	0.015
$< \log P_{256} >$	0.223	0.018	0.100	0.015	0.459	0.021
$< \log F_T >$	−6.213	0.025	−5.526	0.015	−4.750	0.021
$< \log H_{32} >$	0.744	0.013	0.391	0.010	0.501	0.009
$< \log H_{321} >$	0.534	0.014	0.123	0.010	0.254	0.014
$< \cos\theta >$	−0.027	0.028 (0.9σ)	−0.008	0.023 (0.3σ)	−0.019	0.012 (0.2σ)
$< \sin^2 b - 1/3 >$	−0.0097	0.014 (0.6σ)	−0.019	0.012 (1.5σ)	0.019	0.013 (1.5σ)
$< V/V_{\max} >$	0.475	0.019	0.374	0.014	0.145	0.009

Following Mao and Paczynski (1992) and taking $< V/V_{\max} >$ values for each group from Table I, a value of $Z_{\max}$ has been derived for a model universe, $\Omega_M = 0.3, \Omega_A = 0.7, H_o = 65\,\mathrm{km\,s^{-1}\,Mpc^{-1}}$ and assuming the GRBs to be standard candles with spectral slope $\alpha = 1$ (Mallozzi et al. 1996). They are 0.16, 1 and 10, respectively, which is more reasonable than the high values of $z_{\max}$, 4.06, 3.08 and 45.24, respectively, obtained by Balastegui et al. (2001) before revision. Also in our case class I is the closest one. So class III can have super massive stars as progenitors. Also the three classes are as follows. Class I is the class having shortest duration and maximum hardness but fainter than class II. The duration varies from 0.03 to 6 s. It has most uniform spatial distribution and is most homogeneous and closest to the observer (Fig. 8.3).

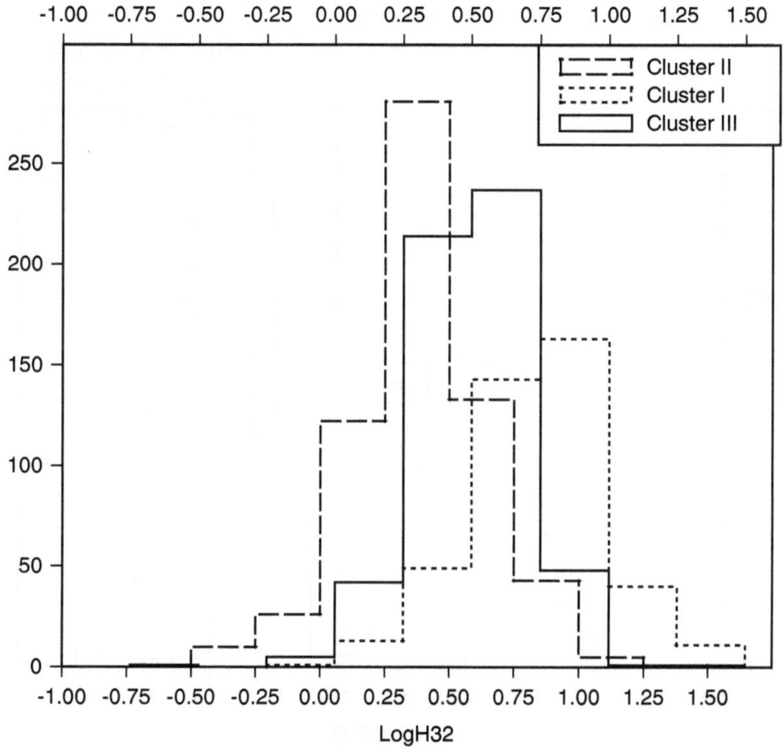

Figure 8.3 Histograms of the hardness (H_{32}) parameter for the three classes found in the cluster analysis

Class II is the intermediate class having duration (T_{90}) varying in the range 10–125 s which is much larger than the intermediate class found by Mukherjee et al. (1998). It is softer but brighter than class I. It also has uniform spatial distribution and more or less homogeneous as found from $< V/V_{max} >$ value which is close to 0.5. Class III is the class having longest duration from 5 to 673 s, it is most soft and most inhomogeneous in nature. The duration of three classes is more or less non overlapping (Fig. 8.4) unlike the previous classifications though hardness parameters do not differ significantly (Fig. 8.3), i.e. they follow overlapping zones. So for the present study we have taken six variables like most of the authors but found similar type of classes as found by Balastegui et al. (2001) where they have considered nine variables. Also in the present classification the durations are well separated unlike other works. In Table 8.3 correlation matrices have been computed for the above three classes. Unlike class I, for classes II and III fluence and durations have very little correlations. It is minimum for class II. Peak flux and durations have anti correlations in classes II and III, respectively. The effect is more pronounced in longest duration class. Hardness has practically no correlation with duration for classes I and II but a fare relation for class III.

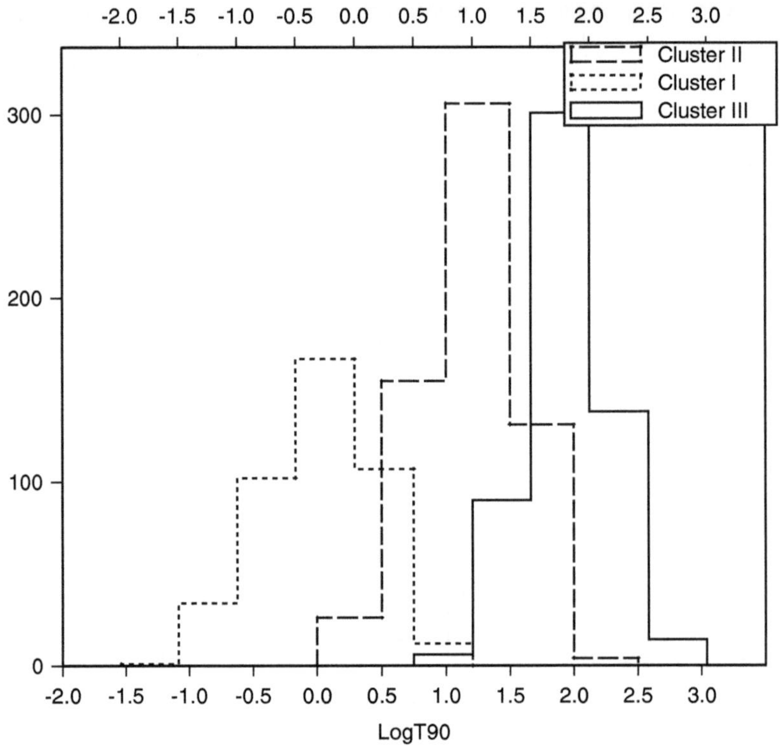

Figure 8.4 Histograms of the duration (T_{90}) parameter for the three classes found in the cluster analysis

Assuming the actual number of clusters is three we have applied discriminant analysis to compute the miss classification probabilities. The results obtained are shown in Table 8.4. From Table 8.4 we see that proportion of correct classification is 0.954. Hence it may be inferred that the choice of three class classification is quite realistic. While looking at the miss classification probabilities (not listed in the paper as the sample size is quite large) we see that for most of the miss classified GRBs the differences in classification probabilities for the true clusters and the identified clusters are not very significant.

8.5.2 Cluster Analysis of HETE 2 and Swift Samples

Among six parameters used in k-means clustering for BATSE data only two, viz., duration (T_{90}) and total fluence (F_t) are available for GRBs in HETE 2 swift catalogs. Also the total fluence ($F_t = F_1 + F_2 + F_3 + F^4$) uses in the BATSE data has the range from 30 KeV and beyond 400 keV (F_4), while for HETE 2 it is 30–400 keV and for Swift it is 15–150 keV, respectively.

Table 8.3 Correlation matrix for the composite parameters in clusters I, II and III, respectively

Parameters	$\log T_{50}$	$\log T_{90}$	$\log P_{256}$	$\log F_T$	$\log H_{32}$	$\log H_{321}$
Cluster I						
$\log T_{50}$	1					
$\log T_{90}$	0.84	1				
$\log P_{256}$	−0.003	0.071	1			
$\log F_T$	0.26	0.27	0.69	1		
$\log H_{32}$	−0.008	−0.044	0.035	0.27	1	
$\log H_{321}$	−0.048	−0.097	0.051	0.214	0.93	1
Cluster II						
$\log T_{50}$	1					
$\log T_{90}$	0.87	1				
$\log T_{256}$	−0.44	0.37	1			
$\log F_T$	0.016	0.051	0.51	1		
$\log H_{32}$	−0.086	−0.094	0.075	0.12	1	
$\log H_{321}$	−0.078	−0.1	0.092	0.14	0.96	1
Cluster III						
$\log T_{50}$	1					
$\log T_{90}$	0.82	1				
$\log T_{256}$	−0.53	0.45	1			
$\log F_T$	0.26	−0.15	0.71	1		
$\log H_{32}$	−0.290	−0.26	0.24	0.33	1	
$\log H_{321}$	−0.30	−0.27	0.26	0.34	0.97	1

Table 8.4 Results of discriminant analysis for the classification

	True groups (clusters)		
Put into clusters	Cluster 1	Cluster 2	Cluster 3
Cluster 1	578	6	23
Cluster 2	28	417	0
Cluster 3	21	0	526
Total	622	423	549

This variation influence may introduce an error or at most 10 %. The GRBs with known redshifts are selected from the catalogs. They are 14 and 36 in HEHE 2 and Swift, respectively. Since in the previous analysis the optimum number of classes is three, we have carried out k-means clustering of BATSE data taking two parameters (T_{90} and F_1) instead of six and assuming there are three classes. Then a discriminant analysis is performed with the same BATSE data but now with respect to above two parameters and the proportion constant is calculated (at 95 % confidence level). It is found that the proportion constant is comparatively lower than the previous one but as high as 0.879. This means that though the other parameters have a fare contribution but the contribution of duration and fluence are very important in the classification. This is done to make a calibration among the BATSE, HETE 2 and Swift GRBs. Now again a discriminant analysis is performed with HETE 2 and Swift samples to calculate the probabilities of different GRBs of falling into above three classes. In this way though the other parameters are not available for HETE 2 and Swift data we have become able to include them into any of the above three classes. Having the redshifts of these GRBs are known their cosmological distances are calculated following Kim et al. (1997) and hence their luminosities (L) are known ($L = 4\pi D^2 F_t$). The BATSE GRBs (since there is only one GRB in the shortest class it is excluded) having known redshifts are also included in the combined sample of GRBs from BATSE, HETE 2 and Swift having known redshifts. It is seen from the figures that three classes are well separated in luminosities. The distinction will be more pronounced if all the six parameters are available for HETE 2 and Swift data. Also the present separation suffers from the observational error in the sense that two GRBs, one with less luminosity, closer to the observer and other with higher luminosity, farther from the observer seem to have same fluence and may fall into the same class. This may be the reason for some of the GRBs with high redshifts which should have fallen in class III with higher luminosities have fallen in class II. But still it is clear from the figure that the classification is merely a luminosity classification at various distances. In this context it is to be mentioned that for BATSE data longest duration class has highest value of Z_{max} and shortest class has lowest value of Z_{max}. But here though shortest class (class I) GRBs have comparatively lower redshifts but intermediate and longest classes have redshifts of various ranges. This may be due to the fact that we have found Z_{max} of the three classes of BATSE sample assuming a model universe where GRBs are taken as standard candles of spectral index 1, i.e. all of them have the same luminosities which is not the situation here and it is the luminosity which acts as a differentiating maker among the classes. More authenticity of the result will be increased when the sample size is quite large. To see the effect of fluence on the classification we have carried out the cluster analysis with only T_{90} and T_{50}, the optimum number of classes found is 2 which is consistent with the two class classification on basis of duration only.

R Code for k means clustering and Classification is given below.

```
data3 ← read.table ("C:\\Users\\Tanuka \\Desktop\\ grb2007 \\.txt",
header = TRUE)
data3
library(MASS)
cor(data3) kmeans(data3, 3)
c1 ← kmeans(data3, 3)
c1
clusmem ← cbind (data3, c1$cluster)
clusmem
group ← c (rep (1, 546), rep (2, 426), rep(3, 622))
discr ← lda (data3, group)
x ← predict (discr) $class
tab ← table (predict (discr)$ class)
tab cbind (data3, c1$cluster)
tab
tab1← table (predict (discr) $ class, group)
tab1
```

The data file grb2007.txt is given in the Appendix.

8.6 Clustering for Large Data Sets: Data Mining

8.6.1 Subspace Clustering

Clustering is a technique used to place data elements into related groups without advance knowledge of the group definitions. Clustering algorithms are attractive for the task of identification in coherent groups for existing data sets under consideration. However, clustering algorithm needs the following requirements when applied to large data sets.

1. Minimal requirements of domain knowledge to determine the input parameters.
2. Discovery of clusters with arbitrary shape and good efficiency on large databases.
3. Automatic determination of the optimum number of homogeneous classes. Popular clustering techniques such as the K-Means Clustering and Expectation Maximization (EM) Clustering fail to give solution to the combination of these requirements. Thus keeping in view the above considerations some new approaches have been developed known as Density Based Clustering Techniques and Subspace Clustering Techniques.

In the Density based approach the main reason why a cluster is recognized is that within each cluster there is a typical density of points which is considerably higher than outside the cluster. Furthermore, the density within the areas of noise is lower than the density in any of the clusters. In other words,the clusters and consequently the classes are easily and readily identifiable because they have an increased density with respect to the points they possess. The single points scattered around the database are outliers, which means they do not belong to any clusters as a result of being in an area with relatively low concentration. Here discussions have been focused on Subspace Clustering Techniques which is a data mining task.

Clustering seeks to find groups of similar objects based on the values of their attributes. Traditional clustering algorithms i.e., the Full Space algorithms use distance on the whole data space to measure similarity between objects.As the number of dimensions in a data set increases, distance measures become increasingly meaningless. For very high dimensional data sets, the objects are almost equidistant from each other. This is known as the curse of high dimensionality.

The concept of subspace clustering has been proposed to cope with this problem by discovering clusters embedded in the subspaces of high dimensional data sets. Subspace Clustering is the task of detecting all clusters in all subspaces. This means that a point might be a member of multiple clusters, each existing in a different subspace.

Subspaces can either be axis parallel or arbitrarily oriented affine subspaces. The two approaches towards clustering differ in how they interpret the overall goal, which is finding clusters in data sets with high dimensionality.

In both cases, the data objects which are grouped into a common subspace cluster are very dense (i.e. the variance is small) when projected onto the hyperplane which is perpendicular to the subspace of the cluster (called the perpendicular space plane). The objects may form a completely arbitrary shape with a high variance when projected onto the hyperplane of the subspace in which the cluster resides (called the cluster subspace plane). This means that the objects of the subspace cluster are all close to the cluster subspace plane. The knowledge that all data objects of a cluster are close to the cluster subspace plane is valuable for many applications.

If the plane is axis-parallel, this means that the values of some of the attributes are more or less constant for all cluster members. The whole group is characterized by this constant attribute value, an item of information which can definitely be important for the interpretation of the cluster. This property may also be used to perform a dedicated dimensionality reduction for the objects of the cluster and may be useful for data compression (because only the higher-variance attributes need to be individually considered for the search) and an index needs only to be constructed for the high-variance attributes.

If the cluster subspace plane is arbitrarily oriented, the knowledge is even more valuable. In this case, it is known that the attributes which define the cluster subspace plane have a complex dependency among each other. This dependency defines a rule, which again characterizes the cluster and which is potentially useful for cluster interpretation.

Many subspace clustering algorithms use a grid based approach to find dense regions. They partition the data space into non-overlapping rectangular cells by discretizing each dimension into s number of bins. A cell is dense if the fraction of total objects contained in the cell is greater than a threshold. Dense cells in all subspaces are identified using a bottom-up strategy and connected dense cells are merged together to form clusters. In the grid based approach, objects around the boundaries of the bins have similar values, but they are put into different bins. As a result, a cluster may be divided into several small clusters.

These methods are popular due to two main reasons. Firstly, conventional (full space) clustering algorithms often fail to find useful clusters when applied to data sets of higher dimensionality, because typically many of the attributes are noisy, some attributes may exhibit high correlations with others and only few of the attributes really contribute to the cluster structure. Secondly, the knowledge gained from a subspace clustering algorithm is much richer than that of a conventional clustering algorithm because it can be used for interpretation, data compression, similarity search, etc.

Arbitrarily Oriented Clustering assumes that the cluster structure is significantly dense in the local neighbourhood of the cluster centres or other points that participate in the cluster.

In the context of high-dimensional data, this locality assumption is rather optimistic. Theoretical considerations show that concepts like local neighbourhood are not meaningful in high-dimensional spaces because distances can no longer be used to differentiate between points. This is a consequence of the well-known curse of dimensionality.

8.6.2 Clustering in Arbitrary Subspace Based on Hough Transform: An Application (Chattopadhyay et al. 2013)

The locality assumption that the clustering structure is dense in the entire feature space and that the Euclidean neighbourhood of points in the cluster, or of cluster centres, does not contain noise is a very strict limitation for high-dimensional real-world data sets. In high-dimensional spaces, the distance to the nearest neighbour and the distance to the farthest neighbour coverage. As a consequence, distances can no longer be used to differentiate between points in high-dimensional spaces and concepts like the neighbourhood of points become meaningless. Usually, although many points share a common hyperplane, they are not close to each other in the original feature space. In those cases, existing approaches will fail to detect meaningful patterns because they cannot learn the correct subspaces of the clusters. In addition, as long as the correct subspace the clusters cannot be determined. Obviously outliers and noise cannot be removed in a preprocessing step. In present work concept of an arbitrarily oriented subspace clustering technique developed by Achtert et al. (2008) has been applied.

In this method development of an original principle to characterize the subspace holding a cluster is based on the idea of the Hough Transform. This transform charts out the points from a two-dimensional data space (also known as picture space) of Euclidean co-ordinates (e.g. pixel of an image) into a parameter space. It is the parameter space that stands for all possible one-dimensional lines in the original two-dimensional data space. In principle, each point of the data space is mapped into an infinite number of points to the parameter space which is, however, not an infinite set but actually a trigonometric function relating to the parameter space. Each function in the parameter space represents all lines in the picture space crossing the corresponding point in data space. The intersection of the dual curves in the parameter space points to a line through the corresponding points alike in the picture space.

The objective of a clustering algorithm is to find intersections of many curves in the parameter space representing lines through many database objects. The key feature of the Hough transform is that the distance of the points in the original data space is not considered any more. Objects can be identified as associated with a common line even if they are far apart in the original feature space. As a consequence, the Hough transform is a promising candidate for developing a principle for subspace analysis that does not require the locality assumption and, thus, enables a global subspace clustering approach. The simplest case of Hough transform is the linear transform for detecting straight lines. In the image space, the straight line can be described as $y = mx + b$ and can be graphically plotted for each pair of image points (x, y). In the Hough transform, a main idea is to consider the characteristics of the straight line not as image points $(x_1, y_1), (x_2, y_2)$, etc., but instead in terms of its parameters, i.e., the slope parameter m and the intercept parameter b. Based on that fact, the straight line $y = mx + b$ can be represented as a point (b, m) in the parameter space. However, one faces the problem that vertical lines give rise to unbounded values of the parameters m and b. For computational reasons, it is therefore better to use a different pair of parameters, denoted r and θ, for the lines in the Hough transform.

The parameter r represents the distance between the line and the origin. while θ is the angle of the vector from the origin to this closest point. Using this parameterization, the equation of the line can be written as $r = x\cos\theta + y\sin\theta$.

It is therefore possible to associate with each line of the image a pair $(r;\theta)$ which is unique if $\theta\epsilon[0;\pi)$ and $r\epsilon R$, or if $\theta\epsilon[0;2\pi)$ and $r \geq 0$. The $(r;\theta)$ plane is sometimes referred to as Hough space for the set of straight lines in two dimensions.

8.6.2.1 Input Parameters

CASH requires the user to specify two input parameters. The first parameter m specifies the minimum number of sinusoidal curves that (minpts) need to intersect a hypercuboid in the parameter space such that this hypercuboid is regarded as a dense area. Obviously, this parameter represents the minimum number of points in a cluster and thus is very intuitive. The second parameter s specifies the maximal number of splits along a search path (split level). CASH is robust with respect to the choice of s. Since CASH does not require parameters that are hard to guess like the number of clusters, the average dimensionality of the subspace clusters, or the size of the Euclidean neighbourhood based on which the similarity of the subspace clusters is learned, it is much more usable.

8.6.2.2 Data Set

In order to evaluate the efficiency of the algorithm CASH, the method has been applied to a data complied and standardized by Hudson et al. (2001) for a sample of 56 low-redshift galaxy clusters containing 699 early-type galaxies. After eliminating the missing observations the sample size has been reduced to 528 and the CASH method has been performed using four parameters (variables), viz. the logarithm of the effective Radius ($logR_c$ in kpc), the surface brightness averaged over half light radius (μ in mag arcsec^{-2}, central velocity dispersion (σ in $\text{km}\,\text{s}^{-1}$) and magnesium index (M_{g2} index).

8.6.2.3 Experimental Evaluation

Initial choice of constraints: Since CASH only needs two constraints viz., m the minpts and s the number of split levels, the constants have been selected by trial and error method. The jitter has been fixed to a preassigned small value 0.15. The value of m has been taken from 100 to 40 and values of s have been varied from 1 to 3. It is expected that the value for m should not be larger than 100 for a sample of size 528 because it is the minimum number of points to be included per cluster. Also the number of split levels should be moderate for a data set of moderate size.

It is seen that the stability has been achieved by taking $m = 60$ and $s = 2$. After that even a decrease in the value of m has not contributed in the result. With the above-mentioned combination of s and m, the number of cluster has been found to be seven.

8.6.2.4 Properties of the Groups

The efficiency of CASH has been checked by several properties.

The average properties of the seven groups are shown in Table 8.5 where N_{gal} represents the size of each clusters and M_{dyn} represents the dynamical mass which can be obtained from the relation

$$M_{\text{dyn}} \approx A\sigma^2 R_e/G, \text{ where A and G are constants.}$$

It is well known that Fundamental Plane (FP) is a relationship between the effective radius, average surface brightness and central velocity dispersion of normal elliptical galaxies and Virial Plane (VP) is the parametric plane constituted by effective radius, surface brightness averaged over effective radius and velocity dispersion when a galaxy is in dynamical equilibrium.

The slopes for $\log R_e$ with respect to $\log M_{dyn}$ are shown in Table 8.6 for seven clusters.

From Table 8.6 it is clear that all the slopes are greater than 0.38. So the galaxies are not formed as a result of pure disk mergers (Robertson et al. 2006). Since the slope of C_4 is more or less close to 0.38, it might be formed due to pure disk merger. For the remaining ones, the slopes are steeper which might be due to merger of non-disky objects or the result of repeated merging of small systems (Shen et al. 2003)

The Mg2 index more or less increases chronologically. So accordingly, higher Mg2 indicates that the galaxies are dynamically more evolved and lower Mg2 value signifies that the galaxies are dynamically less evolved (viz. Table 8.5, column 6).

The Fundamental Plane (FP) is expressed by the relationship,

$$\log_{10} R_e = a \log \sigma + b\mu_e + c$$

where a, b and c are constants to be determined.

The virial Plane (VP) is expressed by the relationship.

$$\log_{10} R_e = 2 \log \sigma + 0.4\mu_e$$

Table 8.5 Average properties of the seven groups of galaxies obtained by CASH method

Clusters	N_{gal}	log_{α}	μ_e	$logR_e$	M_{g^2}	$logM_{dyn}$
c1	21	2.2345 ± 0.021	19382 ± 0.119	2.7791 ± 0.017	0.26286 ± 0.0052	10.170 ± 0.012
c2	60	2.2575 ± 0.236	20.414 ± 0.231	2.8638 ± 0.016	0.28143 ± 0.0035	10.310 ± 0.021
c3	75	2.2526 ± 0.025	19.461 ± 0.172	2.8694 ± 0.027	0.28239 ± 0.0043	10.306 ± 0.013
c4	167	2.2537 ± 0.071	19.437 ± 0.261	2.8257 ± 0.035	0.28837 ± 0.0036	10.264 ± 0.014
c5	82	2.2895 ± 0.013	20.487 ± 0.266	2.8032 ± 0.013	0.26951 ± 0.0023	10.313 ± 0.035
c6	63	2.2752 ± 0.014	19.667 ± 0.143	2.9244 ± 0.029	0.29371 ± 0.0014	10.406 ± 0.023
c7	60	2.3450 ± 0.023	19.346 ± 0.035	2.9127 ± 0.013	0.30400 ± 0.0028	10.534 ± 0.034

Table 8.6 Slopes of seven different clusters

Clusters	N_{gal}	Slope
C1	21	0.447
C2	82	0.443
C3	60	0.578
C4	75	0.417
C5	167	0.417
C6	63	0.663
C7	60	0.662

Table 8.7 The values of the M_{g2} index, Tilt and slope seven clusters

Clusters	N_{gal}	M_{g2}	Tilt	Slope
C1	21	0.26286	0.3487	0.447
C2	82	0.28143	0.744	0.443
C3	60	0.28239	0.645	0.587
C4	75	0.28837	0.649	0.417
C5	167	0.26951	0.728	0.471
C6	63	0.29371	0.749	0.663
C7	60	0.30400	0.732	0.662

The ratio of the slopes of the FP with VP is defined as the tilt. Hence a small value of tilt indicates that the FP is farther from the corresponding VP. The tilt values for the seven groups are shown in Table 8.7.

It is also clear from Table 8.7 that tilts are almost increasing in parity with the $Mg2$ index and also the tilts are approximately increasing from $C1$ to $C7$ indicating that the galaxies in the later groups are more dynamically evolved (hence closer to their corresponding virial planes). This is also consistent with the fact that the magnesium indices are also increasing for the groups indicating that in a dynamically evolved galaxy the metal content is higher. Since none of the tilt values are close to 1, it can be concluded that these galaxies have been formed by dissipational Mergers (Robertson et al. 2006). So the groups can be considered as evolutionary tree with respect to FP, VP and $Mg2$ indices as

$$C1 \rightarrow C2 \rightarrow C3 \rightarrow C4 \rightarrow C7 \rightarrow C6(\text{excluding } C5)$$

which is irrespective of the scatter of the FP giving rise to several controversial arguments so far.

References

Achtert, E., et al. 2008. *Statistical Analysis and Data Mining* 1:111.

Balastegui, A., P. Ruiz-Lapuente, and R. Canal. 2001. *Monthly Notices of Royal Astronomical Society* 328:283.

Chattopadhyay, T., et al. 2007. *The Astrophysical Journal* 667:1017.

Chattopadhyay A.K., et al. 2013. *Astrostatistical Challenges for the New Astronomy*, Springer series in Astrostatistics. Edited by Hilbe J, New York, Springer.

Hartigan, J.A. 1975. *Clustering algorithms.* New York: Wiley.

Hakkila, J., et al. 2000. *Astrophysical Journal* 538:165.

Hudson, M.J., et al. 2001. *Monthly Notices of Royal Astronomical Society* 327:265.

Kim, et al. 1997. Publ. of Astro. Soc. of Australia 14:119.

MacQueen, J. 1967. In *Fifth Berkeley symposium on mathematical statistics and probability*, vol. 1, 281.

Mallozzi, R.S., G.N. Pendleton, and W.S. Paciesas. 1996. *The Astrophysical Journal* 471:636.

Mao, S., and B. Paczynski. 1992. *The Astrophysical Journal Letters* 389:L13.

Milligan, G.W. 1980. *Psychometrika* 45:325.

Mukherjee, S., et al. 1998. *The Astrophysical Journal* 508:314.

Robertson, B., et al. 2006. *The Astrophysical Journal* 641:21.

Schmidt, et al. 1988. *The Astrophysical Journal* 329:L85.

Shen, S., et al. 2003. *Monthly Notices of the Royal Astronomical Society* 343:978.

Sugar, A.S., and G.M. James. 2003. *Journal of the American Statistical Association* 98:750.

Chapter - 9

Time Series Analysis

9.1 Introduction

A time series is a sequential collection of data indexed over time. In most cases the data are continuous but they are recorded at a discrete and finite set of equally spaced points. If a time series has N-observations $(x_0, x_1, \ldots, x_N)$, then the time indexed distance between any two successive observations is referred to as the sampling interval. Time series lets one to explore, analyse and forecast univariate time series. Moreover, autocorrelations and partial auto correlations of the series indicate how and to what degree each point in the series is correlated with earlier values in the series. The analysis of time series is of immense significance not only for economist and businessman but also for scientist, geologist, biologist, research worker, etc. for various reasons. For example, by observing data over a period of time one can easily understand what changes have taken place in the past, i.e. such analysis is extremely helpful for forecasting purpose. It also helps in planning future operations, provided the various features can be extracted from the given time series after a successful analysis. It also facilitates comparison, i.e. different time series are often compared and important conclusions are drawn from there. So, in other words we can say that time series helps in modelling the physical phenomena responsible for the observed variability in various parameters of an object. Below some examples of time series (Figs. 9.1 and 9.2) are shown. Figure 9.1 shows a trend whereas Fig. 9.2 shows a combination of trend and seasonal fluctuations and this is typical of sales data. So from the time series data it is clear that a time series may have several components like trend, seasonal, cyclical, other irregular fluctuations and a purely random part, called the "White noise." At first trend, cyclic variations are removed from the time series and we are left with a series of residuals that may or may not be "random". In the subsequent part we will see that there are various sophisticated techniques for analysing time series of this type to examine whether any cyclic variation is still left in the residuals or whether irregular variation may be explained in terms of probability models, e.g. moving average (MA) or autoregressive models (AR), etc.

A.K. Chattopadhyay, T. Chattopadhyay, *Statistical Methods for Astronomical Data Analysis*, Springer Series in Astrostatistics 3,
DOI 10.1007/978-1-4939-1507-1_9

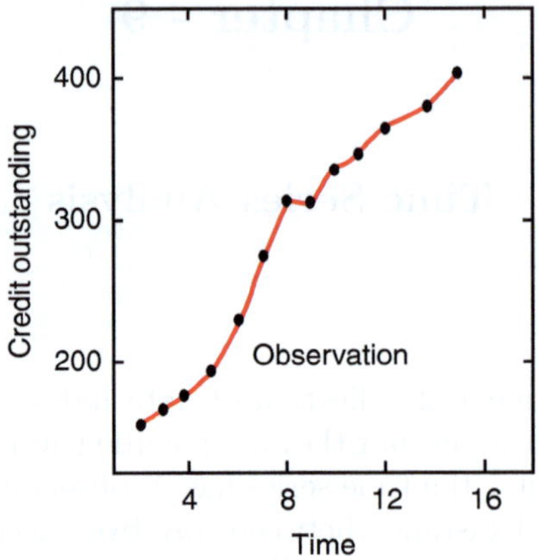

Figure 9.1 Time series showing trend

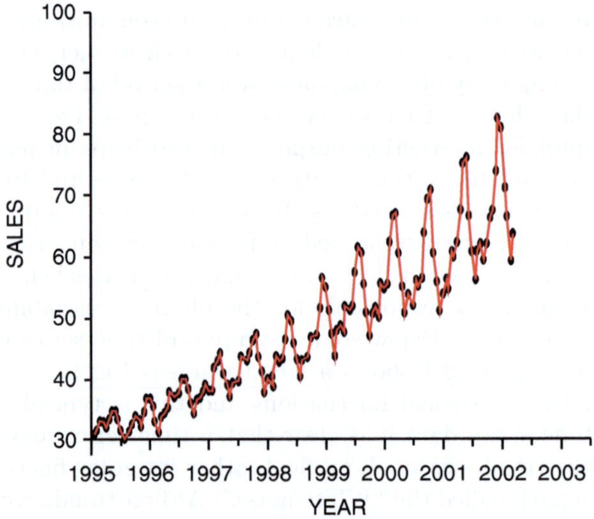

Figure 9.2 Time series showing trend with seasonal fluctuations

9.2 Several Components of a Time Series

A time series has generally the following components. (1) Trend, (2) Seasonal variation, (3) other cyclic variations and (4) irregular fluctuations.

Trend: When there is a long-term change in the mean level of the time series, we say a trend is present (viz. Fig. 9.2). The term "long term" is variable sensitive. In case of climatic variables, sometimes there are variations over long time, e.g. 50 years. So observations over 20 years will show a trend. Therefore, analyser should be very much attentive of variable under concern.

Seasonal variation: Many time series have variation over the year. For example, the temperature is higher in summer than winter or the rainfall is maximum during June–September in India. This yearly variation is called seasonal variation.

Other cyclic variations: Sometimes, there is variation over a fixed period apart from yearly variation. For example, number of sun spots generally have a period of 22 years but besides that there are many short cycles to account for. Economic data are sometimes influenced by business cycles varying from 3/4 years to more than 10 years.

Irregular fluctuations: If trend, seasonal fluctuations, other cyclic variations are removed from a given time series (provided they are present), one is left with a series of residuals which may or may not be random. So in that case one has to see whether any cyclic variation is still left or whether the residual variations can be modelled by probability theory (e.g. MA or AR models), i.e. the residuals are close to "stationary series". This will be discussed in the subsequent sections.

9.3 How to Remove Various Deterministic Components from a Time Series

Trend

If one is likely to remove the trend from the given time series (provided it is present there), he/she can use a filter of the form $p_t = \sum_{r=-i}^{+j} k_r q_{t+r}$ such that the time series $\{q_t\}$ is transformed to $\{p_t\}$ with a set of weights $\{k_r\}$. If $\sum k_r = 1$ it is called "moving average" method and if $i = j, k_r = k_{-r}$ the method is called "symmetric moving average method". If $k_r = \frac{1}{2i+1}$, then $p_t = \frac{1}{2i+1} \sum_{r=-i}^{i} q_{t+r}$. This is the simplest form of moving average method. Sometimes Spencer's 15-point moving average or "Henderson moving average" methods (Chatfield 2004) are also used for detrending the data. Another kind of filtering "trend" from the data is "Differencing", i.e. one has to follow differencing until the data is stationary. Generally first difference is sufficient to remove the trend, i.e. $p_t = q_t - q_{t-1} = \nabla q_t, t = 2, 3, \ldots N$ (Kleibergen 1996) and occasionally second differencing is required, e.g.

$$\nabla^2 q_t = \nabla q_t - \nabla q_{t-1} = q_t - 2q_{t-1} + q_{t-2}$$

Seasonal variation: For a time series containing trend the seasonal fluctuations for monthly data can be eliminated by

$$p_t = \frac{\frac{1}{2}q_{t-6} + q_{t-5} + q_{t-4} + \ldots + q_{t+5} + \frac{1}{2}q_{t+6}}{12}$$

For quarterly data the relation comes as

$$p_t = \frac{\frac{1}{2}q_{t-2} + q_{t-1} + q_t + q_{t+1} + \frac{1}{2}q_{t+2}}{4}$$

For 4-weekly data one can use moving average over 13 successive observations. For monthly data, seasonal variation can also be removed by $\nabla_{12}q_t = q_t - q_{t-12}$. Further details on seasonal variation can be found in Hylleberg (1992), and Gómez and Maravall (2001).

9.4 Stationary Time Series and Its Significance

A time series is said to be stationary if it has constant mean and variance, i.e. if there is no trend, no seasonal and other cyclic variations in the data. In other words one can say that one section of data is identical with any other section of the data. We are primarily interested in the stationary part of the time series because most of the probability models regarding time series are based on the assumption of stationary time series. For this reason one has to transform non stationary time series containing trend and/or seasonal and cyclic fluctuations into a stationary time series to apply several probability models to the residual time series thus obtained.

9.5 Autocorrelations and Correlogram

Sample autocorrelation coefficients are a series of quantiles defined in the following way. Let $\{q_1, q_2, \ldots, q_N\}$ represent a time series. Then $\{q_{1+k}, q_{2+k}, \ldots, q_N\}$ are its $(N-k)$ values k-steps apart. Then the auto covariance coefficients c_k of lag k between these two series is given as

$$c_k = \frac{1}{N}\sum_{t=1}^{N-k}(q_t - \overline{q})(q_{t+k} - \overline{q})$$

The auto correlation coefficient r_k of lag k is defined as

$$r_k = c_k/c_o, k = 1, 2, \ldots, m, \; m < N.$$

Correlogram is the plot of r_k versus $k, k = 0, 1, \ldots, m$. Hence it is clear that for a stationary series $r_k \simeq 0$ for all nonzero values of k.

Figures 9.3 and 9.4 show correlograms of two time series, the first one having a trend and other one being stationary.

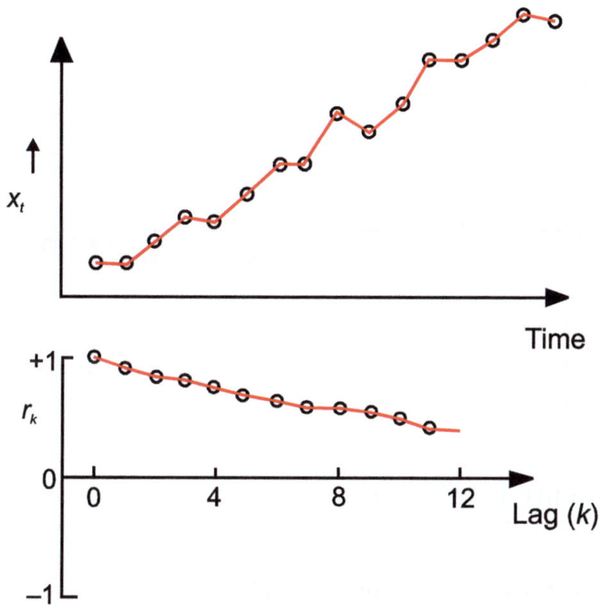

Figure 9.3 Observed data (Top) and Correlogram (Bottom) of a time series having trend

9.6 Stochastic Process and Stationary Process

A stochastic process is a process that evolves in time following some probabilistic laws. Mathematically it is a collection of random variables indexed over time. It is denoted by $X(t)$ if time is continuous ($0 < t < \infty$) and X_t if time is discrete ($t = 0, \pm 1, \pm 2, \ldots$ etc.).

Stationary process is a class of stochastic process if the joint distribution of $X(t_1), \ldots, X(t_k)$ is the same as that of $X(t_1+\tau), \ldots, X(t_k+\tau)\ \forall t_1, t_2, \ldots, \tau$. In this case the stationary process is also called **strictly stationary.** τ is called the time lag. The auto covariance function is defined as $\gamma(\tau) =$ Cov $[X(t), X(t+\tau)]$ and auto correlation function is defined by $\rho(\tau) = \gamma(\tau)/\gamma(0)$. For strictly stationary process, the first two population moments are constant, i.e.

$$\mu(t) = \mu$$
$$\sigma^2(t) = \sigma^2$$

The autocovariance function is defined as $\gamma(\tau) =$ Cov $[X(t)X(t+\tau)]$. In case of **weakly stationary process or second order stationary process**

$$E[X(t)] = \mu,$$
$$\text{Cov}\ [X(t)X(t+\tau)] = \gamma(\tau).$$

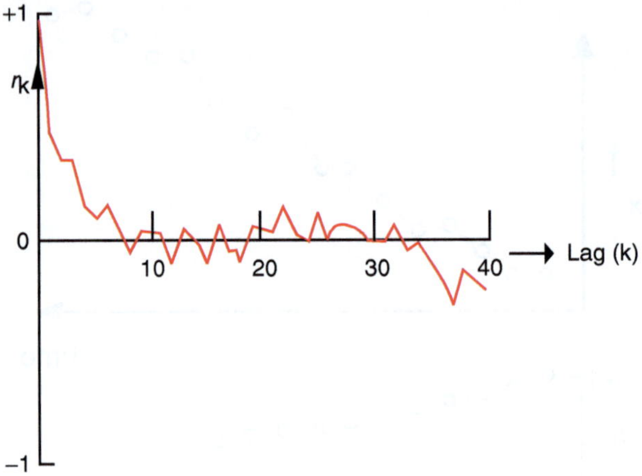

Figure 9.4 Correlogram of a stationary time series

Properties of Autocorrelation Function

Let $\{X(t)\}$ be a stationary stochastic process with mean μ and variance σ^2, then autocorrelation function

$$\rho(\tau) = \gamma(\tau)/\gamma(0)$$

Since, $\gamma(0) = \sigma^2$ hence $\rho(0) = 1$

(i) $\gamma(\tau) = \gamma(-\tau)$.

Proof. $\gamma(\tau) = \text{Cov}\ [X(t)X(t+\tau)]$
$= \text{Cov}\ [X(t-\tau)X(t)]$, since $\{X(t)\}$ is stationary
$= \gamma(-\tau)$.

(ii) $|\ \rho(\tau)| \leq 1$.

Proof. $\text{Var}\ [\delta_1 X(t) + \delta_2 X(t+\tau)] \geq 0$.
(for any constants δ_1, δ_2.)

or $\delta_1^2\ \text{Var}[X(t)] + \delta_2^2\ \text{Var}[X(t+\tau)] + 2\delta_1\delta_2\ \text{Cov}[X(t)X(t+\tau)] \geq 0$.

or $(\delta_1^2 + \delta_2^2)\sigma^2 + 2\delta_1\delta_2\gamma(\tau) \geq 0$.

When $\delta_1 = \delta_2 = 1, \gamma(\tau) \geq -\sigma^2$ i.e. $\rho(\tau) \geq -1$

When $\delta_1 = 1, \delta_2 = -1, \sigma^2 \geq \gamma(\tau)$ i.e., $\rho(\tau) \leq +1$

Hence, $|\rho(\tau)| \leq 1$.

9.7 Different Stochastic Process Used for Modelling

9.7.1 Linear Stationary Models

Purely random process: A discrete stochastic process is called a **purely random process,** $\{Z_t\}$ which are mutually independent and identically distributed (iid). It is assumed that

$$\gamma(k) = \text{Cov}(Z_t, Z_t + k) = \begin{cases} \sigma_Z^2 & k = 0 \\ 0 & k = \pm 1, \pm 2, \ldots. \end{cases}$$

$$\text{Hence,} \qquad \rho(k) = \begin{cases} 1 & k = 0 \\ 0 & k = \pm 1, \pm 2, \ldots. \end{cases}$$

Purely random process sometimes is called **"white noise".**

Autoregressive process: Let $\{Z_t\}$ be a purely random process. Then $\{X_t\}$ is to be an autoregressive process of order p (AR(p)) if

$$X_t = \alpha_1 X_{t-1} + \alpha_2 X_{t-2} + \ldots + \alpha_p X_{t-p} + Z_t \tag{9.1}$$

For, $p = 1, X_t = \alpha X_{t-1} + Z_t$, is called first order autoregressive process.

Then X_t can be written as using the recurrence relations,

$$X_t = Z_t + \alpha Z_{t-1} + \alpha^2 Z_{t-2} + \ldots, -1 < \alpha < 1, \tag{9.2}$$

which is an infinite order moving average (MA) process. Thus there exists a duality between AR and MA process. This can be seen also using the shift operator B as, $BX_t = X_{t-1}$.

Applying it to first order AR model,

$$(1 - \alpha B)X_t = Z_t.$$

$$\begin{aligned} or\ X_t &= Z_t/(1 - \alpha B) \\ &= (1 + \alpha B + \alpha^2 B^2 + \ldots)Z_t \\ &= Z_t + \alpha Z_{t-1} + \alpha^2 Z_{t-2} + \ldots \end{aligned}$$

This is identical to Eq. (9.2).

It can be seen easily that

$$\begin{aligned} E(X_t) &= 0 \\ \text{Var }(X_t) &= \sigma_z^2(1+\alpha^2+\alpha^4+\ldots), \\ &= \sigma_z^2/(1-\alpha^2), |\alpha| \leq 1, \\ \gamma(k) = E[X_t X_{t+k}] &= E[(\sum \alpha^i Z_{t-i}(\sum \alpha^j Z_{t+k-j})] \\ &= \sigma_z^2 \sum_{i=o}^{\infty} \alpha^i \alpha^{k+i} \\ &= \alpha^k \sigma z^2/(1-\alpha^2), |\alpha| < 1 \\ &= \alpha^k \sigma_X^2 \end{aligned}$$

For $k < 0$, we find $\gamma(k) = \gamma(-k)$. Here $\gamma(k)$ does not depend upon t. So AR process of order 1 is a second order stationary process provided $|\alpha| < 1$. The auto correlation function is given by

$$\rho(k) = \alpha^k, k = 0, 1, 2, \ldots$$

Figure 9.5 shows some autocorrelation function for 1st order AR process for various values of α. In each case $|\alpha| < 1$ and hence the series are stationary.

AR process can also be reduced to a MA process in general situation using backward shift operator. Let us consider AR (p) process.
Then, $X_t = \alpha_1 X_{t-1} + \alpha_2 X_{t-2} + \ldots + \alpha_p X_{t-p} + Z_t$.

i.e., $(1 - \alpha_1 B - \alpha_2 B^2 - \ldots\ldots - \alpha_p B^p)X_t = Z_t$
or $\Phi(B)X_t = Z_t$, where $\Phi(B) = 1 - \alpha_1 B - \alpha_2 B^2 - \ldots - \alpha_p B^p$
or $X_t = (1 - \alpha_1 B - \alpha_2 B^2 - \ldots - \alpha_p)^{-1} Z_t$
$= (1 + \beta_1 B + \beta_2 B^2 + \ldots)Z_t$

Now, $E(X_t) = 0$, and $\text{Var}(X_t) = \sum_{i=0}^{p} \beta_i^2$ provided the sum converges which is necessary condition for stationarity. The auto covariance function,

$$\gamma(k) = \sigma_Z^2 \sum_{i=0}^{\infty} \beta_i \beta_{i+k}, \ \beta_0 = 1$$

Sufficient condition for convergence is $\sum |\beta_i|$ converges. In principle, this is the way to find $\gamma(k)$. But $\{\beta_i\}$ are hard to find. So alternatively, the autocorrelation function $\rho(k)$ is found as follows. Multiplying Eq. (9.1) by X_{t-k}, taking expectations, dividing by σ_X^2, assuming $\{X_t\}$ is stationary, using the fact $\rho(k) = \rho(-k) \ \forall k > 0$, $\rho(k)$'s reduce to

$$\rho(k) = \alpha_1 \rho(k-1) + \ldots + \alpha_p \rho(k-p)$$

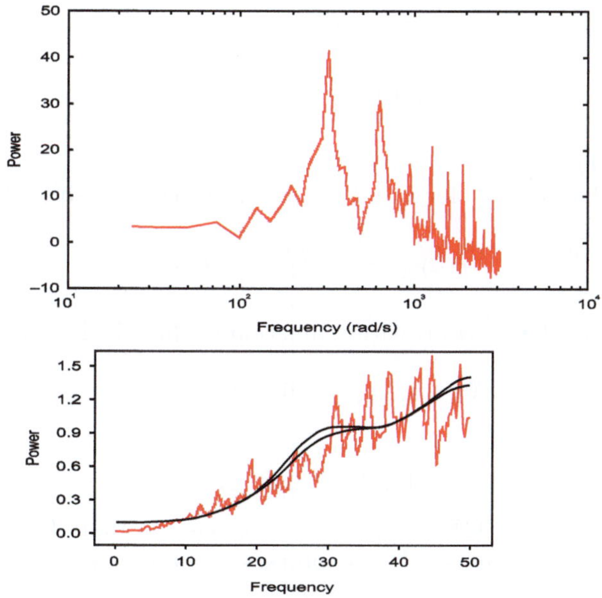

Figure 9.5 Autocorrelation function for AR(1) process

These are the set of difference equations known as Yule Walker equation and has the solution,

$$\rho(k) = A_1\pi_1^{|k|} + \ldots + A_p\pi_p^{|k|}$$

where $\{\pi_i\}$ are the roots of

$$x^p - \alpha_1 x^{p-1} - \ldots - \alpha_p = 0.$$

Depending on initial conditions, $\rho(0) = 1$, it follows $\sum A_i = 1$. It shows that $\rho(k) \to 0$ as k increases provided $|\pi_i| < 1 \ \forall i$, which is the necessary and sufficient condition for stationarity. It can be shown that an equivalent way of expressing the stationarity is that the roots of $1 - \alpha_1 B - \ldots - \alpha_p B^p = 0$ must lie outside the unit circle.

Prob. 1 Consider the AR(2) process given by $X_t = X_{t-1} - \frac{1}{3}X_{t-2} + Z_t$ Show that the process is stationary. Find the auto correlation coefficients.

Solution: $(1 - B + \frac{1}{3}B^2)X_t = Z_t$

So, the equation $1 - B + \frac{1}{3}B^2 = 0$ has the roots $\frac{3 \pm i\sqrt{3}}{2}$ the modulus of which is$\sqrt{3} = 1.732 > 1$. So it is a stationary process.

The Yule Walker equation for $k = 1$ is

$$\rho(1) = \rho(0) - \frac{1}{3}\rho(-1) = 1 - \frac{1}{3}\rho(1)$$

This gives $\rho(1) = \frac{3}{4}$.
The other values of $\rho(k)$ for $k \geq 2$ are given by the recursive relation,

$$\rho(k) = \rho(k-1) - \frac{1}{3}\rho(k-3)$$

Prob. 2 Write the autocorrelation function of the AR(2) process,

$$X_t = 1.2X_{t-1} - 0.32X_{t-2} + Z_t.$$

Solution:

$$(1 - 1.2B + 0.32B^2)X_t = Z_t.$$

So the equation $0.32B^2 - 1.2B + 1 = 0$ has the roots (2.5,1.5). As the roots lie outside the unit circle, the process is stationary.

Hence

$$\rho_k = A_1\pi_1^k + A_2\pi_2^k$$

where π_1, π_2 are roots of the auxiliary equation,

$$y^2 - 1.2y + 0.32 = 0$$

Giving,

$$\pi_1 = 0.4, \pi_2 = 0.8$$

Hence

$$\rho_k = A_1 0.4^k + A_2 0.8^k$$

From the Yule Walker equation, $\rho(k) = 1.2\rho(k-1) - 0.32\rho(k-2)$,

$$\rho(1) = 1.2\rho(0) - 0.32\rho(-1)$$

$$= 1.2 - 0.32\rho(1)$$

$$\text{or}\rho(1) = 1.2/1.32 = 0.91$$

$$\text{Also, }\rho(0) = 1.\text{ Hence}A_1 + A_2 = 1$$

$$\rho(1) = 0.91.\text{ Hence}0.91 = 0.4A_1 + 0.8A_2$$

$$\text{So, }A_2 = 0.51/0.4, A_1 = -0.11/0.4$$

$$\text{So, }\rho_k = -\frac{0.11}{0.4}0.4^k + \frac{0.51}{0.4}0.8k$$

k	0	1	2	3	4	5	6	7	8
ρ_k	1	0.91	0.77	0.63	0.51	0.41	0.33	0.27	0.21

Mixed Auto Regressive Moving Average Model (ARMA)

Let $\{Z_t\}$ be a purely random process then$\{X_t\}$ is said to be ARMA of order (p,q) process if $X_t = \alpha_1 X_{t-1} + \alpha^2 X_{t-2} + \ldots + \alpha_p X_{t-p} + Z_t + \beta_1 Z_{t-1} + \ldots + \beta_q Z_{t-q}$ i.e. using backward shift operator, $\phi(B)X_t = \theta(B)Z_t$

where $\phi(B) = 1 - \alpha_1 B - \alpha_2 B^2 - \ldots - \alpha_p B^p$ and $\theta(B) = 1 + \beta_1 B + \beta_2 B^2 + \ldots + \beta_q B^q$.

For stationary process roots of $\phi(B) = 0$ and $\theta(B) = 0$ lie outside the unit circle. It is laborious to calculate to a.c.f in the above manner. Hence sometimes ARMA is converted to a purely MA process $X_t = \psi(B)X_t$ to make the computations simpler.

Prob 3 For the ARMA (2, 1) process, $X_t - 0.8X_{t-1} - 0.1X_{t=2} = Z_t + 0.3Z_{t-1}$, show that it is stationary. Find the ψ weights of process.

Soln: Here, $\phi(B) = 1 - 0.8B - 0.1B^2$, $\theta(B) = 1 + 0.3B$

$\psi(B) = \theta(B)[\phi(B)]^{-1}$

Now, $\phi(B) = 0$ gives roots , 1.09902 , -9.09902

and $\theta(B) = 0$ gives root, 10/3.

All the roots lie outside the unit circle. So the process is stationary.

$$\begin{aligned}
\psi(B) &= (1 + 0.3B)(1 - 0.8B - 0.1B^2)^{-1} \\
&= (1 + 0.3B)(1 + 0.8B + 0.1B^2 + \ldots) \\
&= 1 + 1.1B + 0.34B^2 + 0.03B^3 + \ldots \\
\psi(B) &= \sum_{i=0}^{\infty} \psi_i B_i \\
So, \psi_1 &= 1.1, \psi_2 = 0.34, \psi_3 = 0.03 \text{ etc.}
\end{aligned}$$

9.7.2 Linear Non Stationary Model

Integrated Autoregressive Moving Average Process (ARIMA)

We have discussed earlier that all probability models are based on stationary time series. In most cases the time series are non stationary. So, one has to remove the non stationary sources of variation, e.g. trend, seasonal fluctuations or any other cyclical fluctuations, etc. If the non stationary source is due to trend, then one can difference the series, i.e. X_t is replaced by $\nabla^d X_t$, etc. This type of model is called an "integrated" model because the stationary model, i.e. fitted to the differenced data has to be "integrated" to provide the model for the original non stationary data.

$$\text{Thus, if } Y_t = \nabla^d X_t$$

and $Y_t = \alpha_1 Y_{t-1} + \ldots + \alpha_p Y_{t-p} + Z_t + \ldots \beta_q Z_{t-q}$

then, writing, $\phi(B)Y_t = \theta(B)Z_t$ we have to replace Y_t by

X_t, i.e. $\phi(B)(1-B)^d X_t = \theta(B)Z_t$

This is an Autoregressive Integrated Moving Average (ARIMA) process of order (p, d, q).

9.8 Fitting Models and Estimation of Parameters

Autocovariance and Autocorrelations Functions

Theoretical autocovariance and autocorrelations functions are very useful in checking the stationarity of a time series. It can be shown that $lim_{N\to\infty} E(c_k) = \gamma(k)$, i.e. c_k is an asymptotically unbiased estimator of $\gamma(k)$.

The sample auto correlation function of lag k is,

$$r_k = c_k / c_0$$

and $E(r_k) \simeq -1/N$

Var $(r_k) \simeq 1/N$ (Kendall et al. 1983)

So the confidence limits for r_k is $= 1/N \pm 2/\sqrt{N} \simeq \pm 2/\sqrt{N}$

Hence values of r_k falling outside this limit is significantly different from zero at 5 % level of significance.

The following points might be noted for modelling with respect to $\mathbf{r_k}$

1. When r_k does not converge to zero quickly, it implies nonstationarity. In that case the time series is to be differenced unless it becomes stationary.
2. When r_k cuts at lag q, then it is preferable to use MA(q) model.
3. When r_k falls slowly to zero, AR/ARMA models are appropriate (Chatfield 2004).

An Ar(p) Process

For AR(p) process, (with mean μ)

$$X_t - \mu = \alpha_1(X_{t-1} - \mu) + \ldots + \alpha_p(X_{t-p} - \mu) + Z_t.$$

Then the parameters can be fitted by least square method.
In particular for AR(1) and AR(2) process,

$$\hat{\mu} = \overline{x}, \hat{\alpha}_1 = r_1$$

and,

$$\hat{\mu} = \overline{x},\ \hat{\alpha}_1 \simeq \frac{r_1(1-r_2)}{1-r_1^2},\ \hat{\alpha}_2 \simeq \frac{(r_2 - r_1^2)}{1-r_1^2}$$

For determining the order of an AR process, AR process of several orders, $p = 1, 2\ldots$ etc. are fitted and residual sum of squares are plotted against $p = 1, 2, \ldots$ etc. When $(p+1)$th term does not improve residual sum of squares, take then p as the order of the process.

Fitting for MA(q) process: For this process parameters cannot be found by explicit least square estimation. So the following procedure is followed. Some suitable starting values of the parameters are considered and the residual sum of squares is computed. The same is calculated for other values of the parameters in a parameter space grid. Then those values of parameters are chosen for which residual sum of squares is minimum.

Fitting ARMA model: The same procedure of minimizing residual sum of squares is followed for this process also.

Fitting Non Stationary Models

ARIMA Model

For nonstationary model it is transformed to stationary model by first order or occasionally second order difference scheme and then linear models are fitted. Then the resulting undifferenced series is the fitted ARIMA model. For seasonal data seasonal differencing is required.

SARIMA Model

When the time series contains seasonal periodic component of periods (e.g. for monthly data s = 12) then the generalized ARIMA model to deal with seasonality is defined as

$$\phi_p(B)\Phi_P(B^s)W_t = \theta_q(B)\Theta_Q(B^s)Z_t$$

where B is the shift operator, ϕ_p, Φ_P, θ_q and Θ_Q are polynomials of orders p, P, q and Q, respectively, and Z_t is purely random process. $W_t = \nabla^d \nabla_s^D X_t$ denotes the differenced series.

The above model is called SARIMA model of order$(p, d, q) \times (P, D, Q)_s$.

For example SARIMA model of order $(1, 0, 0) \times (0, 1, 1)_{12}$ reduces to

$$X_t = X_{t-12} + \alpha(X_{t-1} - X_{t-13}) + Z_t + \Theta Z_{t-12}$$

where $W_t = \nabla_{12} X_t$ and α, Θ are parameters.

The model parameters are estimated by iterative process and the orders p, P, q, Q are determined from the auto covariance function of the stationary series W_t found by differencing.

9.9 Forecasting

Suppose we have an observed time series $x_1, x_2, \ldots, x_N$. Then forecasting is the prediction of x_{N+1} after time step h. It is denoted as $x_N(h)$. Forecasting might be univariate, i.e. $x_N(h)$ depends on $x_1, x_2, \ldots, x_N$, i.e. on a single time series or it might be multivariate, i.e. the variable depends on more than one time series. It can be "point forecasting" or "prediction interval". When the future value is described by some distribution instead of "point" or "interval", then it is called density forecasting.

Point Forecasting for Univariate Methods

(i) Simple Exponential Smoothing (SES):

This is used for non seasonal and non trend time series and is given by $\hat{x}_{N+1} = c_0 x_N + c_1 x_{N-1} + c_2 x_{N-2} + \ldots$

where $c_i = \alpha(1-\alpha)^i, i = 0, 1, \ldots$ so that $\sum c_i = 1, 0 < \alpha < 1$.

Then,

$$\begin{aligned} \hat{x}_{N+1} &= \alpha x_N + \alpha(1-\alpha) x_{N-1} + \ldots \\ &= \alpha x_N + (1-\alpha)\hat{x}_{N-1} \end{aligned} \tag{9.3}$$

If we assume $\hat{x}_1 = x_1$, then the above equation can be used for successive forecasting. It is to be noted that

$$\begin{aligned}\hat{x}_{N+1} &= \alpha[x_N - \hat{x}_{N-1}] + \hat{x}_{N-1} \\ &= \alpha e_N + \hat{x}_{N-1}\end{aligned}$$

where e_N is the prediction error.

Estimate of α

Since we have assumed $\quad \hat{\alpha}_1 = x_1$

$$e_2 = x_2 - \hat{x}_2$$

$$\hat{x}_2 = \alpha e_2 + \hat{x}_1$$

$$e_3 = x_3 - \hat{x}_2$$

$$e_N = x_N - \hat{x}_{N-1}$$

We compute $\sum_{i=2}^{N} e_i^2$ and repeat the procedure for other value of α between [0,1] and consider that value of α for which $\sum e_i^2$ is minimum.

(ii) The Holt and Holt-Winters Point Forecasting: This is used for non-seasonal time series containing trend. In Eq. (9.3) replacing $\hat{x}_{N+1}$ by L_t, i.e. local level at time t,

$$L_t = \alpha x_t + (1-\alpha)L_{t-1}$$

If there is a trend T_t in addition then we have

$$L_t = \alpha x_t + (1-\alpha)(L_{t-1} + T_{t-1})$$

$$T_t = \gamma(L_t - L_{t-1} + (1-\gamma)T_{t-1}$$

Hence,

$$\hat{x}_{t+h} = L_t + h\ T_t,\ h = 1, 2, 3, \ldots.$$

If there is seasonal variation in addition and I_t is the seasonal component at time t, then we have

$$\begin{aligned}L_t &= \alpha(x_t/I_{t-12}) + (1-\alpha)(L_{t-1} + T_{t-1}) \\ T_t &= \gamma(L_t - L_{t-1}) + (1-\gamma)T_{t-1} \\ I_t &= \delta(x_t/L_t) + (1-\delta)I_{t-12}\end{aligned}$$

Hence, $\hat{x}_{t+h} = (L_t + hT_t)I_{t-12+h}, h = 1, 2, \ldots, 12$

(iii) The Box–Jonskins Procedure:

Here the forecasting procedure is based on ARMA or SARIMA model for non-seasonal/seasonal data. This has been already discussed in Sect. 9.8.

9.10 Spectrum and Spectral Analysis

When a time series is transformed from time domain to frequency domain, then the corresponding analysis is called "Spectral Analysis". The advantage of frequency approach is widely accepted in various fields, e.g. electrical engineering, geophysics, astrophysics and meteorology. Spectral analysis is based on the assumption that the time series is made up of sine and cosine waves with different frequencies and the device which uses this idea is called the "periodogram" (Schuster 1898). To study "Periodogram" it is necessary to have an idea of what is "spectral distribution function" and "spectral density function".

Let us consider that a time series, apparently random in nature, contain a mixture of several cyclical fluctuations of different frequencies. Then it can be expressed as

$$X_t = \sum_{i=1}^{m} A_i \cos\,(w_i t + \phi_i) + Z_t$$

where, A_i, w_i, ϕ_i denote the amplitude, angular frequency and phase at frequency w_i. To make "stationary", we assume $\{A_i\}$ are uncorrelated random variables with mean zero or $\{\phi_i\}$ are random variables uniformly distributed over $[0,2\pi]$

or, $X_t = \sum_{i=1}^{m} (b_i cos\ w_i t + d_i sin\ w_i t) + Z_t$

where $b_i = A_i cos\ \phi_i, d_i = -A_i sin\ \phi_i$.

Letting

$$k \to \infty, X_t = \int_0^{\pi} \cos wt\ du(w) + \int_0^{\pi} \sin wt\ dv(w), \tag{9.4}$$

where $u(w)$ and $v(w)$ are uncorrelated continuous process (Cox and Miller 1968).

The upper limit is taken π instead of ∞ because

$$\cos\,[(w+m\pi)t] = \text{Cos } wt,\ m,t \text{ are integers, } m, is \text{ even}$$
$$= \text{Cos } (\pi - w)t,\ m,t \text{ are integers, } m, is\ odd.$$

So the highest frequency is π, called the "Nyquist frequency" and all the frequencies lie between $[0,\pi]$.

Instead of $u(w)$ and $v(w)$ we use a function $F(w)$, so that auto covariance function of lag k can be expressed as

$$\gamma(k) = \int_0^{\pi} \cos\, wkdF(w) \text{ (Wierner–Khintchine theorem)} \tag{9.5}$$

F(ω) is called **"power spectral distribution function".**

If we assume $F(w) = 0, w < 0$, then $F(w)$ is monotonic in $[0,\ \pi]$

and $\gamma(0) = \sigma_X^2 = F(\pi) = Var(X_t)$

If $F(w)$ contains any deterministic sinusoidal component at $w = w_0$ (say), then there is a step increase at $w = w_0$ and for indeterministic part it is continuous. So we can write $F(w) = F_1(w) + F_2(w)$, where the first one is completely "statistical" and the second one is due to the deterministic component. We are primarily interested in the "indeterministic" part and for the present case put $F_2(w) = 0$.

Since $F_2(w) = 0, F(w)$ is continuous and hence derivable.

So we define $f(w) = \frac{dF(w)}{dw}$, where $f(w)$ is called **"Power spectral density function".** Then Eq. (9.5) reduces to

$$\gamma(k) = \int_0^{\pi} \cos\, wk\ f(w)dw$$

So that$f(w) = \frac{1}{\pi} \sum_{k=-\infty}^{\infty} \gamma(k)e^{-iwk}$, i.e. the "Discrete Fourier Transform" of $\gamma(k)$. Since $\gamma(k)$ is an even function (i.e. $\gamma(k) = \gamma(-k)), f(w) = \frac{1}{\pi}[\gamma(0) + 2\sum_{k=1}^{\infty} \gamma(k) \cos\, wk]$

The Periodogram: An Estimate of the Power Spectrum

Let $(x_1, x_2, \ldots, x_N)$ be N observations of a time series and we suspect that it is a mixture of sine and cosine functions with "hidden" periodicities. Then if N is odd,

$$X_t = \alpha_0 + \sum_{i=1}^{m} (\alpha_i cos\ w_i t + \beta_i sin\ w_i t) + Z_t$$

where $w_i = 2\pi f_i$, and $f_i = i/N$

Then, $\alpha_0 = \sum_{t=1}^{N} x_t/N = \overline{x}$

$$\alpha_i = \tfrac{2}{N} \sum_{t=1}^{N} x_t cos\ w_i t$$

$$\beta_i = \tfrac{2}{N} \sum_{t=1}^{N} x_t sin\ w_i t, \ \ i = 1.2, \ldots, m.$$

Then the "Periodogram" consists of $m = (N-1)/2$ values

$$I(w_i) = \frac{N}{4\pi}(\alpha_i^2 + \beta_i^2), i = 1, 2, \ldots m.$$

Significance of "Periodogram"

Now, $\frac{(\alpha_i^2+\beta_i^2)}{2}$ is the contribution of variance in the range, $(w_i \pm \pi/N)$, so we can plot a histogram whose width is $\frac{2\pi}{n}$.

Then, $\frac{\alpha_i^2+\beta_i^2}{2}$ = area of the histogram rectangle

$= \text{height} \times \frac{2\pi}{N}$

So, height $= \frac{N(\alpha_i^2+\beta_i^2)}{4\pi}$
$= I(w_i)$

$$\text{Then, } I(w_i) = \left\{ (\sum_{t=1}^{N} x_t cos[\frac{2\pi it}{N}])^2 + (\sum_{t=1}^{N} x_t sin[\frac{2\pi it}{N}])^2 \right\} /N\pi.$$

Since, $\sum cos\ w_i t = 0 = \sum sin\ w_i t$

$$I(w_i)=[\left\{(\sum_{t=1}^{N} x_t - \overline{x})cos\ w_i t\right\}^2 + \left\{\sum_{t=1}^{N}(x_t - \overline{x})sin\ w_i t\right\}]^2/N\pi$$

$$= \sum_{t,l=1}^{N}(x_t - \overline{x})(x_l - \overline{x})(cos\ w_i t\ cos\ w_l s + sin\ w_i t\ sin w_l s)/N\pi$$

or, $I(w_i) = (c_0 + 2\sum_{k=1}^{N-1} c_k cos\ w_i k)/\pi$

$$= (\sum_{k=-(N-1)}^{N-1} c_k e^{-iw_i k})/\pi.$$

So it appears to be the estimate of the power spectrum

$$f(w) = (\gamma_0 + 2\sum_{k=1}^{\propto} \gamma_k coswk)/\pi$$

9.11 Cross-Correlation Function ($\mathbf{w_{cross}}(\theta)$)

The cross-correlation function $w_{cross}(\theta)$ between two populations 1 and 2 (say), in terms of angular scale θ, is defined as the fractional excess in the probability, relative to a random unclustered distributions, of finding an object in population 1 in a solid angle $\delta\Omega_1$ and an other object in population 2 in a solid angle $\delta\Omega_2$, separated by angle θ and defined as

$$\delta \underline{P} = \sum_1 \sum_2 [1 + w_{cross}(\theta)]\delta\Omega_1\delta\Omega_2$$

where $\sum_1$ and $\sum_2$ are surface densities of populations 1 and 2, respectively (Peebles 1980). The cross correlation function $w_{cross}(\theta)$ is measured in terms of the pair counts between two populations 1 and 2 and is denoted by$D_1D_2(\theta)$. Suppose populations 1 and 2 contain M and N objects $(p_1, p_2, \ldots, p_M)$ and $(q_1, q_2, \ldots, q_N)$ (say), respectively. Let $p_i(i = 1, 2, \ldots, M)$ has the position co-ordinates $(p_{i1}, p_{i2}, \cdots, p_{ik})$ (in k-dimensional space) and $q_j(j = 1, 2, \ldots, N)$ has the position co-ordinates $(q_{j1}, q_{j2}, \cdots, q_{jk})$. Then the separations of every object of population 1 with that of population 2 are $d_{ij}, (i = 1, 2, \ldots, M, j = 1, 2, \ldots, N)$. The entire range of separations are binned into a finite number of intervals $(\theta, \theta + \delta\theta)$ and the number of separations falling in $(\theta, \theta + \delta\theta)$ is denoted by $D_1D_2(\theta)$.

Let R_1, R_2 be the random realizations of populations 1 and 2, respectively. Similar procedure is followed for D_1R_2, D_2R_1 and R_1R_2 and R_1R_2, respectively. Then the various estimates of the cross-correlation functions $w_{cross}(\theta)$ are (Landy and Szalay 1993; Hamilton 1984; Bernstein 1994; Bernardeau et al. 2002)

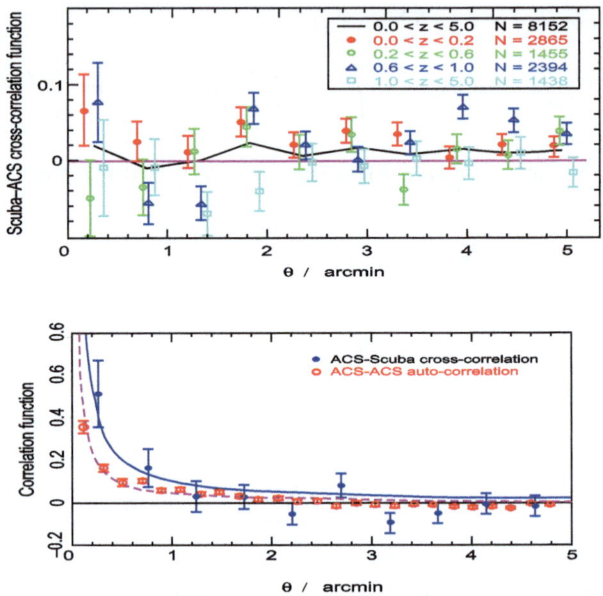

Figure 9.6 Cross correlation functions between two galaxy samples

$$W_{cross}(\theta) = \frac{D_1 D_2}{D_2 R_1} - 1 \tag{9.6}$$

$$W_{cross}(\theta) = \frac{D_1 D_2}{D_1 R_2} - 1 \tag{9.7}$$

$$W_{cross}(\theta) = \frac{D_1 D_2 x R_1 R_2}{D_1 R_2 x D_2 R_1} - 1 \tag{9.8}$$

and

$$W_{cross}(\theta) = \frac{D_1 D_2 - D_1 R_2 - D_2 R_1 + R_1 R_2}{R_1 R_2} \tag{9.9}$$

Among the above four estimates, last two are the modified versions of those originally suggested for the auto-correlation function by Hamilton (1984), Landy and Szalay (1993). Figure 9.6 shows the cross-correlation function between two galaxy populations (Blake et al. 2006). Here it is to be noted that the position co-ordinates might be replaced by various parameters for objects of populations 1 and 2, respectively.

Exercise

1. Let $X_t = Z_t + \theta Z_{t-1}$ be a MA (2) process where $\{Z_t\}$ is white noise $\{0, \sigma_Z^2\}$,

   (i) Find the autocovariance and autocorrelation function for the process when $\theta = 0.8$

   (ii) Complete the variance of the sample mean when $\theta = 0.8$

2. The Wolf sun spot numbers$\{X_t, t = 1, 2, \ldots, 100\}$ have sample autocovariance $\hat{\gamma}_0 = 1382.2, \hat{\gamma}_2 = 591.82$ and $\hat{\gamma}_3 = 96.201$. Find the Yule Walker estimates of ϕ_1, ϕ_2 and σ_Z^2 in the model $X_t = \phi_1 X_{t-1} + \phi_2 X_{t-2} + Z_t$,

   where $\{Z_t\}$ is white noise $(0, \sigma_z^2)$ for the mean correlated series $X_t = Y_t - 46.93, t = 1, 2, \ldots, 100$.

3. Find approximate values for the mean and variance of the periodogram ordinate $I_{200}(\pi/4)$ of the AR (1) process,

   $X_t = 0.5X_{t-1} + Z_t$.

4. Rewrite the following time series models using the backward shift operator. Classify each of them as ARIMA (p, d, q) process. State whether the following models are (i) stationary and/or (ii) invertable.

   (a) $X_t = 0.5X_(t-1) + Z_t$

   (b) $X_t - 1.5X_{t-1} + 0.6X_{t-2} = Z_t$

   (c) $(X_t - 0.2) - 1.2(X_{t-1-0.2} + 0.2(X_{t-2} - 0.2) = Z_t - 0.5Z_{t-1}$

5. Calculate ρ_1, ρ_2 for the MA processes.

   (i) $X_t = Z_t - \beta Z_{t-1}$

   (ii) $X_t = (1 + 2.4B + 0.8B^2)Z_t$

6. Describe the key difference between the correlograms of a stationary process and a MA process of the same order.

7. Generate samples of an AR (1) process with $\phi = 0.7$ using a computer program as follows:

   (i) Generate a vector a_t of 150 random normal variable (0.1)

   (ii) Take $Z_1 = a_1$

   (iii) For $t = 2, 3, \ldots, 150$ calculate $Z_t = 0.7a_{t-1} + a_t$

   (iv) To avoid effect of the initial conditions, eliminate first 50 observations and take the remaining 100 values of the AR (1) process.

8. Obtain the theoretical autocorrelation function of the process, $X_t = 0.9X_{t-1} + 0.18X_{t-2} + Z_t$ where $\{Z_t\}$ is white noise $(0, \sigma_Z^2)$. Generate a realization of the process using a computer and compare the sample function with the theoretical one.

9. Prove that the MA (1) processes

   $X_t = Z_t + 0.5Z_{t=1}$ and $X_t = Z_t + 2Z_{t-1}$ have the same autocorrelation structure but that one is invertible and the other is not.

10. Given the $MA(2), X_t = Z_t + 1.2Z_{t-1} - 0.35Z_{t-2}$. (a) check whether it is invertible, (b) calculate its autocorrelation structure and (c) write it as an $AR(\infty)$ process.

11. Calculate predictions for t = 100,101 and 102 and the final prediction equation of the $MA(2)$ process $X_t = 5 + Z_t - 0.5Z_{t-1}$, knowing that the predictions carried out with information up to $t = 97$ have been $X_{97}(1) = 5.1$ and $X_{96}(1) = 5.3$ and that we have observed $X_{98} = 4.9$ and $X_{99} = 5.5$.

12. Explain the structure of the forecasts generated by the model

    $\nabla X_t = 3 + (1 - 0.7B)Z_t$

13. Let $\{Z_t\}$ be a stationary process with mean zero and let a and b be constants.

   (i) If $X_t = a+bt+s_t+Z_t$ where $\{s_t\}$ is a seasonal component with period 12 and $\{Z_t\}$ is stationary, show that $\nabla\nabla_{12}X_t = (1-B)(1-B^{12})X_t$ is stationary.

   (ii) If $X_t = (a+bt)s_t+Z_t$ where $\{s_t\}$ is a seasonal component with period 12 and $\{Z_t\}$ is stationary, show that $\nabla_{12}^2 X_t = (1-B^{12}(1-B^{12})X_t$ is stationary.

14. Which, if any, of the following functions defined on the integers is the autocovariance function of a stationary time series?

   (i)
$$f(h) = 1 + \text{ Cos } (\pi h/2) + \text{ Cos } (\pi h/4)$$

   (ii)
$$f(h) = 1 + Cos(\pi h/2)Cos(\pi h/4)$$

   (iii)
$$\begin{aligned} f(h) &= 1, h = 0 \\ &= 0.6, h = \pm 1 \\ &= 0, otherwise \end{aligned}$$

15. Let $\{S_t : t = 0, 1, 2, \ldots\}$ be the random walk with common drift μ, defined by

$$S_0 = 0$$

$$S_t = \mu + S_{t-1} + Z_t, t = 1, 2, 3, \ldots$$

where $\{Z_t\} \sim IID(o, \sigma_Z^2)$, show that ∇s is stationary and compute its mean autocovariance function.

References

Blake, C., A. Pope, D. Scott, and B. Mobashar. 2006. *Monthly Notices of Royal Astronomical Society* 368:732.

Bernstein, G. M. 1994. *The Astrophysical Journal* 424:569.

Bernardeau, F., S. Colombi, E. Gaztañaga, and R. Scoccimarro. 2002. *Physical Review* 367:1.

Chatfield, C. 2004. *The analysis of time series: an introduction*, Sixth Edition. Chapman & Hall: CRC Press.

Cox, D.R. and H.D. Miller. 1968. *The Theory of Stochastic Processes.* New York: Wiley.

Gómez, V., and A. Maravall. 2001. In *A course in time series analysis*, ed. Pe'na, D., G.C. Tao, and R.S. Tsay. New York: Wiley.

Hamilton, J.D. 1984. *Time Series Analysis.* Princeton NJ: Princeton University Press.

Hylleberg, S. 1992. *Modeling seasonality.* Oxford: Oxford University Press.

Kendall, M.G., A. Stuart and J.K. Ord. 1983, *The Advanced Theory of Statistics*, Fourth Edition. London: Griffin.

Kleibergen, F. 1996. In *Equality restricted random variables: densities and sampling algorithms.* Econometrica Institute Reports, No 9662/A.

Landy, S.D., A.S. Szalay. 1993. *The Astrophysical Journal* 412:64.

Peebles, P.J.E. 1980. Large-Scale Structure of the Universe, eds. Sam B. Treiman, Princeton Series in Physics.

Schuster, S.A., 1898. *Terrestrial Magnetism* 3:13.

Chapter - 10

Monte Carlo Simulation

The name "Monte Carlo" emerged from the name of the city "Monte Carlo", famous for its Casino. In the casino there was a roulette where a small button was fixed at the centre of a wheel and the numbers 0–9 were marked at the end of the ten spokes of the wheel. When the button was pressed, the wheel starts rotating. When the wheel stops, a number within 0–9 was marked by a marker. Depending upon the number near the marker, one had to play the game. Thus this mechanical device was the device, first constructed to generate random numbers. The method was theoretically came through the work "The Monte Carlo Method" published by American mathematicians John Von Neumann and Stanslav Ulam in 1949. In spite of the theoretical background this method could not be used in significant scale until the invention of electronic computers. There are two aspects of this method. First it is used to simulate any process whose growth is affected by random factors and second to solve mathematical problems not affected by random factors but can be connected to artificially constructed probabilistic model giving rise to solution, otherwise unavailable. For example, suppose we are to calculate the area A within a unit square in Fig. 10.1.

Let us select N random points within the unit square of which N' is that within A. Then we can say easily that the area of A is proportional to N'/N and the accuracy increases when N is large. It can be shown that the area calculated in this way is within 5–15 % of accuracy compared to the true value. There are two features for this experiment. First it is easy to construct only one algorithm to generate a random point within the limit and the results of all trials are averaged, though the result is not very accurate. So, Monte Carlo methods are easy to compute, not subject to high degree of accuracy but can be applied to problems otherwise whose solutions are not available, e.g. multidimensional computation of volume of arbitrary shape. The second feature is that the points generated in the above experiment should be uniformly scattered. If the points are generated by a qualified gunman aiming to target at the centre, then most of the points will cluster around the centre, as a result of which area A will be increased a lot from its true value. So the "random points" are not just random points but also are to be "uniformly scattered". The precise meaning of the above conjecture is understood

A.K. Chattopadhyay, T. Chattopadhyay, *Statistical Methods for Astronomical Data Analysis*, Springer Series in Astrostatistics 3,
DOI 10.1007/978-1-4939-1507-1_10

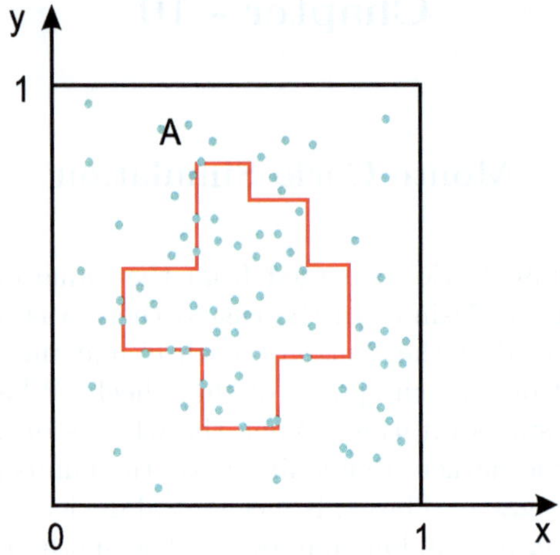

Figure 10.1 Area calculation by Monte Carlo method

through the concept of "random numbers". So "random numbers" are numbers that occur in a sequence such that two conditions are satisfied. (1) The values are uniformly distributed over a predefined interval or set and (2) it is impossible to predict future values based on previous or present values.

10.1 Generation of Random Numbers

The generation of random numbers can be through some formula but the quality of randomness can be checked by means of statistical test which will be discussed below. The random numbers generated by using any formula are hence called pseudorandom numbers. This is called "simulation" and it means that these numbers satisfy a set of tests as if they represent the values of a random variable.

The first algorithm for generating random number was suggested by J. Neumann. It is known as the "mid-square method". Let us consider the following example. Suppose we have a three digit number $x_0 = 345$. Squaring x_0 we get 119,025. Since it has odd number of digits, we can take for $x_1 = 902$ or 190. Let us take $x_1 = 902$. Squaring we have 813,604. So $x_2 = 360$ or 136 and so on. However this algorithm did not become a very successful one as fraction of smaller values is higher than is necessary. The most commonly used method of generating pseudorandom sequence is due to Lehmer (1951) which is generated by the recurrence relation

$$x_i \equiv ax_{i-1} \text{ (modulo m)} \tag{10.1}$$

i.e. if ax_{i-1} is divided by m then the remainder is x_i. This congruential relation is generalized by Vyssotslay et al. (1961) as

$$x_i \equiv ax_{i-1} + c \text{ (modulo m)} \tag{10.2}$$

where a, c, x_i, x_{i-1} are integers and m is an integer whose value depends on the design of the computer.

If $m = 17, a = 4, c = 1$ and $x_0 = 2$, then the sequence of x_i's generated using (10.2) is 2, 9, 3, 13, 2, 9, ... so the period is 4. Our intention is that the period should be longer than the number of random numbers required. So m should be sufficiently large. The full period of m can be achieved using (10.2) under the following criteria:

1. c and m have no common divisor,
2. $a \equiv 1$ (modulo p) for every prime factor p of m, i.e. if p is a prime factor that divides m, then p divides $a - 1$.
3. $a \equiv 1$ (modulo 4) if m is multiple of 4, i.e. if 4 divides m, then 4 divides $a - 1$.

For the formula (10.1), the period is always less than m and if $m = 2^\gamma {q_1}^{\delta_1} {q_2}^{\delta_2}$ $\ldots {q_r}^{\delta_r}$

where q_i's are distinct odd primes, the maximum possible period is

$$\begin{aligned} \lambda(m) &= lcm\{\lambda(2^\gamma), \lambda({q_1}^{\delta_1}), \ldots, \lambda({q_r}^{\delta_r})\} \\ \text{where } \lambda(q^\delta) &= q^{\delta-1}(q-1) \text{ if } q \text{ is odd} \\ \lambda(2^\gamma) &= 1(\gamma = 0, 1), \\ &= 2(\gamma = 2), \\ &= 2^{\gamma-1}(\gamma > 2). \end{aligned}$$

The maximum period is achieved provided that

1. $a^n \not\equiv 1(\text{modulo } {q_j}^{\delta_j})$ for $0 < n < \lambda({q_j}^{\delta_j})$,
2.

$$\begin{aligned} \text{a}^n \equiv 1(modulo\ 2)\ if\ \gamma &= 1 \\ \equiv 3(modulo\ 4)\ if\ \gamma &= 2 \\ \equiv 3 \text{ or } 5(modulo\ 8)\ if\ \gamma &> 2 \end{aligned}$$

3. x_0 is prime relative to m.

All the above criteria are due to Frisch (1962).

The widely used values of m is $m = 2^\gamma$. So for 16 bit computer $m = 2^{15}$.

Usually random numbers are converted to random fractions between (0, 1) by using $\xi_i = x_i/m$ in order to correspond them to probabilities.

Example 1: Write a program in C to generate random fractions using linear congruential series. The values are generated by $r[i] = (ar[i-1]+c) \bmod m$. Given the values of multiplier a, increment c, seed $r[0]$ and a large integer m (usually taken as max int).

Solution.

```
# include < stdio.h >

# include < math.h >

main( )

{

int i, n, m;

float r[100], a, c, u[100];

printf ("Enter the number of random number required :");

scanf ("%d", &n);

printf ("Enter the value of a&c : \n\n");

scanf ("%f\n\n%f", &a, &c);

printf ("Enter the value of the seed r[0] : \n\n");

scanf ("%f", &r[0]);

printf ("Enter a large value for m:");

scanf ("%d", &m);

for (i = 0; i < n; i + +);
```

```
{
r[i] = fmod(a * r[i - 1] + c, m);
u[i] = r[i]/m;
printf ("%f\n\n", u[i]);
}
}
```

10.2 Test for Randomness

In this section we are giving a simple widely used test for checking randomness of a data set. Suppose the random numbers generated in the data set lie between 0 and 9, and the total numbers generated is N (say). Then one can use the following equidistribution test to verify the randomness of the generated values. For this test we initially develop a table which contains in the first column the ten possible values, 0–9, in the second column the corresponding observed frequencies and in the last column the expected frequencies, each of which will be equal to $N \times \frac{1}{10}$ as there are ten values and we assign the equal probability namely $\frac{1}{10}$ to each of them for being selected and according to the concept of binomial distribution the expected frequency will be N times the success probability. Then we apply the standard chisquare test on the observed and expected frequencies. Here the null hypothesis is that the numbers are independent, i.e. they are randomly generated against the alternative that they are not random. Hence if the value of the chisquare statistic given by $\mathcal{X}^2 = \sum_{i=1}^{10} \frac{(O_i - E_i)^2}{E_i}$ where O_i and E_i's are observed and expected frequencies for the ith number is too small, we will accept the null hypothesis, i.e. according to statistical terminology we will reject the null hypothesis if observed value of $\mathcal{X}^2 > \mathcal{X}^2{}_{\alpha/2,N-1}$ or $\mathcal{X}^2 < \mathcal{X}^2{}_{1-\frac{\alpha}{2},N-1}$ where α is the level of significance and $\mathcal{X}^2{}_{\alpha/2,N-1}$ and $\mathcal{X}^2{}_{1-\frac{\alpha}{2},N-1}$ are the tabulated values given in biometric table. In other words one can accept the null hypothesis of randomness if the p value (already defined in Chap. 4) is large (at 5 % level of significance, i.e. > 0.05, say) and reject it otherwise.

10.3 Generation of Random Numbers from Various Distributions

(i) Simulation from Exponential Distribution

The p.d.f. of exponential distribution is given by

$$f(x) = \lambda e^{-\lambda x}, x \geq 0, \lambda > 0$$

Then the c.d.f is

$$\begin{aligned} F(x) &= \int_0^x f(x)dx \\ &= \int_0^x \lambda e^{-\lambda x} dx \\ &= 1 - e^{-\lambda x} \end{aligned}$$

Since $F(x)$ follows uniform distribution over [0, 1], so here $F(x)$ is a random fraction between [0, 1] and let us denote it by $r, r \in [0, 1]$.

Then $r = 1 - e^{-\lambda x}$

$$\text{i.e. } x = -\left(\frac{1}{\lambda}\right) \ln(1 - r) \tag{10.3}$$

(ii) Simulation from Standard Gaussian Distribution (Box–Muller Transformation)

The p.d.f of a univariate standard Gaussian distribution is

$$f(x) = \frac{1}{\sqrt{2\pi}} e^{-\frac{x^2}{2}}, -\infty < x < \infty$$

So, p.d.f. of a bivariate standard Gaussian distribution is

$$f(x, y) = \frac{1}{2\pi} e^{-(\frac{x^2+y^2}{2})}, -\infty < x < \infty, -\infty < y < \infty$$

Substituting $x = r\cos\theta, y = r\sin\theta$

$$\begin{aligned} f(r, \theta) &= \frac{1}{2\pi} e^{-r^2/2} \left| J\left(\frac{x, y}{r, \theta}\right) \right| \\ &= \frac{1}{2\pi} e^{-r^2/2} r, \text{ since, } J\left(\frac{x, y}{r, \theta}\right) = r \end{aligned}$$

substituting $z_1 = \frac{r^2}{2}, z_2 = \frac{\theta}{2\pi}$,

$$\begin{aligned} f(z_1, z_2) &= \frac{1}{2\pi} e^{-z_1} \sqrt{2z_1} J \left(\frac{r, \theta}{z_1, z_2} \right) \\ &= \frac{1}{2\pi} e^{-z_1} \sqrt{2z_1} \pi \sqrt{\frac{2}{z_1}} \\ &= e^{-z_1} \end{aligned}$$

which is the p.d.f. of an exponential distribution with $\lambda = 1$.

Let r$''$ be a random fraction, i.e. $r'' \in [0, 1]$ then,

$$\begin{aligned} \text{or } z_1 &= \ln(1 - r'') \\ \text{or } z_1 &= \ln r', (r' \in [0, 1]) \\ \text{or } \frac{r^2}{2} &= -\ln r' \text{ and } \theta = 2\pi\xi_2,\ \xi_2 \in [0, 1] \\ \text{or } x &= r\cos\theta = \sqrt{-2\ln r'} \cos(2\pi\xi_2) \end{aligned} \tag{10.4}$$

where $\xi_2 \in [0, 1]$.

(iii) Simulation from Binomial Distribution

The p.m.f for the binomial distribution with parameters n and p is given by

$$f(x) = \binom{n}{x} p^x q^{n-x}$$

where n is the number of trials, p is the probability of success and q is the probability of failure. Hence, $q = 1 - p$.

Let us define n new set of random variables $Z_i, i = 1, 2, \ldots n$ as

$$\begin{aligned} \mathrm{Z}_i &= 1 \text{ if } r_i \le p, i = 1, 2, \ldots n. \\ &= 0 \text{ otherwise} \end{aligned}$$

where $r_i \in [0, 1]$.

Then, $x = \sum_{i=1}^{n} z_i$ where z_i's are the values of the random variables

$$Z_i's, i = 1, 2, \ldots n \tag{10.5}$$

(iv) Simulation from a Poisson Distribution

Consider an interval of length t. Suppose within this interval $N(t)$ events occur at random and the random time interval between $(i+1)$th and ith occurrences be X_i. Then according to the definition of Poisson process

$$P[N(t) = y] = e^{-\lambda t}\frac{(\lambda t)^y}{y!}, y = 0, 1, \ldots \infty$$

where λ is the rate of occurrence. Further from Poisson process it follows that X_i follows an exponential distribution with parameter λ, i.e.

$$f(x_i) = \lambda e^{-\lambda x_i}, x_i > 0$$

Let us define $S_y = X_1 + X_2 + \ldots + X_y$

Then the event $[N(t) = y] \equiv [S_y \leq t]$

Now if we take $t = 1$, then

$$P[N(t) = y] = \frac{e^{-\lambda}\lambda^y}{y!}, y = 0, 1, \ldots, \infty \tag{10.6}$$

Here the problem is to generate the values of y. From Eq. (10.6) this can be done as follows:

Generate samples $x_1, x_2 \ldots$ from $\exp(\lambda)$. At each step check whether $\sum_{i=1}^{n} x_i \leq 1$ and $\sum_{i=1}^{n+1} x_i > 1$. If so, then the first Poisson sample is $y = n$.

(v) Simulation from Cauchy Distribution

The p.d.f of Cauchy (x_0, γ) distribution is given by

$$f(x) = \frac{1}{\pi\gamma\left[1 + \left(\frac{x - x_0}{\gamma}\right)^2\right]}, \quad -\infty < x < \infty$$

and the corresponding c.d.f is

$$F(x) = \frac{1}{\pi}\arctan\left(\frac{x - x_0}{\gamma}\right) + \frac{1}{2}$$

Let r be a random fraction.

Then by using the relation

$$\begin{aligned} F(x) &= r \\ \text{or } x &= F^{-1}(r) \end{aligned}$$

we can generate a sample from Cauchy distribution as

$$x = x_0 + \gamma \tan\left[\pi\left(r - \frac{1}{2}\right)\right]$$

when the values of the parameters x_0 and γ are given.

(vi) Simulation from Gamma Distribution

Simulation from Gamma distribution $G(\alpha, n)$ is as follows where n is an integer. When n is not an integer we follow "rejection sampling".

We know that if $X_1, X_2, \ldots, X_n$ are independent and identically distributed (*iid*) as $\exp(\alpha)$, then $Y = X_1 + X_2 + \ldots + X_n \sim Gamma(\alpha, n)$. So to generate the first sample from Gamma, we generate n samples from $\exp(\alpha)$. Hence, $y = x_1 + x_2 + \ldots + x_n$ will be the first sample generated from $G(\alpha, n)$.

For R code of generation of random numbers from several distributions see Chap. 11, p. 301.

(vii) Rejection Sampling

This method is used when it is difficult to generate sample from a distribution with p.d.f $f(x)$. Here we choose some enveloping distribution with p.d.f $h(x)$ such that $h(x)$ has the same ranges as $f(x)$ but from $h(x)$ it is relatively easy to simulate.

It is possible to choose $h(x)$ to be roughly of similar shape as that of $f(x)$ and then to envelope $f(x)$ by $h(x)$ we would obtain the desired scatter of points under $f(x)$, first by obtaining a scatter of points under $h(x)$ and then rejecting just those which were under $h(x)$ but not under $f(x)$. While it is often possible to choose an appropriate $h(x)$ to be of similar shape to $f(x)$, it is clearly not possible to envelope $f(x)$ by $h(x)$ so that for all $x, f(x) \leq h(x)$ since $f(x)$ and $h(x)$ are both density functions and $\int f(x)dx = \int h(x)dx = 1$. This can be obtained by uniformly stretching the scatter of points under h(x) in a direction at right angle to the x axis until $h(x) \geq f(x)$ for all x. This is achieved by choosing a suitable scaler $k > 1$ and generating samples from $g(x) = kh(x)$. So having sample points from $g(x)$ we will accept only those which are under both $f(x)$ and $g(x)$ and reject those which are under $g(x)$ but not under $f(x)$.

Hence probability of rejection,

$$\frac{\int_{\infty}^{\infty}(g(x)-f(x))dx}{\int_{-\infty}^{\infty}g(x)dx}=\frac{\int_{-\infty}^{\infty}\{kh(x)-f(x)\}dx}{\int_{-\infty}^{\infty}kh(x)dx}=1-\frac{1}{k}$$

So the technique will be efficient when k is small enough in order to increase the probability of acceptance.

Example 2: Write a program in C to generate a sample of size n (say) from an exponential distribution, given the parameter λ.

Solution:

```
# include < stdio.h >

# include < math.h >

# include < time.h >

# include < stdlib.h >

main ( )

{

int i, n;

float r, m[100], lambda;

printf ("Enter the parameter for the exponential distribution");

scanf ("%", & lambda);

printf ("Enter the sample size\n");

scanf ("% d", & n);

printf ("\n");

printf ("The sample is :")

randomize ( );

for (i = 0; i < n; i + +)
```

```
{
r = rand ( ) / (float) RAND_MAX;
m[i] = -(1/lambda) * log(1 - r);
printf ("\n%f", m[i]);
}
}
```

Example 3: Write a program in C to generate a sample of size n from a standard normal distribution using pointer, randomize function and Box Muller transformation. Also compute the mean value for values greater than 2.5.

Solution:

```
# include < stdio.h >
# include < stdlib.h >
# include < math.h >
# include < time.h >
# include < malloc.h >
main ( )
{
int i, n;
long int s;
s = 15000;
float *x, r1, r2, sum, mean;
FILE *fp ;
x = (float*) malloc (s* size of (long int));
```

```
fp = fopen ("normal2.dat", "w");
printf ("\n Enter the sample size :");
scanf ("%d", &n);
randomize ( );
for (i = 0; i < n; i++)
{
r1 = rand ( ) / (float) RAND_MAX;
r2 = rand ( ) / (float) RAND_MAX;
x[i] = sqrt(- 2 * log r1) * cos(2 * 4 * (atan(1)) * r2;
fprintf (fp, "f\n", x[i]);
printf ("%f\n", x[i]);
}
sum = 0;
for (i = 0; i < n; i++)
{
if (x[i] > 2.5)
sum = sum + x[i];
}
mean = sum / n;
```

```
fprintf (fp, "Value of mean = % f", mean);

f close (fp);

}
```

Example 4: Write a program in C to generate a sample of size m from a Binomial distribution, given the values of the parameters n and p.

Solution:

```
# include < stdio.h >

# include < math.h >

# include < stdlib.h >

# include < time.h >

main ( )

{

int i, n, m, j, x;

float r, p;

printf ("Enter the parameters for Binomial distribution : ");

scanf ("%d%f", &n, &p);

printf ("\n Enter the sample size :");

scanf ("%d", &m);

printf ("The sample generated of size n is :");

randomize ( );

for (i = 0; i < m; i + +)

{
```

```
x = 0;

for (j = 0; j < n; j++)

{

r = rand ( ) / (float) RAND_MAX;

if (r <= p)

x = x + 1;

}

printf("\n%d", x);

}

}
```

Example 5: Write a program in C to generate a sample from a Poisson distribution, given the parameter λ.

Solution:

```
# include < stdio.h >

# include < math.h >

# include < stdlib.h >

# include < time.h >

main ( )

{

int i, n, c;

float r, m, s, lambda;

printf ("Enter the parameter for Poisson distribution : \n");
```

```
scanf ("% f", & lambda);

printf ("Enter the sample size : \n");

scanf ("%d", &n);

randomize ( );

for (i = 0; i < n; i + +)

{

s = 0;

c = 0;

while (s <= 1)

{

r = rand ( ) / (float) RAND_MAX;

m = - (1 / lambda) * log(1 − r);

s = s + m

c = c + 1;

}

printf ("\n%d", c − 1);

}

}
```

Example 6: Generate a sample from half normal distribution with p.d.f, $f(x) = \sqrt{\frac{2}{\pi}} e^{-x^2/2}, 0 \leq x < \infty$ using rejection sampling.

Solution: We choose $h(x) = e^{-x}, x \geq 0$ so that $g(x) = ke^{-x}, x \geq 0$. One way of choosing k is to consider condition for equal roots arising from the equation, $ke^{-x} = \sqrt{\frac{2}{\pi}}e^{-x^2/2}$.

If the equation has no real roots, then k is very large, if the equation has two distinct roots, then k is too small. If roots are equal, k has the smallest possible values, where the two curves touch each other. Taking logarithm of the above equation,

$$x^2 - 2x + 2\log_e(k\sqrt{\pi/2}) = 0$$

It has two equal roots iff $4 = 8\log_e(k\sqrt{\pi/2})$ i.e. $k = \frac{2e}{\pi}$.

Algorithm:

1. Simulate x, from e^{-x}, i.e. $x = -\ln(r_1)$ where r_1 is a random fraction.
2. Compute $f(x))$ and $g(x)$.
3. Take another random fraction r_2 independent of r_1.
4. Accept x if $r_2 < \frac{f(x)}{g(x)}$ and reject otherwise and go to step 1 for a new sample x.

10.4 Monte Carlo Method

Monte Carlo method consists of a system of computational algorithms through repeated random sampling to compute any specific problem. These are widely used in physical and mathematical systems, e.g. calculation of risk in business or evaluating the multidimensional integrals which is otherwise not possible by any deterministic way. Two major classes of numerical problems that arise in statistical inference are optimization and integration. Suppose we have to estimate

$$E_f[h(x)] = \int h(x)f(x)dx \tag{10.7}$$

We can estimate it by using a sample generated from the density function $f(x)$. Let $x_1, x_2, \ldots, x_m$ be the points. Then we can approximate (10.7) by the empirical average as

$$\overline{h_m} = \frac{1}{m}\sum_{j=1}^{m} h(x_j)$$

since $\overline{h_m}$ converges almost surely to $E_f[h(x)]$ by the strong law of large number (SLLN). Again when h^2 has a finite expectation under f, the speed of convergence of $\overline{h_m}$ can be assessed since the variance

$$v_m = \frac{1}{m^2}\sum_{j=1}^{m}\left[h(x_j) - \overline{h_m}\right]^2$$

For large m, $\frac{\overline{h_m}-E_f[h(x)]}{\sqrt{v_m}}$ follows $N(0,1)$, by Central Limit Theorem which gives convergence test and confidence bound on the approximation of $E_f[h(x)]$

Estimating by Monte Carlo Simulation is not unique.

Suppose we are to estimate $p = \int_2^\infty \frac{1}{\pi(1+x^2)}dx$.

Then by definition (13.7) the above integral can be written as $p = \int_{-\infty}^{\infty} I_{(x>2)} \frac{1}{\pi(1+x^2)}dx$ where $I_{(x>2)} = 1$ if $x > 2$ and $= 0$ otherwise. If $h(x) = I_{(x>2)}$ and $f(x) = \frac{1}{\pi(1+x^2)}, -\infty < x < \infty$ is the Cauchy p.d.f., then by (10.7) the estimator of p is $\hat{p}_1$ (say) and is given by

$$\hat{p}_1 = \frac{1}{m}[\text{Number of } x_j > 2] = \frac{\sum_{j=1}^{m} Ix_{j>2}}{m}$$

of an independent and identically distributed (*iid*) sample $x_1, x_2, \ldots, x_m \sim$ Cauchy(0, 1). The variance of the estimator is $v(\hat{p}_1) = \frac{mp(1-p)}{m^2} = \frac{p(1-p)}{m}$.

So, for, $p = 0.15, v(\hat{p}_1) = 0.127/m$.

Since Cauchy distribution is symmetric,

$$\begin{aligned} p &= \int_{-\infty}^{-2} \frac{1}{\pi(1+x^2)}dx \\ \text{so, } 2p &= \int_{-\infty}^{-2} \frac{dx}{\pi(1+x^2)} + \int_2^\infty \frac{dx}{\pi(1+x^2)} \\ \text{so, } p &= \frac{1}{2}\left[\int_{-\infty}^{-2} \frac{dx}{\pi(1+x^2)} + \int_2^\infty \frac{dx}{\pi(1+x^2)}\right] \end{aligned}$$

So, the estimator from this expression $\hat{p}_2$ (say) is given as, $\hat{p}_2 = \frac{1}{2}\left[\frac{1}{m}\sum_{j=1}^{m} I_{|x_j|>2}\right]$

$$\text{So, } var(\hat{p}_2) = \frac{p(1-2p)}{2m}$$

This is as follows:

$$p = P[X_j > 2]$$

$$\text{Also, } p = P[X_j < -2]$$

$$\begin{aligned} P[|X_j| > 2] &= 1 - p[|X_j| < 2] \\ &= 1 - p[-2 < X_j < 2] \\ &= 1 - (1 - 2p) \\ &= 2p \end{aligned}$$

$$var(\hat{p}_2) = m\frac{2p(1-2p)}{(2m)^2} = \frac{p(1-2p)}{2m}$$

So, for $p = .15, var(\hat{p}_2) = 0.052/m$, i.e. the variance is reduced, so $var(\hat{p}_2)$ is a better estimate than $var(\hat{p}_1)$ due to generation of values outside the domain of interest $[2, \infty]$.

10.5 Importance Sampling

Let us use Monte Carlo method in evaluating $\int_0^1 g(x)dx$ for the function $g(x)$ as shown in Fig. 10.2. We find that $g(x) = 0$ for $x < 0$ and $x > 1$.

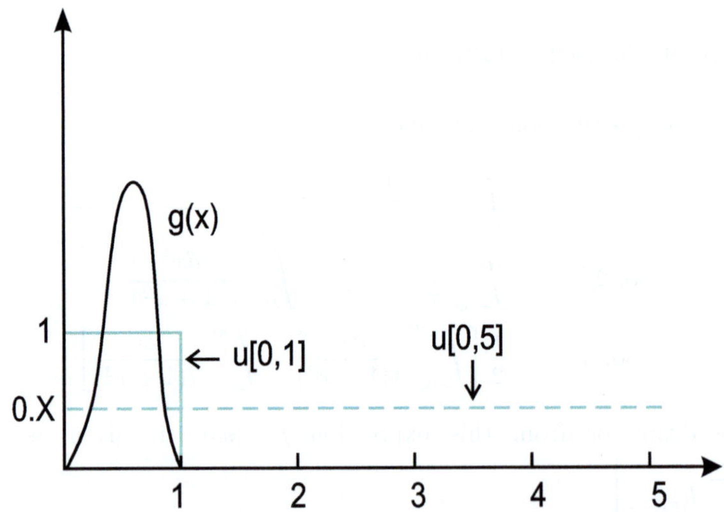

Figure 10.2 Monte Carlo method applied to two different functions

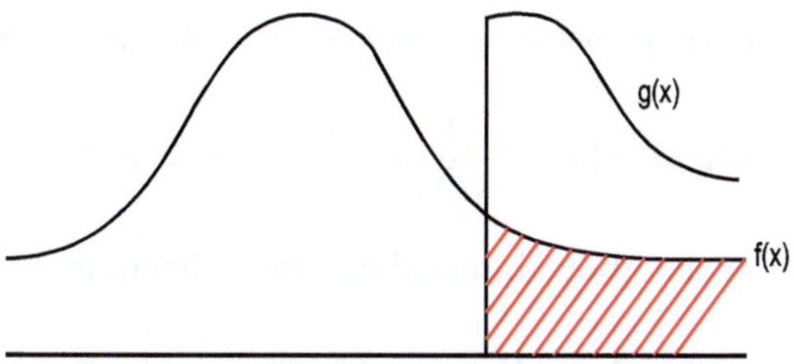

Figure 10.3 Importance sampling

Let us have the density function $f(x)$ as $U(0,1)$. Then we have $\int_0^1 g(x)dx = \int_0^1 g(x).1dx$ following (10.7) where $h(x) = g(x)$ and $f(x) = 1, x \leftarrow [0,1]$. Then value of the integral is $\sum_{i=0}^{1} g(x_i)$ where x_i's are generated from the uniform distribution over [0, 1]. So $\int_0^1 g(x)dx = E_U[g(x)]$ reasonably works well. Now instead of $U[0,1]$ if we use $U[0,5]$ in place of $f(x)$, then $f(x) = \frac{1}{5}, x \in [0,5]$. and, $\int_0^1 g(x)dx = 5\int_0^1 g(x).\frac{1}{5}dx = 5\int_0^1 g(x)f(x)dx = \frac{5}{n}\sum_{i=1}^{n} g(x_i)$ and it will make no sense as, on average 80 % of the random points will lie in $1 < x_i < 5$ for which $g(x_i) = 0$. So it is clear that one's choice of distribution from which to draw the random sample will affect the quality of their Monte Carlo estimator. Here comes the idea of "importance sampling". The objective in importance sampling is to concentrate the distribution of sample points in that parts of the interval that are of "most importance" instead of spreading out evenly, mathematically

$$\begin{aligned} E_f[h(x)] &= \int h(x)f(x)dx \text{ can be written as,} \\ &= \int h(x)\frac{f(x)}{g(x)}g(x)dx \end{aligned}$$

so that the random sample is now drawn from the new density function $g(x)$ which covers completely the interval of X of interest (Fig. 10.3). $g(x)$ is called the instrumental density function.

$$\text{so, } \overline{h_n} = \frac{1}{n}\sum_{j=1} h(x_j)\frac{f(x_j)}{g(x_j)}$$

Importance sampling is of considerable interest since it puts very little restriction on the choice of the instrumental distribution $g(x)$, which can be chosen from the distributions that are easy to simulate. Moreover the same sample, generated from $g(x)$, can be used repeatedly not only for different functions $h(x)$ but also for densities $f(x)$. Here also, $\overline{h_n}$ converges completely to $E_f[h(x)]$.

Example 7: Suppose we want to estimate $\int_{-\infty}^{t} \phi(x)dx = \Phi(t)$,

$$\text{where, } \phi(x) = \frac{1}{\sqrt{2\pi}} e^{-x^2/2}, -\infty < x < \infty.$$

In normal situation the above integral can be written as

$$\int I(x)\phi(x)dx \text{ where } I(x) = 1 \text{ if } x \leq t \;\; and = 0 \text{ otherwise}$$

Then following (10.7), $h(x) = I(x)$ and $f(x) = \phi(x)$.

So, by Monte Carlo method, the estimated value of $\Phi(t)$ is

$$\begin{aligned} \hat{\Phi}(t) &= \frac{1}{n}\sum_{i=1}^{n} I(x_i \leq t) \\ &= \frac{(number \;\; of \;\; x_i \leq t)}{n} = \frac{N(x_i)}{n} \text{ (say)} \end{aligned}$$

with estimated variance $[\hat{\Phi}(t)(1 - \hat{\Phi}(t)]/n$.

The above method breaks down if t is very small or large say -4.5 or 4.5. Then,

$P[t > 4.5] = \frac{1}{n}\sum_{i=1}^{n} I_{t_i > 4.5}$ and for, $n = 10{,}000$ the above summation produces usually zero.

If we use importance sampling with choice of $g(x)$ as

$$g(x) = e^{-(x-4.5)}$$

which is an exponential p.d.f truncated at 4.5 with scale 1, then

$$P(t > 4.5) = \int_{4.5}^{\infty} \phi(x)dx = \int_{-\infty}^{\infty} I(x > 4.5)\frac{\phi(x)}{g(x)}.g(x)dx$$

$$\begin{aligned} \text{where, } I(x > 4.5) &= 1 \text{ for } x > 4.5 \\ &= 0 \text{ otherwise} \end{aligned}$$

Then, by (10.7), $h(x) = I(x > 4.5)\frac{\phi(x)}{g(x)}$ and $f(x) = g(x)$.

$$\begin{aligned} \text{So, } P(t > 4.5) &\simeq \frac{1}{n}\sum_{i=1}^{n} I(x_i > 4.5)\frac{\phi(x_i)}{g(x_i)} \\ &= 0.000003377 \end{aligned}$$

An alternative estimator (which is slightly biased) is

$$\frac{\sum_{i=1}^{n} h(x_i)\frac{f(x_i)}{g(x_i)}}{\sum_{i=1}^{n} \frac{f(x_i)}{g(x_i)}}$$

which is better since the variance is reduced.

Example 8: Run an experiment to estimate the variance of the estimator of $P[t > 4.5]$ to some accuracy obtained from an importance sampler using the instrument p.d.f as exp(1) left truncated at 4.5.

10.6 Markov Chain Monte Carlo (MCMC)

It is not always necessary to simulate directly a sample from the distribution $f(x)$ to approximate the integral $I = \int h(x)f(x)dx$, since the other approaches like importance sampling can be used. Here we shall show that it is possible to obtain a sample $x_1, x_2, \ldots, x_n$ approximately distributed from $f(s)$ without directly simulating form $f(x)$. Here we use an ergodic Markov Chain with stationary distribution $f(x)$.

A MCMC method for the simulation of a distribution $f(x)$ is any method producing an ergodic Markov Chain $(X^{(t)})$ whose stationary distribution is $f(x)$.

10.7 Metropolis–Hastings Method

Let $P = \{p_{ij}\}$ be the transition matrix of an irreducible Markov Chain with states $0, 1 \ldots s$. Then if $X(t)$ denotes the state occupied by the process at time t, then

$$P\{X(t+1) = j | X(t) = i\} = p_{ij}$$

If $\pi = (\pi_0, \pi_1, \ldots, \pi_s)$ be a probability distribution with $\pi_i > 0$ for all i and if $h(\cdot)$ is a function defined on the states and we wish to estimate $I = E_\pi(h) = \sum_{i=0}^{S} h(i)\pi_i$.

we proceed as follows.

In order to use this algorithm for a given distribution $\pi = (\pi_0, \pi_1, \ldots, \pi_s)$ we must construct a Markov Chain with π as its stationary (or target) distribution.

Here, $\pi_i p_{ij} = \pi_j p_{ji}$ for all i and j and this implies $\sum_i \pi_i p_{ij} = \pi_j$ for all j.

Proof: Here we assume that p_{ij} has the form

$$\begin{aligned} p_{ij} &= q_{ij}\alpha_{ij}(i \neq j) \\ \text{with, } p_n &= 1 - \sum_{i \neq j} p_{ij} \end{aligned}$$

where $Q = \{q_{ij}\}$ is the transition matrix of an arbitrary Markov Chain on the states 0, 1, ..., s and α_{ij} is given by

$$\alpha_{ij} = \frac{s_{ij}}{1 + \frac{\pi_i}{\pi_j}\frac{q_{ij}}{q_{ji}}}$$

where s_{ij} is a symmetric function of i and j chosen so that $0 \leq \alpha_{ij} \leq 1$ for all i and j. Q is called the instrumental or proposal distribution. With this form for p_{ij}.

$$\begin{aligned} \pi_i p_{ij} &= \pi_i q_{ij} \frac{s_{ij}}{1 + \frac{\pi_i}{\pi_j}\frac{q_{ij}}{q_{ji}}} \\ &= \pi_i q_{ij} \frac{s_{ij}}{\left[\frac{\pi_j q_{ji} + \pi_i q_{ij}}{\pi_j q_{ji}}\right]} \\ &= \pi_i q_{ij} s_{ij} \times \frac{\pi_j q_{ji}}{(\pi_j q_{ji} + \pi_i q_{ij})} \\ &= \frac{\pi_i \pi_j q_{ij} q_{ji} s_{ij}}{\pi_j q_{ji} + \pi_i q_{ij}} \end{aligned}$$

Similarly, $\pi_j p_{ji} = \frac{\pi_i \pi_j q_{ij} q_{ji} s_{ij}}{\pi_i q_{ij} + \pi_j q_{ji}}$, so, $\pi_i p_{ij} = \pi_j p_{ji}$.

Simulation Process:

Let us assume that (1) $X(t) = i$ and select a state j using the distribution given by the ith row of Q.

(2) $X(t+1) = j$ with probability α_{ij}

$= i$ with probability $1 - \alpha_{ij}$

Simple choice of s_{ij} is given by

$$
{s_{ij}}^{(M)} = \begin{cases} 1 + \frac{\pi_i q_{ij}}{\pi_j q_{ji}} & (if \frac{\pi_j q_{ji}}{\pi_i q_{ij}} \geq 1) \\ 1 + \frac{\pi_j q_{ji}}{\pi_i q_{ij}} & (if \frac{\pi_j q_{ji}}{\pi_i q_{ij}} < 1) \end{cases}
$$

With $s_{ij} = {s_{ij}}^{(M)}$ we have Metropolis method and with $s_{ij} = {s_{ij}}^{(B)}$ we have Baker's method. So we have

$$
{\alpha_{ij}}^{(M)} = \begin{cases} 1 & if \frac{\pi_j q_{ji}}{\pi_i q_{ij}} \geq 1 \\ (\pi_j q_{ji})/(\pi_i q_{ij}) & if \frac{\pi_j q_{ji}}{\pi_i q_{ij}} < 1 \end{cases}
$$

Thus $i/\pi_i = \pi_j$, we take $X(t+1) = j$ with probability 1 for Metropolis method.

The corresponding algorithm for probability density function can be restated as follows:

A Metropolis–Hastings algorithm (MH) starts with the objective (target) density of f. A conditional density $q(y/x)$ is then chosen. The MH algorithm can be implemented in practice when $q(\cdot/x)$ is easy to simulate from and is either explicitly available (up to a multiplicative constant independent of x) or symmetric, i.e. such that $q(x/y) = q(y/x)$. The target density f must be available to some extent. A general requirement is that the ratio $f(y)/q(y/x)$ is known up to a constant independent of x.

The MH algorithm associated with the target density f and the conditional or proposal or instrumental density q produces a Markov Chain $(X^{(t)})$ through the following transition.

1. Given $x^{(t)}$
2. Generate $y_t \sim q(y/x^{(t)})$
3. Choose $x^{(t+1)} = \begin{cases} y_t \text{ with probability } \rho(x^{(t)}, y_t) \\ x_t \text{ with probability } 1 - \rho(x^{(t)}, y_t) \end{cases}$

where $\rho(x, y) = \min \left\{ \frac{f(y)}{f(x)} \cdot \frac{q(x/y)}{q(y/x)}, 1 \right\}$

The probability $\rho(x, y)$ is called the MH acceptance probability.

Example 9: Write a programme in C to perform a MH algorithm using the target distribution as, $\pi_i = \frac{e^{-\lambda}\lambda^i}{i!}(i = 0, 1, \ldots, \infty)$ taking $q_{ij} = 1/2(j = i-1, i+1, i \neq 0), q_{00} = q_{01} = 1/2$.

Solution: Here, $\pi_{i+1}/\pi_i = \frac{\lambda}{(i+1)}, \frac{\pi_{i-1}}{\pi_i} = i/\lambda, i \neq 0$.

(a) **Program using factorial (restricted for i = 15)**

```
# include < stdio.h >

# include < math.h >

# include < stdlib.h >

# include < time.h >

long int fact (long int n);

main ( )

{

float pi1, pi2, pit, pitp1, pif, lamda, r1, r2, alpa, r3, s;

long int i, j;

lamda = 2;

randomize ( );

i = 1;

j = 1;

pit = exp(-lamda);

s1 : if(j = 0)

{

pi1 = exp(- lamda)*pow(lamda, j) / fact(j);

pi2 = exp(- lamda)*pow(lamda, j + 1) / fact(j + 1);

}
```

```
else
{
pi2 = exp(- lamda)*pow(lamda, j + 1) / fact(j + 1);
pi1 = exp(- lamda)*pow(lamda, j − 1) / fact(j − 1);
}
r1 = rand( ) / (float)RAND_MAX;
if (r1 <= 0.5)
{
pitp1 = pi1;
j = j + 1;
s = j − 1;
}
else
{
pitp1 = pi2;
if (j == 0)
goto s5;
else
j = j − 1;
s5 : s = j + 1;
```

```
}
r2 = pitp1 / pit;
if (r2 >= 1)
{
alpa = 1;
}
else
{
pif = pit;
j = s;
}
printf("value = % ld \n", j);
pit = pif;
i = i + 1;
if (i <= 10)
goto s1;
}
longint fact (long int n)
{
float c = 1;
long int i;
```

```
if (n == 0)
goto s4;
else
{
for(i = 1; i <= n; i++)
c = i * c;
}
}
s4: return(c);
}
```

(b) **Program without using factorial (generalized)**

```
# include < stdio.h >
# include < math.h >
# include < stdlib.h >
# include < time.h >
main ( )
{
float r1, r2, r3, alpa, lamda;
int i, j, s, j1;
i = 1;
```

```
j = 3;
lambda = 3;
randomize ( );
while (i <= 100)
{
r1 = rand ( ) / (float) RAND_MAX;
if (r1 <= 0.5)
{
if (j == 0)
s = j;
else
s = j - 1;
}
else
s = j + 1;
if (s == j + 1)
{
alpa = lamda / s;
if (alpa > = 1)
j1 = s;
else
```

```
{
r2 = rand ( ) / (float) RAND_MAX;
if (r2 < = alpa)
j1 = s;
else
j1 = j;
}
}
else
{
alpa = s / lamda;
if (alpa > = 1)
j1 = s;
else
{
r3 = rand ( ) / (float) RAND_MAX;
if (r3 < = alpa)
j1 = s;
else
j1 = j;
```

```
}

}

print f ("value = %d\n", j1);

j = j1;

i = i + 1;

}

}
```

Monte Carlo Method for evaluating double and multiple integral

Suppose we are to integrate $I = \int_{a_1}^{b_1}\int_{a_2}^{b_2}\dots\int_{a_n}^{b_n} f(x_1, x_2, \dots, x_n)dV$ where $dV = dx_1 dx_2 \dots dx_n$. Then,

$$I \approx V < f > \pm V\sqrt{\frac{< f2 > - < f >^2}{N}}$$

where $< f >= \frac{1}{N}\sum_{i=1}^{N} f(x_{1i}, x_{2i}, \dots, x_{ni})$
and $(x_{1i}, x_{2i}, \dots, x_{ni}), i = 1, 2, \dots, N$ are the N, n-dimensional points uniformly distributed over the multidimensional volume $V = (b_1 - a_1)$ $(b_2 - a_2)\dots(b_n - a_n)$ and, $< f^2 >= \frac{1}{N}\sum_{i=1}^{N} f^2(x_{1i}, x_{2i}, \dots, x_{ni})$.

Ex. Evaluate $\int_{x=1}^{2}\int_{y=3}^{4} x^2y^3dxdy$ using Monte Carlo Method.

```
# include < stdio.h >

# include < math.h >

# include < stdlib.h >

# include < time.h >

main ( )
```

```
double x, y, x1, y1, sum, vol

int n

vol = (2 - 1) * (4 - 3);

sum = 0;

for (i = 0; i < 1000; i++)

{

x1 = rand ( ) / (float) RAND_MAX;

y1 = rand ( ) / (float) RAND_MAX;

x = 1 + (2 − 1) * x1;

y = 3 + (4 − 3) * y1;

sum = sum + (x * x) * (y * y * y),

}

sum = vol * sum / 1000;

printf ("The value of the integral = %d", sum);

}
```

When the region of integration is not easy to sample randomly

Suppose we are to evaluate $I = \int_w f(x, y, z)dxdz$ where the region of integration is the intersection of a torus with the edge of a large box (Fig. 10.4). Then W can be represented by

$$z^2 + (\sqrt{x^2 + y^2} - 3)^2 \leq 1$$

$$x \geq 1, y \geq -3$$

Then we enclose the torus by a volume of a rectangular box, $1 \leq x \leq 4, -3 \leq y \leq 4, -1 \leq z \leq 1$.

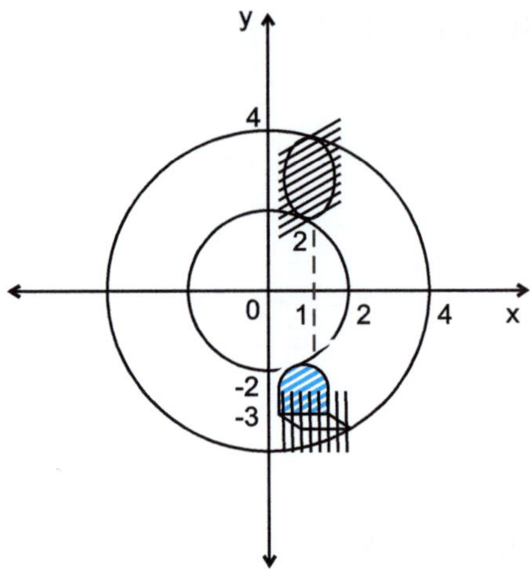

Figure 10.4 Evaluation of double integral by Monte Carlo method

Then $V = (4-1)*(4+3)*(1+1)$ is the volume of the box. Now we are to choose N random points $(x_i, y_i, z_i), i = 1, 2, \ldots, N$ from the volume V so that they fall within the region W. Then value of the integral $I \cong V.\frac{1}{N}\sum_{i=1}^{N} f(x_i, y_i, z_i) =< f >$ and the corresponding error is $= \pm V \sqrt{\frac{<f^2>-<f>^2}{N}}$ where $< f^2 >= \frac{1}{N}\sum[f(x_i, y_i, z_i)]^2$.

Computer Program: $f(x, y) = \sqrt{x^2 + y^2 + z^2}$ (say)

```
# include < stdio.h >

# include < math.h >

# include < stdlib.h >

# include < time.h >

main ( )

{

double x, y, z, x1, y1, z1, sum, var, vol, sum1;

int i, n;
```

```
sum = 0;
var = 0;
vol = (4 - 1) * (4 + 3) * (1 + 1);
scanf ("% d", & n);
for (i = 0; i < n; i++)
{
x1 = rand ( ) / (float) RAND_MAX;
y1 = rand ( ) / (float) RAND_MAX;
z1 = rand ( ) / (float) RAND_MAX;
x = 1 + 3 * x1;
y = -3 + 7 * y1;
z = -1 + 2 * z1;
if (z * z + pow(sqrt(x * x + y * y) - 3.0, 2) < 1.0)
{
sum = sum + sqrt(x * x + y * y + z * z);
var = var + (x * x + y * y + z * z);
}
}
sum1 = vol * sum / n;
var = vol * sqrt((var / n - (sum / n) * (sum / n)) / n);
```

```
printf("The value of integral = %d\,", sum1);

printf("The value of error = %d", var);

}
```

Exercise

1. Write a C program to generate 100 random samples from two Gaussian distributions with means and standard deviations (5, 10) and (6, 10), respectively, using pointer variables and draw histograms using 20 bins with fitted curves. Hence compute the means and standard deviations of the generated samples.
2. Write a program to generate 100 random samples from a standard normal, exponential and uniform distributions, respectively, using pointer variables and draw the histograms with fitted curves.
3. Using Metropolis–Hastings algorithm draw the histogram of the sample generated when the target density is standard normal and proposed density is exponential. Sample size is 1,000. Number of bins is 100.
4. Do a Monte Carlo simulation to evaluate the integral $I = \int_0^1 \frac{e^x - 1}{e-1} dx$ where x is the value of a random variable following uniform distribution. Find an estimate of the standard error of the integral taking 100 and 16 points, respectively.
5. Do a Monte Carlo simulation taking sample size 500 to estimate the first and second moments of the distribution with the probability density function $f(x) = \frac{1}{\sqrt{2\pi}\sigma} e^{-\frac{(x-\mu)^2}{2\sigma^2}} \; -\infty < x < \infty$ where $\mu = 5$ and $\sigma = 10$. Compare with the original values and find the errors.
6. Do a Monte Carlo simulation taking sample size 500 to estimate the first and second moments of the distribution with probability density function,

$$\begin{aligned} f(x) &= \lambda e^{-\lambda x}, x \geq 0 \\ &= 0, x < 0, \lambda = 5 \end{aligned}$$

   Compare with the original values and find the errors.
7. Generate 100 samples from a power law distribution whose pdf is $f(x) = Ax^{-\alpha}, a \leq x \leq b$ where $a = 0.01$ and $b = 100$. A is a constant to be determined from the normalization condition of the pdf.

References

Frisch, H.L. 1962. *Physical Review* 126:949.

Lehmer, D.H. 1951. Mathematical methods in large-scale computing units. In *Proceedings of a second symposium on large-scale digital calculating machinery*, 141–146. Cambridge: Harvard University Press.

Chapter - 11

Use of Softwares

11.1 Introduction

With the advent of large telescopes ample virtual databases have come into the field with a great challenge of its proper usage, preservation and interpretation. For the past century, the discipline of statistics has primarily focused on the analysis of date from different areas such as biology, medical science, social science and engineering. Recently considerable advances have been made in the application of statistical analysis to fields in the physical sciences, e.g. geology. Astronomers and astrophysicists have also sought to understand astronomical data using statistical techniques and a new field "Astrostatistics" has been emerged during the last few decades. But much of these analyses have been at a rather elementary level. The majority of the astronomers have not kept up with the most recent advancement in statistical modelling and have therefore not been able to take advantage of the insights afforded by using these sophisticated statistical techniques to the huge database already existing in various data archives. These are various software packages to meet up the above requirements for astrophysicists but a proper training is necessary for using these softwares. Among the various softwares some are priced but some are command based and are freely available in the Internet.

In the present chapter some preliminary descriptions are given on these freely available softwares for the beginners. The main software to be discussed is "R". "R" is suitable for handling large data sets and also for graphical representation and the precious thing for "R" is that every time new statistical techniques are being uploaded by several authors which can be at once used for multipurpose applications.

11.2 Preliminaries on R

Initially scientists like Rick Beeker, John Chambers and Allan Wilks of Bell Laboratory developed a language known as S language and the software related to it is S-Plus. Then some of the scientists carried over some changes

A.K. Chattopadhyay, T. Chattopadhyay, *Statistical Methods for Astronomical Data Analysis*, Springer Series in Astrostatistics 3,
DOI 10.1007/978-1-4939-1507-1_11

over "S" and claimed that the language is superior to S-Plus in computational speed and hence gave the name "R" (a letter appearing before S in the alphabetical list).

11.3 Advantages of R Programming

"R" has the following facilities for data manipulation, calculation and graphical display.

(1) It has effective data handling and storage facility.

(2) It is suitable for calculation on arrays in particular matrices.

(3) It includes conditional loops, user defined recursive functions and I/O facilities.

(4) Give minimal output and store results in a fit object for subsequent use by further R functions.

How to Get "R" Under "Windows" Operating System

R-software is freely available at `http://www.r-project.org/`. The users may try themselves, but beginners may get confused to choose proper working file or even after downloading he/she may not get the actual executing file for installation. So in order to face no disturbances beginners are recommended to follow the download instructions below.

1. Click `http://www.r-project.org/`.
   Then a window describing the introduction of R appears. On the left side several downloading options appear.
2. Download **CRAN**
3. A CRAN mirror window appears where a comprehensive R-archive network is available at the listed address.
4. Choose a location nearest to you.
5. Click on the address given below that country.
6. Three versions of R, namely Linux, Mackintosh, Windows are available. Click on **Download R for Windows**.
7. Click on **base** system among **base contributory** and **R tools**.
8. Click on the executing file and ask for downloading.
9. Click on for **Download R2.15.1 for Windows** at the top. It is to be noted that every time "R" is going under improvement. So user has suggested to download the latest version available on site.
10. Double click on **R2.15.1** for installation and clicking Next → Next → etc. responding to subsequent instructions.

11.4 How to Get R Under Ubuntu Operating System

The following steps should be followed for installing "R".

(1) Open the terminal.

(2) Type: sudo apt-get install r-base.

(3) Type: password.

(4) Follow the commands subsequently.

11.5 Basic Operations

11.5.1 Computation

R can be used as an ordinary calculator.

Examples:

(i)	> 2 + (7 ∗ 5) / (4 - 9)	# Use of brackets is preferred.
	> log (10)	# Natural logarithm.
	> log 10 (432)	# Logarithm with base 10.
	> 4 ∧ 2	# 4 raised to power 2.
	> 3 / 2	# division.
	> sqrt(16)	# square root.
	> abs (3 - 7)	# Absolute value of 3 - 7.
	> pi	# π
	> exp (2)	# e^2
	> 15% / %4	# integer division of 15 / 4.
	> #	# A command line.
	> $x \leftarrow 5 + \log(10)$	# a variable x, given the value $5 + \log(10)$.
	> floor (x)	# largest integer $\leq x$.
	> ceiling (x)	# smallest integer $\geq x$.
	> $x \leftarrow 3 + 2i$	# x is a complex number.
	> Re(x)	# gives real part of x.

$> \mathrm{Im}(x)$	# gives imaginary part of x.
$> y \leftarrow -1 + 1i$	
$> x + y$	# gives $2 + 3i$.
$> x * y$	# gives $-5 + 1i$.

11.5.2 Vector Operations

Vectors in R can be created by the concatenated function, c, which combines all values given as arguments to the function into a vector.

$> x \leftarrow c(1, 4, 2, 7);$	# creates a vector x with 4 components 1, 4, 2, 7.
$> x$	
$>$ length (x)	# gives length of x as 4.
$> y \leftarrow c(1 : 4)$	# Creates a vector y with consecutive integers 1, 2, 3, 4.
$> x + y$	# gives addition of vectors x and y. Here, (2, 6, 5, 11).
$> y \wedge 2$	# raises each component of y to power 2. Here, (1, 4, 9, 16).
$> 2 \wedge y$	# raises 2 to powers of y. Here, (2, 4, 8, 16).
$> x[2 : 4]$	# Make a subset from 2nd to 4th element of x.
$> x[-c(2, 7)]$	# All elements of x except 2, 7. Here, (1, 4).
$> x[x > 3]$	# All elements of x greater than 3. Here, (4, 7).

> colour ← c ("green", "blue", "orange")	# Defines a character vector of three colours. " " denotes a string.
> x colour ← $c(x,$ colour);	# The vector is a character vector with strings "1", "2", "7" for "green", "blue", "orange".

11.5.3 Matrix Operations

A matrix of objects can be created in "R" using "matrix" function. The general syntax is matrix (data, nrow, ncol, byrow = T). The last argument specifies that the matrix is to be filled by "row by row" or "column by column", with the latter being the default.

Example:

> $m1 \leftarrow$ matrix $(c(1, 4, 7, 5, -1, 1, 3, 9)$, ncol = 3, byrow = T);

> $m1$.

The output is:

	[, 1]	[, 2]	[, 3]
[1,]	1	4	7
[2,]	5	−1	1
[3,]	1	3	9

Again a matrix can be created by combining vectors of equal length and using the function c bind ( ), meaning "column bind".

Let > $x \leftarrow c(1, 3, 2, 10, 5)$
$y \leftarrow c(1, 2, 3, 7, 9)$
be two vectors of same length. Then,

> $m2 \leftarrow$ cbind(x, y);

$m2$

Output:

	x	y
[1,]	1	1
[2,]	3	2
[3,]	2	3
[4,]	10	7
[5,]	5	9

Transpose of the matrix $m2$

$> t(m2)$

Output:

	[,1]	[,2]	[,3]	[,4]	[,5]
x	1	3	2	10	5
y	1	2	3	7	9

Dimension of the matrix $m2$

> dim ($m2$) # Gives 5 2

A matrix can also be created by using the function "rbind" meaning "row-bind".

Example:

$> m3 \leftarrow$ rbind (x, y);

$m3$

Output:

	[,1]	[,2]	[,3]	[,4]	[,5]
x	1	3	2	10	5
y	1	2	3	7	9

To Extract a Particular Element/Particular Row/Particular Column

$> m1[2,3]$ # Gives element at 2nd row, 3rd column. Here it is '1'.

$> m1[2,]$ # Gives entire 2nd row. Here it is 5 -1 1

$> m1[,3]$ # Gives entire 3rd column. Here it is 7 1 9

$> m1[c(-1),c(-1)]$ # Gives a sub matrix removing first row and first column.
Here it is

	[,1]	[,2]
[1,]	−1	1
[2,]	3	9

Arithmetic Operations with Matrix

$> 2 * m1$ # Scalar multiplication by 2.

$> m1 + m2$ # matrix addition of same orders.

$> m1 * m2$ # multiplication of two matrices of same order component wise.

> sqrt $(m1)$ # Constitute a matrix whose elements are square roots of elements of original matrix $m1$.

$> m1\% * \%m2$ # Gives usual matrix multiplication of two matrices of order $m \times n$ and $n \times p$.

> solve $(m1)$ # Gives inverse of $m1$ which is a square matrix.

Inverse of Any Matrix

For generalized inverse of any matrix (square or non square), one should include MASS package prior to writing "ginv" function as follows.

> library (MASS)
> ginv $(m1)$ # Gives generalized inverse of $m1$.

Example:

> $m1 \leftarrow$ matrix $(c(1,3,5,2,-1,2,3,3,9)$,
ncol = 3, byrow = T);
> $m1$
> $m4 \leftarrow m1[-1]$; # Gives $m1$, without first row.

> $m4$ # Gives

	[,1]	[,2]	[,3]
[1,]	2	−1	2
[2,]	3	3	9

> library (MASS)
> ginv ($m4$)

Output:

	[,1]	[,2]
[1,]	−0.30	−0.03333333
[2,]	−0.36	0.10666667
[3,]	0.02	0.08666667

Eigen Values and Eigen Vectors of a Matrix

> eigen ($m1$) # Gives eigen values of the matrix $m1$.
$ values Here it is 11.4082607 −2.6097848 0.2015242

> eigen ($m1$) # Gives a matrix whose columns are eigen vectors of
$ vectors matrix $m1$. Here it is

	[,1]	[,2]	[,3]
[1,]	−0.4724087	0.59234748	−0.7401019
[2,]	−0.2139582	−0.80382721	−0.5179304
[3,]	−0.8550157	0.05464694	0.4289490

Note: To use any R function which is not in the library the following steps should be carried out under windows operating system.

1. Press "Package" option at top and press "install package".
2. A "CRAN" mirror will appear. From that select any country name.
3. A list of functions will appear and select the required function and press "ok".
4. In the "R Console" it is shown that "package successfully unpacked and MDS sum checked". Then press "Packages" and "Load package".
5. The list of functions with the new function will be shown. Select the required one and press "ok".

The corresponding command in Linux is
install.packages ("package name") # should be typed in R Console.
R Site Search ("* * *") # For searching for any program.

11.5.4 Graphics in "R"

R has extensive graphical facilities for constructing a variety of plots and diagrams from very simple to complex one. The simplest is the "plot" function.

Example:

plot (height, weight)	# scatter plot of height along x axis and weight along y axis. Fig. (11.1)
text (height, weight, labels = as.character (x $ sex))	# to add text as male or female on the above plot. Fig. (11.2) # read file x as height, weight and sex

The effect of different options in plotting parameters are shown in Tables 11.1 and 11.2, respectively.

Multiple Graphs

The plotting area can be divided to incorporate graphs in panel by using "par" function.

Example: To incorporate four graphs, scatter plot, histogram, sine function and another function (Fig. 11.3).

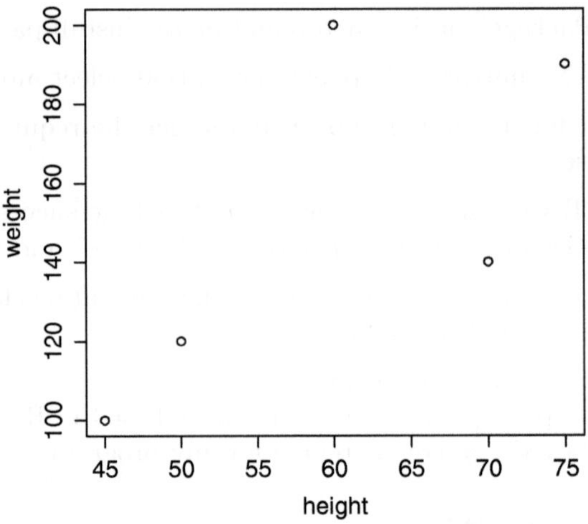

Figure 11.1 Scatter plot of heights and weights

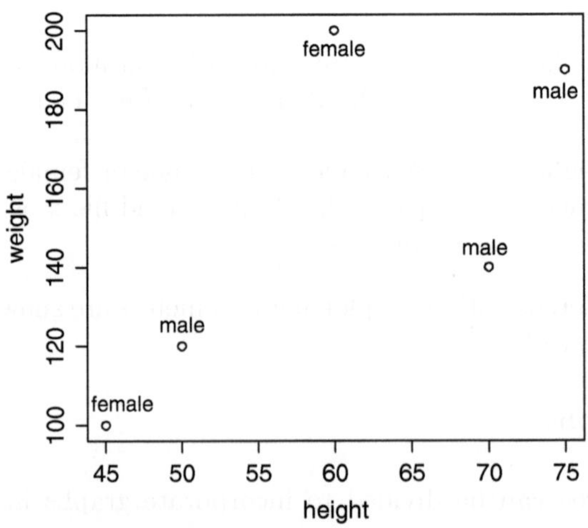

Figure 11.2 Scatter plot of heights and weights with labels

Table 11.1 Various options for "plot" function

Parameter	Purpose
type = "p" / "l" / "b" / "h" / "s" / "n"	Points / lines / both / vertical bars / steps / nothing
axis = T / F	With / Without axis
main = "..."	Main title
sub = "..."	Subtitle
x lab = "..."	Label for x-axis
y lab = "..."	Label for y-axis
x lim, y lim = c (min, max)	x/y axis range
pch = 1 / 2 / 3 etc or pch = "+" / "." etc	Plot characters
lty = 1 / 2 / 3 etc.	Line style
col = 1 / 2 / 3 etc.	Colour option, by default 1 for black.
lwd = 1 / 2 / 3 etc.	Line width (1 for default)

> par (mf row = c(2, 2))	# Splits the area into 2 by 2 rows and columns
> var1 < − c(2,5,7,9,2,1,1,4,3,11)	
> var2 < − c(5,7,8,2,4,6,6,3,2,1)	
> plot (var1, var2, main ="Scatter Plot")	# Scatter plot of var1 and var2 with legend "Scatter plot" in (1, 1) panel.
> hist (rnorm (1000, 0, 1), main = "Histogram of rnorm (1000, 0, 1))	# Draw histogram of 1000 random number drawn from N(0, 1) with a legend in panel(1, 2)
> plot (sin, - pi, 2 ∗ pi, main = "sine function")	# Plot of sin function in the range $(-\pi, 2\pi)$ with a legend in panel (2, 1)
> plot (var1 + var2, main = "plot of var1 + var2")	# Plot of var1 + var2 in panel (2, 2)

Table 11.2 Addition to existing graph

Function	Description
points (x, y)	Points at co-ordinates x and y
text $(x, y$, text)	Texts at specified co-ordinates
lines (x, y)	Lines to connect the points given by x and y
abline (a, b)	Line with intercept a and slope b
abline (h = 10)	Horizontal line at height $y = 10$
abline (v = 10)	Vertical line at distance $x = 10$
legend $(x, y, c$ ("var1", "var2", "varN"), lty = 1 : N, col = 1 : N)	To put legend at co-ordinates (x, y)
legend (locator(1), c("var1", ..., "varN"), lty = 1 : N, col = 1 : N)	To put legend at any convenient place
title ("title", "subtitle")	To write title at top of figure

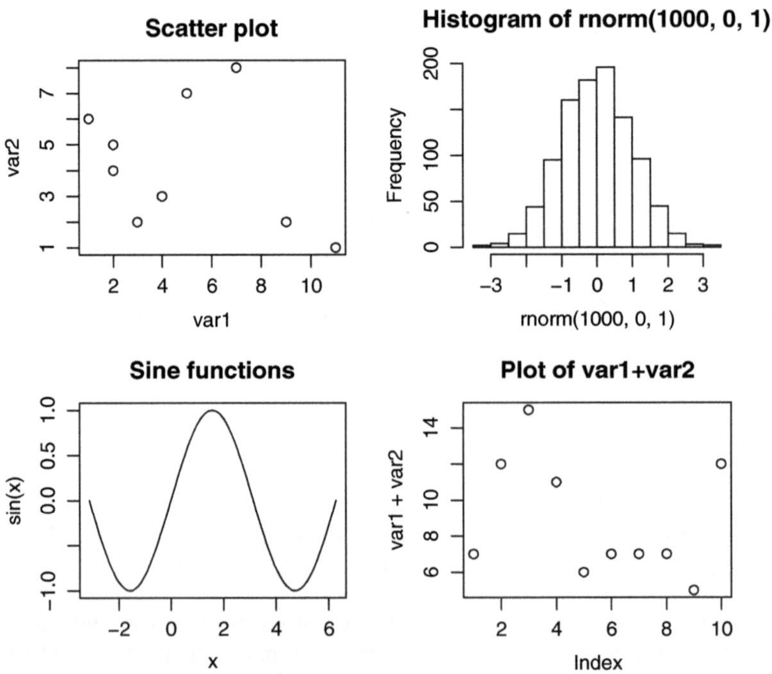

Figure 11.3 Plot in panel

To Plot with Error Bars

$x \leftarrow c$ (1, 2, 3, 4, 5, 6, 7) # values of x

$y \leftarrow c$(10, 50, 120, 134, 145, 160, 199) # values of y

errors $\leftarrow c$(0.6324555, 3.6523965, 7.7097341,11.5991379, 13.0713427, 14.6853669, 14.8842198) # values of errors of y

Plot (x, y, type = "b", pch = 16, lty = 3, $y$lim = c(0, 200), $y$lab = "label y", $x$lab "label of x", main = "Title of graph")

len = 0.07 # horizontal length of error bar

for (i in 1 : 7)

{

arrows ($x[i], y[i], x[i], y[i]$ + errors[i], angle = 90, length = len)

arrows ($x[i], y[i], x[i], y[i]$ - errors[i], angle = 90, length = len)

}

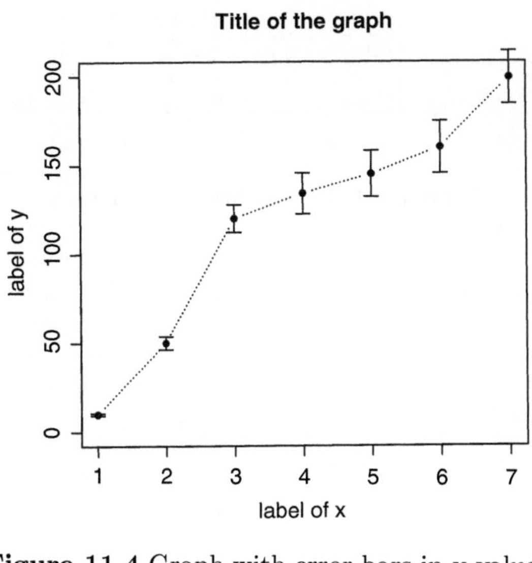

Figure 11.4 Graph with error bars in y values

Drawing of Pie Chart and Bar Chart

> pie (*c*(32, 87, 65, 29), labels = *c* ("india", "America", "England", "China"), density = 10, angle = 15 + 10 ∗ *c*(1 : 4), col = *c*(1 : 4))

pie diagram of four countries with shading line having density 10 and various angles and colours (Fig. 11.5).

> bar plot (*c*(235, 532, 789, 123), names.arg = *c*("A", "B", "C", "D") ylim = c(0, 800))

bar diagram of four characters A, B, C, D and y axis has the range (0, 800). (Fig. 11.5 (bottom)).

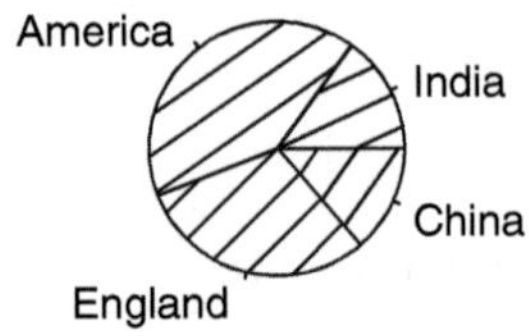

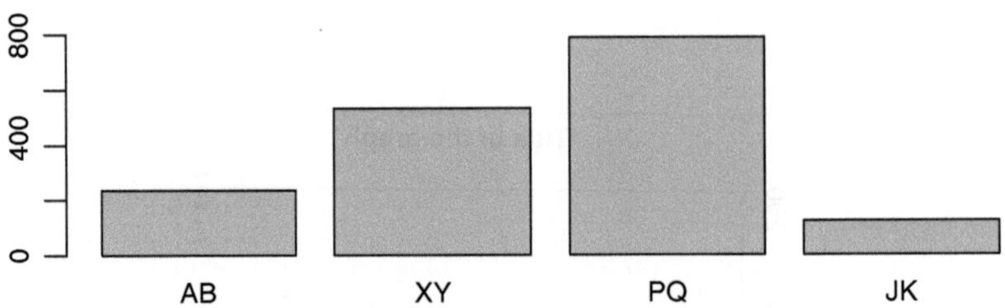

Figure 11.5 Pie chart and bar plot

Drawing Pair Plot

data ← read.table ("C:\\Users\\Tanuka \\Desktop\\UCD1 \\.txt", header = TRUE)	# Read the file UCD1.txt as matrix, data, with its headers.
x1 ← data [,1] y1 ← data [,2] z1 ← data [,3]	# Defines variables, X1, Y1, Z1 as the first, 2nd and 3rd columns of the matrix, data.
Pairs (cbind (x1, y1, z1), upper.panel = NULL, pch = 42)	# Draw the pair plots.

Figure 11.6 shows Matrix scatter plot for three variables x1, y1 and z1.

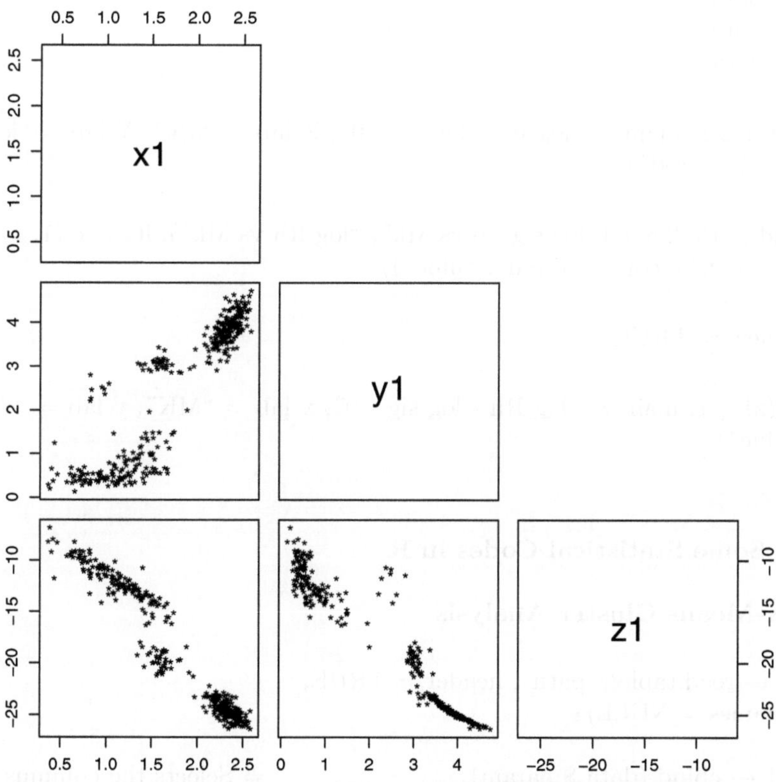

Figure 11.6 Matrix scatter plot of each pair of three variables (say)

For Different Colours

pairs (cbind (x1, y1, z1), upper.panel = NULL, pch = 42, col = (1 : 3))

For Only One Colour

pairs (cbind (x1, y1, z1), upper.panel = NULL, pch = 42, col = "green")

To draw two scatter plots for the same horizontal axis and with legend.

load the data of your choice and then follow the commands below (Fig. 11.7).

x1 ← data [,1]
y1 ← data [,2]
z1 ← data [,3]

plot (z1, x1, main = "log Rh - log sig - 0", X lab = "Mk", Y lab = "log sig - 0", col = "red")

legend (- 15, 2.5, c ("log sig - 0 vs Mk", "log Rh vs Mk"), lty = c (1, 1), lwd = c (2.5, 2.5), col = c ("red", "blue"))

par (new = TRUE)

plot (z1, y1, main = "log Rh - log sig - 0", x lab = "MK", y lab = " ", col = "blue")

11.6 Some Statistical Codes in R

(i) K-Means Cluster Analysis

data ← read.table ("path", header = TRUE,
row.names = NULL)

data1 ← cbind (data $ param1,
data $ param2, data $ param3)

Selects the columns with param1, param2, param3 from the file data.

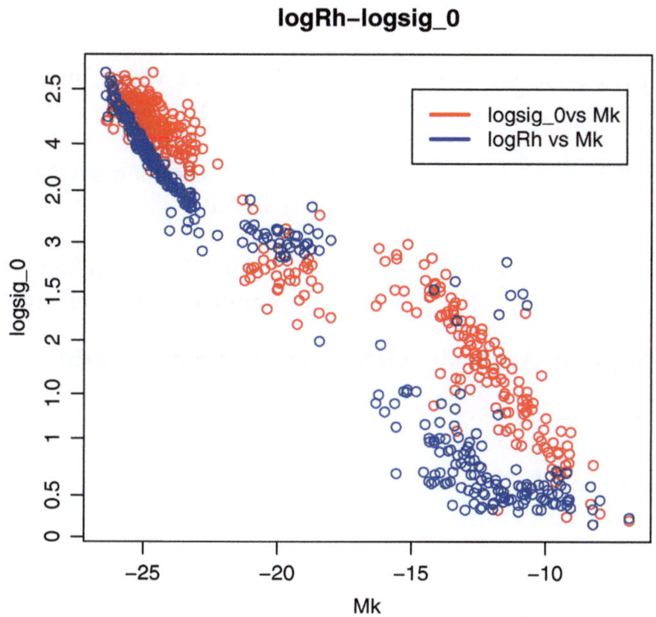

Figure 11.7 Two scatter plots with the same horizontal axis

kmeans (data1, 3)	# Do, k-means cluster analysis assuming k = 3.
c1 ← kmeans (data1, 3)	
c1	# Shows the numbers with cluster membership 1/2/3 here.
plot (data [,1], data [,3], $ col = c1 cluster)	# plot the clusters with param1 and param3 with respect to cluster membership.
clus_mem ← cbind (data1, c1 $ cluster)	# show the parameters with membership column.

clus_mem

t1 ← which (c1 $ cluster = = 1) # shows which members fall in cluster1.

data.frame (t1) # put the membership in a column.

data1 [which (c1 $ cluster = = 1),] # sort the group along with all parameters with cluster number1.

(ii) Discriminant Analysis with Misclassification Table

Construct data1 as before.

c1 ← kmeans (data1, 3)

c1

clus_mem ← cbind (data1, c1$cluster)

clus_mem

group ← c (rep (1, 47), rep (2, 210) rep(3, 132) # in k means groups are 1, 2, 3 with numbers of objects, say, 47, 210, 132, respectively.

library (MASS)

discr = lda (data1, group) $ perform discriminant analysis with respect to the predefined groups in k-means cluster analysis.

x ← predict (discr) $class # predict membership w.r.t discriminant analysis.

tab = table (predict (discr) $ class) # show groups with membership of the result of discriminant analysis.

tab1 = table (predict (discr) $ class, group)

tab1 # gives misclassification table.

```
rate = 1 - sum (diag (tab1)) /
sum(tab1)

rate                                    # % of misclassified data.

data2 ← cbind (clus_mem, x)             # data with k-means cluster
                                        member-ship and predicted
                                        membership columns.

Tr.post ← predict (discr) $ posterior

Tr.post                                 # gives posterior probabilities of
                                        discriminated members.

data3 ← cbind (clus_mem, x, Tr. post)   # gives data, cluster membership,
                                        predicted class and corresponding
                                        probabilities.

data 3
```

To Sort File, Cluster Membership, Predicted Class and Its Probability

```
max ← array (NA, c(389, 1))   # if the sample size is 389 (say)

for (i in 1 : 389)

{

max[i] ← max (Tr.post [i, 1],
Tr.post [i, 2], Tr.post[i, 3])

}

max

data4 ← cbind (clus_mem,
Tr.ld1, max)

data4                         # Gives data with cluster membership
                              & discriminated membership with its
                              probability.
```

(iii) Principal Component Analysis (PCA)

Let us load the file under concern and name it as "data" as before. Then the following commands are to be followed.

cor (data) # Gives correlation matrix of data.

eigen (cor (data)) # Give eigen values and eigen vectors of correlation matrix.

prcomp (data, cor = TRUE) # pca through correlation matrix.

summary (pc.cr ← prcomp (data, cor = TRUE) # Gives cumulative variation.

pc.cr ← prcomp (data, cor = TRUE, scores = TRUE) # naming

x ← pc.cr$rotation # Give loadings of the principal components.

x

t ← pc.cr$scores # Give values of PCs for the observations.

t

plot (t[,1], t[,2]) # Gives the plot of observations for pc1 and pc2.

biplot (pc.cr) # Gives PC values as well as loading (i.e. length of the parameters)

Kolmogrov–Smirnov (KS) Test for Two Samples

Let us load two samples mag_{-K} and mag_{-V} (see Appendix). Then we are to follow the command

ks.test(mag_{-K}[,1],mag_{-V}[,1],exact=TRUE)

Before this we are to load the function "stats" from library.

Kruskal–Wallis Rank Sum Test for Two Samples

```
data ← read.table ("C:\\Users\\Tanuka      # Read the file UCD1.txt as
\\Desktop\\UCD1 \\.txt", header =          matrix, data, with its headers.
TRUE,

x1 ← data [,1]                             # Define variables, x1, y1
y1 ← data [,2]                             as the first and 2nd columns

kruskal.test(x1,y1)                        # performs the test
```

Output:
Kruskal–Wallis rank sum test
data: x1 and y1 Kruskal–Wallis
chi-squared = 384.6389, df = 347,
p-value = 0.07998

Wilcoxon Rank-Sum Test for Two Samples

```
data ← read.table ("C:\\Users\\Tanuka      # Read the file UCD1.txt as
\\Desktop\\UCD1 \\.txt", header =          matrix, data, with its headers.
TRUE,

x1 ← data [,1]                             # Define variables, x1, y1
y1 ← data [,2]                             as the first and 2nd columns
wilcox.test(x1, y1 = NULL, alternative     # any one from "two.sided",
= c("two.sided"))                          # "less" or "greater"
```

Output:
data: x1 V = 75855, p-value ¡ 2.2e-16
alternative hypothesis: true location is
not equal to 0

Test for Gaussianity for One Parameter

Load any data file, mag_{-K} (say). Then the following command should be followed.

ks.test(mag_{-K}[,1],"pnorm",exact=TRUE)

Test for Gaussianity in a Multivariate Set-Up

For this one has to install the R-function **"mvnorm test"**. For this the following commands are to be followed in **"window"** set-up.

step1. Click "Cran Mirror" under "Packages" on topmost part of R-console and select a country.

step2. Click "install package" under "Packages" and select "mvnorm test" from the dialog box which appears.

step3. Click "load package" under "packages" and load "mvnorm test" from the library which appears.

To Install Package Under Linux

step1. install.packages ("package name") # Type in R console.

step2. Load "mvnorm test".

Load the file, "data" (say). Then follow the command,

mshapiro.test(t(data))

If One Wants to Test for a Number of Columns, 2 (Say)

Construct a new data file, data 1 (say) (e.g. from UCD1.txt), as follows:

data 1 ← as.matrix (cbind (data[,1], data[,2], 389, replace = FALSE)	# 389 is the sample size (say)

mshapiro.test(t(data 1))

Regression Analysis

Ordinary Linear Regression

To load data file, "data" (say)

Then follow the commands.

lm(data$y ~ data$x)	# Fit y on x, where y and x are two headers of "data".
summary(lm(data$y ~ data$x))	# Gives summary of the fitting.
c1 ← lm(data$y ~ data$x)	
plot(c1)	# Plot residual vs fitted values.

More than One Variables as Predictors (glm)

lm(data$y ~ data$x + data$z) # Fit y on (x, z).

Ridge Regression for Two Predictors

data ← read.table ("C:\\Users\\Tanuka \\Desktop\\ UCD1 \\.txt", header = TRUE,	# Read the file UCD1.txt as matrix, data, with its headers.
x1 ← data [,1] y1 ← data [,2] z1 ← data [,3] library(MASS)	# Define variables, x1, y1,z1 as the first, 2nd and 3rd columns
model.ridge← lm.ridge(y1~ x1+z1, lambda = 1) model.ridge$Inter	# y1 response, x1, z1 predictors
model.ridge$coef	# Gives coefficients of predictors
summary(model.ridge)	# Gives summary of the test

Output:

model.ridge$Inter
[1] 1
model.ridge$coef
x1 z1
-3.050482 -3.046672
summary(model.ridge)

	Length	Class	Mode	
coef	2	-none	- numeric	scales
scales	2	-none	- numeric	Inter
inter	1	-none	- numeric	lambda
lamda	1	-none	- numeric	
ym	1	-none	numeric	
xm	2	-none	numeric	
GCV	1	-none	- numeric	
kHKB	1	-none	- numeric	
kLW	1	-none	- numeric	

Stepwise Regression Model

library (MASS)

fit ← lm (y ∼ data $x + data$z + data$p + . . .)   # All initial predictor variables.

step ← step(fit, direction = "both",criterion='AIC')
step$anova   # Displays results.

summary(step)   # Gives summary.

To Increase the Size of the Print Screen on R (Type in R Console)

options (max.print = 5.5E5)   # can print output of size 5.5 $\times 10^5$ (say)

sink ("data1.txt")   # print the output in the file data 1 in text format.

data1

Fitting a Distribution with Parameters e.g. Gamma Distribution to a Data Set x1

library(MASS)
fitdistr(x1,"gamma")

Fitting Any Distribution to a Data Set x1

1. Install (rriskDistributions)
2. Load (rriskDistributions)
3. Read data as matrix:

x1< −as.matrix(read.table("path"))

useFitdist(data2fit=x1[,1], show.output = TRUE, "distributions")

distributions : "norm", "logis", "beta", "exp", "chisq",
"unif", "gamma","lnorm", "Weibull", "f", "t", "cauchy",
"gompertz", "triang"

Generation of Random Numbers (1,000 Say) from Any Distribution and Showing Histogram of the Sample Generated

x←rdistribution (n=1,000, param1= , param2=, ...)
x← hist(x)

List of distributions:

Distribution	Code	Parameters
Beta	beta	shape1, shape2
Binomial	binom	size, prob
Cauchy	cauchy	location, scale
Chi-square	chisq	df
Exponential	exp	1/mean
F	f	df1, df2
Gamma	gamma	shape,1/scale
Geometric	geom	prob
Hypergeometric	hyper	m, n, k
Log-normal	lnorm	mean, sd
Logistic	logis	location, scale
Normal	norm	mean, sd
Poisson	pois	lambda
Student	t	df
Uniform	unif	min, max
Weibull	weibull	shape

Appendix

Kolmogorov–Smirnov Two Sample Test Data Sets (Sample size 180× 1)

File: Mag_{-K}

−23.2074	−24.2015	−24.7079	−25.5826	−24.7811	−25.1276	−24.8011
−23.9370	−25.2397	−24.8089	−26.1979	−24.5410	−24.7249	−24.1075
−26.0539	−23.6367	−24.2693	−23.2323	−25.7012	−24.1509	−26.1449
−25.9689	−25.0855	−24.0194	−24.2857	−24.6511	−24.8092	−24.5812
−23.3918	−25.2042	−25.4348	−25.0648	−23.5343	−24.5418	−24.9718
−24.2809	−23.2797	−24.8669	−25.0720	−24.5944	−24.7167	−25.1317
−23.6899	−23.2316	−24.9971	−25.6536	−25.6193	−24.5588	−24.5719
−24.6807	−23.1846	−23.6369	−26.3391	−25.1475	−25.2033	−25.0586
−24.9286	−24.8506	−24.9465	−26.1826	−25.6414	−25.4523	−23.2752
−24.8370	−25.0072	−24.3658	−24.6577	−25.4488	−25.5579	−25.0294
−26.1232	−25.4694	−25.1920	−24.2621	−23.5923	−24.1142	−24.2020
−25.3757	−24.5041	−24.7747	−25.7152	−24.1143	−25.3687	−24.7354
−24.4115	−24.3380	−25.7742	−24.7011	−24.1123	−25.7985	−24.9438
−24.6560	−24.6532	−25.4850	−25.0423	−25.1852	−25.9123	−26.3992
−23.2716	−23.2178	−25.3023	−25.2550	−25.8977	−24.3428	−26.1957
−24.5214	−23.7626	−24.9774	−25.0164	−25.4323	−25.2627	−25.7207
−24.9423	−24.4995	−25.7128	−24.8642	−25.8406	−23.9685	−23.9120
−25.2993	−25.2526	−24.0529	−24.5669	−25.0076	−24.9662	−24.9538
−24.4216	−23.7023	−24.9559	−24.1416	−25.6263	−26.1305	−25.1000
−25.9275	−24.8766	−26.0467	−24.1687	−24.8133	−24.2621	−24.9971
−24.9496	−25.3179	−24.9445	−23.9591	−24.2294	−25.5327	−24.8643
−24.2765	−24.8809	−24.2355	−23.9229	−23.3174	−23.4615	−24.4003
−24.5944	−22.8810	−25.1942	−22.2361	−24.9120	−25.8170	−25.4085
−24.3570	−25.3671	−23.0617	−24.9171	−25.0693	−24.4197	−26.2672
−24.5719	−25.8708	−24.4439	−25.5150	−25.8977	−24.5485	−24.4043
−26.1178	−24.4003	−23.6574	−24.9904	−20.7443		

A.K. Chattopadhyay, T. Chattopadhyay, *Statistical Methods for Astronomical Data Analysis*, Springer Series in Astrostatistics 3,
DOI 10.1007/978-1-4939-1507-1

File: Mag_{-V}

−20.47	−21.47	−21.94	−22.73	−21.76	−22.14	−21.70	−21.31	−22.25
−22.30	−23.37	−21.73	−21.77	−21.05	−23.09	−20.80	−21.36	−20.37
−22.80	−20.94	−23.23	−23.03	−21.86	−21.41	−21.35	−21.64	−22.19
−21.76	−20.52	−22.15	−22.39	−22.44	−20.77	−21.46	−22.04	−21.37
−20.47	−22.03	−22.22	−21.64	−21.62	−22.29	−21.15	−20.61	−21.89
−22.47	−22.45	−21.87	−21.79	−21.69	−20.37	−20.83	−21.92	−22.27
−22.32	−22.14	−21.85	−21.76	−22.19	−23.18	−22.66	−22.42	−20.55
−21.94	−22.24	−21.39	−21.55	−22.65	−22.14	−22.35	−23.30	−22.83
−22.26	−21.32	−21.14	−21.31	−21.46	−22.29	−21.56	−21.59	−23.06
−21.34	−22.26	−21.74	−21.53	−21.39	−22.79	−21.87	−21.11	−22.92
−22.06	−21.88	−21.59	−22.45	−22.22	−22.44	−23.08	−23.54	−20.40
−20.40	−22.44	−22.34	−23.06	−21.52	−23.30	−21.51	−20.80	−22.51
−21.74	−22.73	−22.50	−22.68	−22.04	−21.45	−22.99	−21.94	−22.95
−21.03	−20.86	−22.28	−22.35	−21.09	−21.78	−22.07	−21.82	−22.04
−21.61	−20.86	−22.11	−21.42	−22.99	−21.87	−22.66	−22.98	−21.81
−23.35	−21.47	−22.14	−21.43	−22.07	−21.97	−22.49	−21.76	−21.07
−21.39	−22.39	−21.79	−21.45	−21.84	−21.40	−21.19	−20.74	−20.53
−21.76	−21.51	−20.12	−21.76	−21.82	−22.05	−22.99	−22.26	−21.26
−22.53	−20.39	−22.10	−22.13	−21.28	−23.27	−21.79	−22.86	−21.26
−22.70	−23.06	−21.43	−21.36	−22.89	−21.76	−20.35	−21.88	−18.51

Data Set for Gaussianity Test in Univariate Set Up (Sample Size 180)

File: Mag_{-K}

Data Set for Step Wise Regression (Sample Size 389 × 4)

File: $UCD1$

$logsig_0$	$logRh$	Mk	$mukh$
1.0607	0.6083	−12.1949	26.9219
1.01703	0.25891	−11.0059	28.544
0.716	0.48076	−9.8997	30.0756
0.716	0.59083	−9.5149	30.6122
0.80618	0.32645	−11.0012	28.5505
0.41497	0.37595	−7.9538	32.8026
0.39794	0.65406	−9.2	31.0523
0.74819	0.51294	−10.89	28.7036
0.90309	0.34397	−10.817	28.8043
0.74036	0.52631	−10.0403	29.88

1.06446 0.52518 −10.9957 28.558
0.75587 0.6911 −11.0854 28.4346
0.5682 0.34397 −9.607 30.4836
1.09342 0.27664 −10.15 29.7276
0.62325 0.32705 −9.1851 31.0731
0.61278 0.67324 −9.5808 30.5202
0.85126 0.54 −10.6954 28.972
0.65321 0.49715 −9.2349 31.0036
0.81954 0.32346 −9.4103 30.7582
0.81291 0.29623 −9.0554 31.2547
1.15534 0.39873 −11.5055 27.8588
0.79239 0.53191 −10.2505 29.5881
0.88081 0.37701 −9.9812 29.9622
0.76343 0.33994 −9.7054 30.3463
0.77085 0.40424 −10.1554 29.72
0.66276 0.36245 −9.1212 31.1626
1.27646 0.2898 −12.42 26.6184
0.65321 0.13211 −8.2151 32.4346
1.25527 0.34909 −11.9159 27.2997
0.82607 0.38385 −10.0654 29.8451
0.38021 0.19531 −6.8555 34.3529
0.91381 0.38437 −10.7455 28.9029
0.4624 0.51565 −8.3055 32.3075
0.73239 0.28063 −10.2954 29.5257
0.9345 0.435 −10.8608 28.7439
0.95424 0.45406 −10.6756
28.9994 0.70757 0.38645 −9.1012
31.1906 0.63347 0.44554 −9.7455
30.2904 0.60206 0.52818 −9.42
30.7447 0.6902 0.62359 −9.8049
30.2077 0.36173 0.28327 −8.2303 32.4133 1.01284 0.45804 −10.6911 28.9781
0.70757 0.446 −9.5206 30.6042 0.91381 0.51139 −10.6959 28.9714 0.74819
0.36952 −9.1755 31.0866 0.43136 1.24613 −11.7754 27.4907 1.39898 0.43297
−10.7605 29.2823 1.06296 0.54033 −11.8605 27.775 1.20817 0.47712
−11.7605 27.911 1.15442 0.43457 −12.2605 27.2333 1.40586 0.55267
−13.6605 25.3735 1.10721 0.33445 −11.9605 27.6392 0.99211 0.45788
−11.5605 28.1837 1.26245 0.47712 −12.8605 26.4286 1.12057 0.3075
−12.3605 27.0985 1.24601 0.50379 −12.5605 26.8297 1.41397 0.32634
−13.4605 25.6351 1.1271 0.38561 −11.2605 28.5943 1.07628 0.47567
−11.6605 28.0473 1.17319 0.4014 −11.6605 28.0473 0.95809 0.43616
−10.4605 29.697 1.09968 0.2833 −11.9605 27.6392 1.06032 0.35984 −11.5605
28.1837 0.93298 0.5832 −11.8605 27.775 0.97864 0.38382 −11.3605 28.4572
1.31175 0.344 −12.6475 26.7131 0.92583 0.5 −10.5035 29.6375 1.31597 0.636
−12.7675 26.5527 0.90902 0.549 −11.4815 28.2917 0.91116 0.376 −11.3735
28.4394 1.00432 0.41 −11.5255 28.2315 1.20412 0.78387 −12.6449 26.3166

1.15229 0.58529 −12.3768 26.6765 1.1271 0.34454 −11.94 27.267 1.44091 0.50709 −13.1305 25.6701 1.29003 1.017 −13.2453 25.5183 1.31806 0.8787 −12.9653 25.8891 1.19312 0.92985 −12.3753 26.6785 1.35984 0.96684 −12.8253 26.0755 1.32222 0.708 −12.3853 26.665 1.53782 0.63837 −14.1153 24.3844 1.1271 0.92009 −12.8853 25.9955 1.21748 0.7238 −12.9653 25.8891 1.24055 0.85642 −12.7953 26.1155 1.17319 0.63837 −12.6453 26.316 1.02531 0.78179 −12.8153 26.0888 1.39094 1.00091 −13.7253 24.889 1.36736 1.13397 −13.3953 25.3207 1.19033 1.05479 −12.5753 26.4098 1.32428 0.88942 −13.5253 25.1501 1.36173 0.75376 −13.1453 25.6504 1.2878 0.84484 −13.0453 25.7829 1.29447 0.57767 −12.3853 26.665 1.20683 0.83294 −13.2653 25.4919 1.23805 0.55539 −12.1153 27.0294 1.14301 0.45273 −12.6453 26.316 1.1038 0.65685 −13.9253 24.6294 1.35218 0.8787 −13.6253 25.0194 1.08636 0.7238 −12.4753 26.544 1.19312 0.45273 −12.3253 26.7459 1.20683 0.27664 −12.9853 25.8625 1.65321 1.27261 −15.9944 22.0622 1.48144 0.97158 −14.3244 24.1166 1.66558 0.64747 −15.5592 22.5812 1.71767 1.4484 −16.2184 21.8004 1.42975 1.47473 −15.1784 23.0455 1.51188 1.35823 −15.6184 22.5098 1.07555 1.46177 −13.1784 25.6067 1.58883 0.54531 −13.7384 24.872 1.55871 1.14737 −14.4884 23.908 1.67025 0.54531 −14.2484 24.2137 1.39094 1.35823 −13.8884 24.6772 1.45025 1.30498 −13.3684 25.3561 1.43457 1.36227 −16.3238 21.6785 1.46687 1.95555 −16.1538 21.8755 1.50651 1.48284 −15.2838 22.9161 1.4014 1.48284 −14.8238 23.4856 1.63849 0.6633 −14.5438 23.8378 1.52763 0.86742 −13.9438 24.6055 1.38202 1.16845 −13.9438 24.6055 1.49415 0.96433 −14.1438 24.3478 1.5092 0.56639 −14.3438 24.0919 1.47712 1.00572 −14.3438 24.0919 1.54531 0.56639 −14.1438 24.3478 1.50515 1.00572 −13.9438 24.6055 1.47422 0.56639 −13.5438 25.126 1.43457 0.6633 −13.4438 25.2571 1.39445 0.6633 −13.4438 25.2571 1.07918 0.74248 −13.3438 25.3885 1.73799 1.49956 −15.1484 23.0824 1.50106 0.98465 −14.1584 24.329 1.68664 1.11934 −15.5684 22.5701 0.94448 2.5188 −14.1733 24.3099 0.82607 2.26636 −11.7499 27.5255 0.96848 2.47718 −10.8609 28.7437 0.81954 2.20377 −13.3067 25.4374 0.83251 2.46228 −11.32 28.1127 0.97772 2.36135 −10.7 28.9658 0.81954 2.79944 −11.4654 27.9137 1.02119 2.60203 −13.38 25.3408 1.88081 1.99264 −18.4405 19.4494 2.31302 3.513 −23.7899 17.2129 2.41196 3.871 −24.8122 17.7639 2.19396 3.46 −23.0821 17.0742 2.31197 3.663 −24.3074 17.4357 2.31091 3.557 −23.8753 17.2422 2.11594 3.347 −23.2594 17.0922 2.30298 3.854 −24.9076 17.8392 2.35908 3.844 −24.656 17.6499 2.25912 3.724 −24.2741 17.418 2.42504 3.417 −23.356 17.1066 2.24993 4.166 −25.5727 18.4917 2.17406 3.385 −23.4486 17.1235 2.09517 3.885 −24.7512 17.7181 2.28103 3.9 −24.7952 17.7509 1.96988 3.382 −23.3221 17.1012 2.58104 4.73 −26.4254 19.6929 2.27207 3.605 −24.0529 17.3124 2.22011 3.514 −23.5756 17.1517 2.33706 3.686 −24.2693 17.4155 2.19396 3.468 −23.2323 17.0888 2.19201 3.726 −24.4206 17.4993 2.17202 3.494 −23.5555 17.1468 2.39199 3.642 −24.2675 17.4145 2.37694 3.566 −23.9002 17.2513 2.52802 4.085 −25.3073 18.2036 2.36903 4.019 −25.0655 17.9735 2.46404 3.928 −24.9197 17.8491 1.716 3.036 −20.5284 17.821 1.95424 3 −21.3127

17.4165 1.81291 2.852 −19.7084 18.3754 1.63548 2.98 −20.2384 18.0028
1.55871 2.949 −19.5384 18.5051 1.59106 3.091 −19.5884 18.4665 1.41664
3.05 −20.3984 17.9005 1.6149 2.988 −18.8784 19.052 1.59439 2.993
−19.3484 18.6557 1.5563 3.068 −19.3584 18.6477 1.51851 3.13 −19.9884
18.1723 1.61805 3.146 −21.0984 17.5137 2.22789 3.364 −23.2074 17.0858
2.14114 3.644 −24.2015 17.3811 2.18808 3.86 −24.7079 17.6865 2.46702
4.227 −25.5826 18.5032 2.40295 3.887 −24.7811 17.7403 2.38507 4.057
−25.1276 18.0298 2.48799 3.823 −24.8011 17.7554 2.34596 3.553 −23.937
17.2652 2.36493 4.092 −25.2397 18.1363 2.34792 3.884 −24.8089 17.7613
2.50907 4.457 −26.1979 19.3287 2.22011 3.856 −24.541 17.5732 2.29491
3.843 −24.7249 17.6988 2.26007 3.571 −24.1075 17.3366 2.54605 4.465
−26.0539 19.1152 2.233 3.543 −23.6367 17.1673 2.33706 3.686 −24.2693
17.4155 2.19312 3.468 −23.2323 17.0888 2.54195 4.258 −25.7012 18.6451
2.28601 3.67 −24.1509 17.3567 2.41397 4.477 −26.1449 19.2486 2.51706
4.392 −25.9689 18.9952 2.30103 3.969 −25.0855 17.9915 2.31492 3.117
−24.0194 17.2982 2.32305 3.688 −24.2857 17.4241 2.58104 3.789 −24.6511
17.6466 2.47407 3.857 −24.8092 17.7616 2.37199 3.785 −24.5812 17.5994
2.20412 3.429 −23.3918 17.1127 2.35295 3.969 −25.2042 18.1018 2.46195
4.176 −25.4348 18.3373 2.34908 4.005 −25.0648 17.9729 2.20085 3.434
−23.5343 17.1418 2.22994 3.701 −24.5418 17.5737 2.34908 3.995 −24.9718
17.8923 2.06707 3.709 −24.2809 17.4216 2.07188 3.497 −23.2797 17.095
2.34104 3.926 −24.8669 17.8065 2.32593 3.988 −25.072 17.9793 2.28892 3.82
−24.5944 17.6081 2.463 3.798 −24.7167 17.6928 2.43297 4.104 −25.1317
18.0336 2.31091 3.516 −23.6899 17.1821 2.25792 3.504 −23.2316 17.0887
2.25503 3.959 −24.9971 17.9139 2.44793 4.288 −25.6536 18.5872 2.44498
4.257 −25.6193 18.5462 2.29601 3.786 −24.5588 17.5847 2.22608 3.831
−24.5719 17.5933 2.43902 3.849 −24.6807 17.6672 2.15806 3.407 −23.1846
17.0833 2.36493 3.551 −23.6369 17.1674 2.34596 4.279 −26.3391 19.5509
2.44994 3.989 −25.1475 18.0483 2.33506 4.019 −25.2033 18.101 2.37603
3.976 −25.0586 17.9674 2.436 3.959 −24.9286 17.8564 2.34203 3.779
−24.8506 17.7937 2.36903 3.958 −24.9465 17.8712 2.48799 4.588 −26.1826
19.3054 2.53504 4.245 −25.6414 18.5726 2.42095 4.133 −25.4523 18.3563
2.09587 3.448 −23.2752 17.0943 2.43505 3.894 −24.837 17.7831 2.35793
4.031 −25.0072 17.9225 2.33905 3.696 −24.3658 17.4678 2.27807 3.82
−24.6577 17.6511 2.39794 4.123 −25.4488 18.3525 2.28511 4.289 −25.5579
18.4747 2.37694 3.99 −25.0294 17.9418 2.46494 4.478 −26.1232 19.2164
2.45102 4.13 −25.4694 18.3751 2.34301 3.941 −25.192 18.0902 2.39393 3.649
−24.2621 17.4117 2.31006 3.511 −23.5923 17.1558 2.20085 3.639 −24.1142
17.3396 2.24601 3.79 −24.202 17.3813 2.44902 4.131 −25.3757 18.2743
2.44295 3.77 −24.5041 17.5499 2.41095 3.883 −24.7747 17.7355 2.45102
4.298 −25.7152 18.6624 2.25503 3.673 −24.1143 17.3397 2.41797 4.17
−25.3687 18.2669 2.38507 3.91 −24.7354 17.7064 2.33304 3.721 −24.4115
17.494 2.41597 3.689 −24.338 17.4523 2.33506 4.262 −25.7742 18.7364
2.32408 3.818 −24.7011 17.6817 2.29003 3.585 −24.1123 17.3388 2.50705
4.326 −25.7985 18.7674 2.40106 3.94 −24.9438 17.8689 2.35908 3.844

−24.656 17.6499 2.34596 3.765 −24.6532 17.648 2.46805 4.14 −25.485 18.3923 2.39393 3.956 −25.0423 17.953 2.45803 3.98 −25.1852 18.0837 2.45803 4.335 −25.9123 18.9176 2.34203 4.495 −26.3992 19.6492 2.38093 3.403 −23.2716 17.0938 2.29601 3.353 −23.2178 17.087 2.21801 4.005 −25.3023 18.1986 2.405 3.993 −25.255 18.1513 2.41296 4.345 −25.8977 18.8979 2.2971 3.642 −24.3428 17.455 2.38899 4.64 −26.1957 19.3254 2.34005 3.825 −24.5214 17.5607 2.28307 3.605 −23.7626 17.2041 2.18808 3.89 −24.9774 17.8971 2.37401 3.957 −25.0164 17.9304 2.31492 4.066 −25.4323 18.3346 2.2971 4.007 −25.2627 18.159 2.56301 4.296 −25.7207 18.6692 2.43902 3.957 −24.9423 17.8677 2.45803 3.777 −24.4995 17.547 2.43996 4.352 −25.7128 18.6594 2.35005 3.894 −24.8642 17.8044 2.50393 4.394 −25.8406 18.8222 2.19893 3.559 −23.9685 17.2775 2.19005 3.6 −23.912 17.2557 2.47799 4.11 −25.2993 18.1955 2.41497 3.95 −25.2526 18.149 2.27207 3.605 −24.0529 17.3124 2.26293 3.795 −24.5669 17.59 2.34104 3.966 −25.0076 17.9229 2.39794 3.952 −24.9662 17.8877 2.37603 3.96 −24.9538 17.8772 2.33203 3.726 −24.4216 17.4999 2.17898 3.548 −23.7023 17.1858 2.42504 3.947 −24.9559 17.879 2.29601 3.629 −24.1416 17.3523 2.48302 4.252 −25.6263 18.5545 2.29994 4.567 −26.1305 19.2271 2.39707 4.073 −25.1 18.0046 2.43505 4.402 −25.9275 18.9383 2.27207 3.978 −24.8766 17.8142 2.41697 4.482 −26.0467 19.1049 2.18298 3.651 −24.1687 17.3652 2.25912 3.962 −24.8133 17.7647 2.233 3.705 −24.2621 17.4118 2.42406 3.982 −24.9971 17.9139 2.41896 3.936 −24.9496 17.8737 2.306 4.101 −25.3179 18.2144 2.36493 3.914 −24.9445 17.8695 2.30492 3.302 −23.9591 17.2738 2.25912 3.621 −24.2294 17.395 2.45606 4.28 −25.5327 18.4459 2.3831 3.939 −24.8643 17.8045 2.34104 3.658 −24.2765 17.4192 2.42095 3.97 −24.8809 17.8177 2.29798 3.663 −24.2355 17.3981 2.266 3.658 −23.9229 17.2598 2.11992 3.208 −23.3174 17.1004 2.32593 3.401 −23.4615 17.1261 2.266 3.748 −24.4003 17.4875 2.44107 3.763 −24.5944 17.6081 2.29003 3.301 −22.881 17.0666 2.22608 3.985 −25.1942 18.0923 2.13001 3.07 −22.2361 17.1276 2.38507 3.944 −24.912 17.8428 2.38399 4.336 −25.817 18.7914 2.47202 4.183 −25.4085 18.309 2.3249 3.73 −24.357 17.4629 2.31492 4.111 −25.3671 18.2653 2.22298 3.368 −23.0617 17.0729 2.3189 3.943 −24.9171 17.847 2.42797 3.985 −25.0693 17.977 2.42991 3.752 −24.4197 17.4988 2.55703 4.612 −26.2672 19.4361 2.22608 3.831 −24.5719 17.5933 2.36996 4.334 −25.8708 18.8619 2.22891 3.761 −24.4439 17.5131 2.436 4.247 −25.515 18.4258 2.41296 4.345 −25.8977 18.8979 2.31994 3.789 −24.5485 17.5781 2.29092 3.783 −24.4043 17.4898 2.38596 4.346 −26.1178 19.2084 2.266 3.748 −24.4003 17.4875 2.15534 3.55 −23.6574 17.173 2.53403 3.922 −24.9904 17.9081 1.59439 3.077 −20.7443 17.6966 1.65896 3.036 −20.5284 17.821 1.52114 3.091 −18.5084 19.3861 1.4609 2.923 −19.1978 18.7791 1.37658 3.023 −18.0211 19.8531 1.65031 2.939 −19.992 18.1698 1.48572 2.886 −19.0122 18.9359 1.51055 3.113 −20.012 18.1558 2.10037 3.09497 −22.9284 17.0672 2.24304 3.12884 −23.3749 17.1098 2.33445 3.23332 −23.0879 17.0747 2.20412 2.90959 −22.797 17.0673 1.39794 2.918 −18.457 19.434 1.62221 3.177 −20.8802 17.6231 1.54407 2.993 −19.9364 18.209

1.34242 3.064 −19.2699 18.7196 1.57978 3.135 −20.0476 18.1312 1.65321 2.98 −19.6494 18.4199 1.64345 2.989 −18.9259 19.0105 1.5563 3.175 −21.2218 17.4565 1.70757 2.855 −19.9046 18.2317 1.90849 2.945 −20.9307 17.5969 1.79239 2.865 −19.7577 18.3386 1.5563 2.931 −18.747 19.1685 1.61278 3.129 −20.975 17.5743 1.61278 3.091 −20.4815 17.8493 1.66276 3.137 −18.8612 19.0671 1.62325 2.967 −19.0081 18.9394 1.5682 3.432 −21.0289 17.5474 1.63347 3.362 −18.7218 19.1911

Data Set for Ridge Regression (Sample Size 389 × 4)

File: *UCD*1

Data Set for Wilcoxon Test (Sample Size 389 × 4)

File: *UCD*1

Data Set for Kruskal Wallis Test (Sample Size 389 × 4)

File: *UCD*1

Data Set for PCA (Sample Size 43 × 9)

File: *hbr*1

$ID\ \ logT_{eff}\ \ HB\ \ DT\ \ HBR\ \ HB_{RE}\ \ L_t\ \ (B-V)_{peak}\ \ BT$

NGC0104 7 −0.97999999999999998 0.83999999999999997 2.5 0.80000000000000004 1 3.7559999999999998 NGC0362 5 −0.87 0.65000000000000002 7.5 0.55000000000000004 6.5 4.0789999999999997 NGC1261 5 −0.70999999999999996 0.69999999999999996 10 0.63 8.5 4.0789999999999997 NGC1851 1 −0.32000000000000001 0.72999999999999998 10 0.65000000000000002 9 4.0970000000000004 NGC1904 1 0.89000000000000001 0.14999999999999999 11 0.050000000000000003 9 4.3520000000000003 NGC2808 1 −0.48999999999999999 0.72999999999999998 13.5 0.63 12 4.5679999999999996 NGC3201 4 0.080000000000000002 0.64000000000000001 9 0.34999999999999998 6 4.0789999999999997 NGC4147 3 0.47999999999999998 0.45000000000000001 8.5 0.20000000000000001 5.5 4.0609999999999999 NGC4590 3 0.14999999999999999 0.55000000000000004 7 0.25 4 4.0410000000000004 NGC4833 1 0.94999999999999996 0.20000000000000001 8 0 5 4.3010000000000002 NGC5024 2 0.76000000000000001 0.10000000000000001 3 −0.050000000000000003 1.5 4.0789999999999997 NGC5694 1 NA 0.14999999999999999 5.5 0 2 4.2039999999999997 NGC5824 NA NA NA NA NA NA 4.3799999999999999 NGC5904 3 0.37

0.59999999999999998 10 0.23000000000000001 6 4.1760000000000002 NGC5927 NA NA NA NA NA NA 3.7240000000000002 NGC5946 NA NA NA NA NA NA 4.2789999999999999 NGC5986 NA NA NA NA NA NA 4.415 NGC6093 1 0.92000000000000004 0.32000000000000001 6 0.050000000000000003 4.2000000000000002 4.4770000000000003 NGC6171 6 −0.69999999999999996 0.78000000000000003 9 0.68999999999999995 6.2999999999999998 3.875 NGC6205 1 0.96999999999999997 0.14999999999999999 10 0.050000000000000003 8.5 4.5049999999999999 NGC6218 1 1 0.19 7.5 −0.070000000000000007 4 4.2169999999999996 NGC6235 NA NA NA NA NA NA 4.1139999999999999 NGC6266 1 0.28000000000000003 0.69999999999999996 15 0.13 10 4.4770000000000003 NGC6273 NA NA NA NA NA NA 4.5679999999999996 NGC6284 1 NA 0.20000000000000001 10 0 7 4.2789999999999999 NGC6287 NA NA NA NA NA NA 4.1139999999999999 NGC6342 NA NA NA NA NA NA 3.778 NGC6356 NA NA NA NA NA NA 3.7559999999999998 NGC6362 5 −0.57999999999999996 0.77000000000000002 9 0.55000000000000004 6.5 3.9540000000000002 NGC6544 NA NA NA NA NA NA 4.1760000000000002 NGC6624 NA NA NA NA NA NA 3.7709999999999999 NGC6637 NA NA NA NA NA NA 3.7480000000000002 NGC6638 5 −0.29999999999999999 0.66000000000000003 7.5 0.40000000000000002 5 4.0970000000000004 NGC6652 NA NA NA NA NA NA 4 NGC6681 NA NA NA NA NA NA 4.3010000000000002 NGC6717 NA NA NA NA NA NA 4.1139999999999999 NGC6723 5 −0.080000000000000002 0.68000000000000005 11 0.5 9 4.1299999999999999 NGC6864 6 −0.39000000000000001 0.75 8 0.39000000000000001 6.5 4.1760000000000002 NGC6934 NA NA NA NA NA NA 4.1299999999999999 NGC6981 4 0.17000000000000001 0.59999999999999998 5 0.34999999999999998 3 4 NGC7078 3 0.67000000000000004 0.55000000000000004 14 0.050000000000000003 9 4.4770000000000003 NGC7089 2 0.95999999999999996 0.27000000000000002 9 0.17999999999999999 8 4.4770000000000003 NGC7099 1 0.88 0.25 4.5 0.050000000000000003 2 4.0789999999999997

Data Set for Fast ICA (Sample Size 127 × 15)

File: $NGC5128new$

c $muvh$ $mu0$ $W0$ Rc $Rtid$ Rh $logMtot$ $logrho_0$ $sigmap0$ $logtrh$ $RgcT1$ $[Fe/H]$ $(C-T1)0$

3.86 21.802 14.93 13.7 0.02588 186.21 5.1168 4.33 6.10 2.2387 8.87 10.45 22.663 −1.67 1.137 1.80 19.202 17.87 6.0 2.46604 173.78 5.7544 5.47 2.94 5.7148 9.41 3.53 20.450 −1.62 1.157 3.03 21.327 18.18 7.8 0.94406 1071.52 7.6033 4.88 3.26 3.1189 9.35 21.69 21.363 −2.20 0.955 1.52 22.277 21.08 5.3 6.23735 245.47 12.1619 5.11 1.44 2.5763 9.77 12.67 21.423 −0.27 1.820 2.70 20.652 18.12 7.4 0.91622 501.19 4.6989 4.90 3.42 3.6644 9.06 10.38 20.672

−0.44 1.707 1.67 20.727 19.49 5.7 2.30144 123.03 4.9204 4.74 2.33 2.6546
9.02 11.99 21.574 −1.30 1.281 1.89 19.952 18.48 6.2 2.11349 181.97 5.2723
5.17 2.81 4.2364 9.24 12.34 20.664 −0.91 1.454 2.24 21.077 19.25 6.8 1.84502
346.74 6.0256 4.81 2.54 2.6977 9.18 9.92 21.620 −1.08 1.376 1.89 19.302
17.89 6.2 2.00909 173.78 5.0119 5.31 3.02 5.0933 9.26 11.85 20.202 −1.65
1.146 1.49 18.952 17.85 5.2 1.53815 56.23 2.9309 4.99 3.15 4.5709 8.78 3.84
21.190 −1.84 1.078 0.99 18.927 18.12 3.1 2.18273 30.20 3.1405 5.06 2.88
4.8195 8.85 2.45 20.525 −1.70 1.128 1.89 19.752 18.30 6.2 2.74157 239.88
6.8391 5.45 2.76 5.1404 9.52 5.39 20.061 −1.05 1.389 1.17 19.552 18.65 4.0
3.09030 60.26 4.8865 5.22 2.54 4.5920 9.20 4.05 20.431 −1.18 1.330 1.89
20.252 18.80 6.2 1.25314 109.65 3.1261 4.56 2.88 2.7227 8.65 7.93 22.161
−2.27 0.931 1.71 19.227 17.97 5.8 1.68655 97.72 3.7068 5.09 3.07 4.5499 8.97
10.83 20.907 −1.99 1.025 2.11 19.827 18.12 6.6 1.27350 181.97 3.7497 4.98
3.23 4.1115 8.94 9.17 21.424 −0.63 1.595 2.05 20.102 18.56 6.5 1.65959
204.17 4.6666 4.95 2.86 3.5237 9.06 8.41 21.297 −2.58 0.836 1.32 20.202
19.25 4.6 3.06902 77.62 5.2602 5.01 2.31 3.4754 9.17 9.91 21.043 −2.47 0.871
2.70 19.202 16.72 7.4 0.60674 331.13 3.1117 4.99 4.05 5.0119 8.82 6.93
21.147 −1.00 1.400 1.67 19.577 18.28 5.7 1.86638 100.00 3.9902 5.16 3.02
4.7643 9.06 6.59 21.013 −0.53 1.653 1.05 19.902 19.04 3.4 2.84446 43.65
4.2170 5.00 2.46 3.8726 9.02 6.39 21.156 −0.84 1.484 1.59 21.802 20.57 5.5
3.89942 177.83 7.9250 4.88 1.80 2.4547 9.40 9.73 21.758 −0.44 1.705 0.01
20.527 19.86 0.1 3.96278 30.20 4.8084 4.85 1.98 3.1189 9.05 22.46 21.310
−0.91 1.451 1.94 20.027 18.67 6.3 2.55270 245.47 6.5917 5.27 2.66 4.2756
9.43 8.45 20.669 −4.77 0.270 0.16 21.677 20.97 0.2 6.42688 50.12 7.8343 4.89
1.38 2.5410 9.38 11.67 21.433 −0.60 1.611 0.93 20.527 19.78 2.8 2.10378
26.92 2.9512 4.38 2.26 2.2751 8.54 9.91 22.446 −2.49 0.863 1.89 21.527 20.08
6.2 1.72187 147.91 4.2954 4.34 2.25 1.8072 8.77 9.84 22.330 −1.00 1.400 1.71
21.277 19.91 5.8 3.97192 229.09 8.7297 5.27 2.13 3.6475 9.62 9.74 20.947
−0.14 1.914 1.71 20.602 19.28 5.8 2.99916 173.78 6.5917 5.14 2.37 3.6224
9.37 12.53 20.831 −0.73 1.540 2.79 21.577 18.92 7.5 2.24905 1479.11 12.7938
5.31 2.64 3.6559 9.87 3.29 20.711 −0.78 1.518 1.40 21.077 20.06 4.9 5.08159
154.88 9.1411 5.12 1.75 3.0061 9.57 2.68 20.945 −1.86 1.071 1.35 21.302
20.33 4.7 3.69828 100.00 6.4565 4.73 1.78 2.2803 9.19 3.87 21.959 −1.95
1.040 1.67 20.502 19.31 5.7 4.72063 251.19 10.0925 5.45 2.11 4.2073 9.77
5.17 20.207 −2.62 0.825 1.89 21.652 20.19 6.2 3.89045 338.84 9.7051 5.00
1.85 2.5763 9.56 7.85 21.358 −1.00 1.410 2.17 21.052 19.38 6.7 3.48337
562.34 10.7647 5.29 2.21 3.4834 9.75 5.04 20.909 −2.60 0.829 0.76 21.152
20.41 1.9 5.04661 52.48 6.6527 4.87 1.64 2.6792 9.26 4.97 20.720 −1.00 1.400
1.46 20.727 19.62 5.1 4.43609 151.36 8.2985 5.22 2.01 3.5318 9.55 8.64
20.670 −1.11 1.360 1.75 21.127 19.74 5.9 3.63915 234.42 8.2224 5.26 2.23
3.7239 9.58 9.35 21.057 −0.19 1.875 2.95 22.377 19.37 7.7 1.85353 1778.28
13.1826 4.99 2.52 2.6182 9.76 7.35 21.838 −0.91 1.451 1.75 21.327 20.02 5.9
3.04089 194.98 6.8865 4.79 1.99 2.3714 9.25 8.78 21.534 −1.34 1.264 1.80
21.802 20.41 6.0 4.92040 346.74 11.4815 5.17 1.74 2.8576 9.75 9.95 21.216
−0.58 1.622 1.80 20.852 19.43 6.0 2.50611 177.83 5.8345 5.06 2.51 3.5318

9.27 11.01 21.358 −0.22 1.854 1.13 20.402 19.47 3.8 3.40408 60.26 5.2602
5.08 2.29 3.7844 9.21 12.16 20.961 −0.48 1.679 2.70 20.652 18.21 7.4 2.49459
1348.96 12.8233 5.59 2.81 4.9317 9.99 9.36 20.039 −1.96 1.036 2.31 21.452
19.51 6.9 1.94089 426.58 6.7143 4.90 2.55 2.8708 9.30 11.72 21.837 −0.37
1.750 1.89 18.152 16.76 6.2 1.22744 107.15 3.0620 5.36 3.71 6.9024 8.96
11.93 20.049 −2.46 0.874 2.46 19.102 16.97 7.1 1.41579 436.52 5.6105 5.68
3.70 7.8524 9.50 4.26 19.818 −0.39 1.738 2.00 17.327 15.78 6.4 0.92683
102.33 2.4946 5.53 4.22 9.3756 8.89 3.82 19.561 −1.18 1.329 1.71 19.027
17.70 5.8 2.34423 138.04 5.1523 5.58 3.14 6.8391 9.40 4.12 19.857 −0.59
1.615 2.24 18.502 16.66 6.8 1.34276 257.04 4.3853 5.64 3.79 8.2794 9.32 7.16
19.668 −0.63 1.592 2.00 18.077 16.53 6.4 1.11944 123.03 3.0061 5.38 3.83
7.2111 8.96 7.56 19.991 −1.27 1.294 2.24 17.002 15.21 6.8 0.75683 144.54
2.4717 5.62 4.52 10.7399 8.93 8.96 19.403 −1.54 1.188 1.71 18.352 17.09 5.8
1.96789 114.82 4.3152 5.56 3.35 7.2946 9.27 12.30 19.507 −1.62 1.155 2.17
18.377 16.64 6.7 0.97499 158.49 3.0130 5.29 3.88 6.6222 8.92 10.46 20.244
−0.98 1.420 2.46 18.227 16.17 7.1 0.90782 281.84 3.6058 5.45 4.05 7.5509
9.10 10.92 19.851 −1.63 1.151 2.11 17.077 15.40 6.6 0.87902 125.89 2.5823
5.68 4.41 11.0917 8.98 3.37 19.300 −1.00 1.400 2.46 18.377 16.27 7.1 0.89743
281.84 3.5645 5.42 4.04 7.3451 9.08 3.08 19.974 −1.16 1.338 1.89 18.477
17.02 6.2 1.59221 138.04 3.9628 5.48 3.49 6.9663 9.18 3.25 19.830 −1.13
1.353 2.31 18.377 16.45 6.9 1.03276 229.09 3.5727 5.50 3.97 7.8524 9.12 4.45
19.940 −0.70 1.556 1.75 18.977 17.65 5.9 1.87499 120.23 4.2364 5.32 3.16
5.5847 9.15 2.98 20.055 −1.24 1.306 2.17 17.477 15.78 6.7 0.86099 141.25
2.6607 5.49 4.23 8.8308 8.92 6.85 19.680 −1.70 1.126 2.11 17.252 15.57 6.6
0.74645 104.71 2.1928 5.47 4.42 9.4624 8.79 2.24 19.780 −1.00 1.400 2.24
18.252 16.45 6.8 1.27644 239.88 4.1687 5.62 3.84 8.2604 9.27 2.97 19.580
−1.00 1.400 2.54 17.577 15.36 7.2 0.87700 323.59 3.7670 5.81 4.43 11.2980
9.28 3.38 19.210 −1.00 1.400 2.11 17.227 15.51 6.6 0.68077 95.50 2.0045 5.55
4.61 10.7895 8.77 4.44 19.929 −0.34 1.769 1.84 18.002 16.58 6.1 1.12980
87.10 2.7227 5.36 3.83 7.3282 8.88 3.33 19.990 −1.00 1.400 1.46 17.052 15.95
5.1 1.13240 38.90 2.1184 5.52 4.08 9.8401 8.79 3.79 19.600 −1.00 1.400 2.38
18.527 16.54 7.0 0.86298 223.87 3.1915 5.28 3.97 6.5013 8.95 7.22 20.410
−1.00 1.400 2.70 19.327 16.84 7.4 0.98401 537.03 5.0466 5.38 3.81 6.1518
9.29 6.97 20.824 −0.89 1.461 2.31 18.527 16.64 6.9 1.00925 223.87 3.4914
5.32 3.82 6.4121 9.03 7.25 20.154 −1.56 1.178 2.24 18.052 16.26 6.8 1.23027
234.42 4.0179 5.63 3.90 8.5507 9.25 12.62 19.384 −1.29 1.285 2.54 17.127
14.94 7.2 0.72444 269.15 3.1117 5.78 4.65 11.9674 9.14 4.00 19.007 −1.37
1.254 2.70 17.802 15.34 7.4 0.71450 389.05 3.6728 5.65 4.49 9.8628 9.20 4.58
19.416 −1.44 1.225 2.62 18.627 16.24 7.3 1.11944 501.19 5.2360 5.81 4.09
9.6828 9.50 4.00 19.479 −0.41 1.726 2.79 17.277 14.70 7.5 0.50119 331.13
2.8510 5.64 4.92 11.2980 9.03 8.17 19.452 −2.19 0.959 2.38 17.827 15.83 7.0
0.95719 251.19 3.5400 5.65 4.21 9.4842 9.18 6.79 19.436 −1.00 1.412 2.95
17.702 14.69 7.7 0.49204 467.74 3.4995 5.72 4.97 11.7220 9.20 7.61 19.436
−0.88 1.466 2.46 17.252 15.20 7.1 0.70632 218.78 2.8054 5.63 4.56 10.5196
9.01 8.32 19.384 −2.00 1.023 2.62 16.977 14.67 7.3 0.59156 263.03 2.7669

5.73 4.84 12.1619 9.04 19.68 19.120 −1.61 1.161 2.62 17.602 15.27 7.3
0.87700 389.05 4.0926 5.83 4.44 11.3240 9.34 9.64 18.896 −1.34 1.263 2.87
17.827 15.01 7.6 0.76208 602.56 4.8306 5.93 4.64 12.3880 9.49 2.67 19.040
−1.00 1.400 2.24 17.802 15.95 6.8 1.07895 204.17 3.5237 5.75 4.19 10.4472
9.22 12.56 19.334 −0.56 1.636 3.10 17.552 14.20 7.9 0.49091 645.65 4.4875
5.91 5.11 13.7404 9.44 9.71 18.814 −1.77 1.101 2.17 16.877 15.14 6.7 0.88716
144.54 2.7416 5.78 4.49 12.1899 9.06 15.02 18.971 −1.27 1.294 2.38 17.152
15.17 7.0 0.85507 223.87 3.1623 5.79 4.49 11.6950 9.16 12.35 18.952 −1.35
1.262 2.00 17.827 16.22 6.4 1.40605 154.88 3.7757 5.87 4.02 11.2460 9.32
11.66 19.132 −0.32 1.783 1.94 16.577 15.09 6.3 1.38676 134.90 3.5892 6.13
4.31 15.6315 9.39 10.47 18.068 −1.24 1.304 2.38 16.802 14.82 7.0 0.77983
204.17 2.8840 5.88 4.70 13.6773 9.14 2.75 18.900 −1.00 1.400 1.43 16.502
15.42 5.0 1.68267 53.70 3.0903 6.07 4.12 15.3109 9.26 2.93 18.230 −1.00
1.400 2.11 16.527 14.85 6.6 0.84918 120.23 2.5003 5.90 4.68 14.5211 9.06
9.41 18.828 −0.82 1.498 2.00 15.777 14.27 6.4 0.71945 79.43 1.9364 5.89 4.91
16.1436 8.88 9.50 18.693 −1.69 1.132 2.17 16.827 15.12 6.7 1.07399 173.78
3.3189 5.94 4.40 13.3045 9.25 22.51 18.553 −1.55 1.181 3.32 16.527 10.81 9.8
0.04677 100.00 2.0989 5.92 7.80 23.4963 8.96 5.46 19.031 −0.17 1.893 2.46
16.002 13.88 7.1 1.33660 416.87 5.3088 6.83 4.92 30.3389 9.95 2.10 16.510
−0.55 1.641 2.54 16.127 13.91 7.2 1.14815 426.58 4.9204 6.61 4.89 24.9459
9.80 4.95 16.891 −1.05 1.386 2.95 16.227 13.21 7.7 0.81470 776.25 5.8076
6.79 5.38 31.1889 9.99 10.87 16.681 −0.67 1.572 3.27 16.777 12.51 8.3
0.28840 562.34 4.3752 6.42 6.20 28.1190 9.65 12.47 17.854 −0.31 1.789 2.62
15.327 12.97 7.3 0.45394 204.17 2.1232 6.24 5.69 24.9459 9.09 11.95 17.962
−0.86 1.474 2.54 15.852 13.63 7.2 0.76208 281.84 3.2659 6.38 5.19 23.3346
9.43 1.85 17.640 −1.00 1.400 0.01 14.452 13.80 0.1 0.95499 7.41 1.1614 6.00
4.98 23.7684 8.59 3.56 17.814 −1.27 1.293 1.94 14.077 12.57 6.3 0.39902
38.90 1.0328 6.09 5.89 27.6058 8.56 2.75 17.820 −1.00 1.400 5.23 16.602 6.32
19.5 0.00049 83.18 1.9187 5.69 11.36 18.3231 8.80 10.51 19.301 −0.63 1.596
4.26 16.127 8.37 15.3 0.00378 67.61 1.5849 5.73 9.68 20.0909 8.69 5.93
19.412 −0.56 1.636 3.40 15.277 9.15 11.1 0.01795 45.71 1.3213 6.02 8.78
33.9625 8.70 3.84 18.741 −0.18 1.887 3.51 16.702 10.43 11.9 0.02432 79.43
2.3823 5.72 7.93 17.3780 8.95 10.72 19.140 −1.33 1.267 4.20 15.977 8.35 15.1
0.00649 104.71 2.4660 6.11 9.40 25.0035 9.14 6.60 18.186 −0.83 1.491 2.54
17.802 15.53 7.2 0.84140 316.23 3.6141 5.85 4.52 12.0226 9.28 9.73 19.262
−0.30 1.794 2.87 19.252 16.43 7.6 1.12202 891.25 7.1121 5.70 3.90 7.8163
9.65 7.91 19.710 −1.00 1.400 1.75 18.952 17.64 5.9 2.51768 162.18 5.6885
5.57 3.02 6.4121 9.45 12.11 19.671 −1.49 1.204 2.17 18.402 16.80 6.7 2.02302
331.13 6.2517 5.87 3.50 8.9536 9.64 8.87 19.098 −4.20 0.404 3.03 17.777
14.54 7.8 0.45604 512.86 3.6728 5.84 5.17 13.6144 9.29 7.28 19.261 −0.39
1.736 3.30 17.802 13.37 8.4 0.62806 1288.25 10.6660 6.76 5.49 27.1019 10.37
8.10 17.081 −0.41 1.723 2.00 16.777 15.25 6.4 1.95434 213.80 5.2602 6.37
4.09 16.9824 9.73 10.53 17.498 −1.42 1.234 2.87 16.577 13.77 7.6 1.21619
954.99 7.7090 6.80 4.90 26.8534 10.17 9.18 16.644 −1.21 1.317 3.27 17.602
13.34 8.3 0.56105 1096.48 8.4918 6.66 5.57 26.4850 10.18 11.26 17.358 −0.34

1.767 2.70 16.677 14.19 7.4 1.07399 575.44 5.5081 6.51 4.82 21.5774 9.83
18.31 17.407 −0.94 1.438 2.38 16.702 14.70 7.0 1.32739 346.74 4.9091 6.40
4.53 19.0108 9.71 7.59 17.554 −0.93 1.444 3.20 17.452 13.60 8.1 0.52481
891.25 6.2230 6.43 5.48 22.4388 9.88 8.49 17.965 −0.39 1.735 3.20 17.652
13.80 8.1 0.69984 1174.90 8.2794 6.58 5.26 23.2809 10.13 21.02 17.533 −0.44
1.707 2.54 16.027 13.83 7.2 0.73451 275.42 3.1550 6.22 5.08 19.9067 9.34
12.95 17.944 −1.61 1.161 2.87 16.852 14.00 7.6 0.47424 371.54 2.9992 6.00
5.33 17.1791 9.22 2.77 18.863 −0.53 1.649 2.24 17.427 15.53 6.8 0.66681
125.89 2.1727 5.65 4.71 11.8304 8.88 3.26 19.952 0.05 2.100 2.79 17.252 14.57
7.5 0.84140 549.54 4.7863 6.27 4.88 18.1134 9.64 5.27 18.173 −0.42 1.714
2.46 17.002 14.88 7.1 0.83560 257.04 3.3113 6.01 4.72 15.0314 9.29 8.86
18.727 −0.57 1.627 2.54 17.627 15.35 7.2 1.21619 446.68 5.2119 6.22 4.42
15.4525 9.68 4.35 18.300 −0.35 1.762 2.54 16.827 14.55 7.2 0.70146 263.03
3.0061 6.09 5.00 17.3780 9.26 3.19 18.599 −0.27 1.819 2.87 16.727 13.91 7.6
0.62661 501.19 3.9719 6.20 5.16 18.6209 9.48 5.04 18.032 −0.99 1.414

Data Set for K-Means Clustering Algorithm (Sample Size 1594 × 6)

File: *grb*2007

*logT*50 *logT*90 *logP*256 *LogFt* *LogH*32 *LogH*321

0.25334 0.71466 1.07397 −5.27852 0.19054 −0.02913 0.10721 0.49638
−0.35754 −5.81951 0.32892 0.13784 1.67306 1.95509 0.55871 −4.22996
0.38283 0.17708 2.48624 2.63348 −0.32057 −5.50934 0.58037 0.28732
1.68699 1.87511 0.17493 −4.76470 0.25791 −0.03144 1.52881 1.79206
0.54008 −4.41162 0.44038 0.22228 1.97878 2.26011 −0.18111 −5.24094
0.51085 0.21099 −0.66154 −0.50169 −0.14146 −6.86566 0.51999 0.38107
0.82737 1.70600 1.67732 −3.53472 0.77891 0.53498 2.62222 2.65707 0.14737
−5.67809 0.10422 −0.19273 0.92345 1.23106 0.47712 −5.31327 0.56201
0.35371 1.10938 1.39389 −0.18977 −5.73393 0.24417 −0.05049 0.50515
0.85926 0.67541 −5.95269 0.25873 −0.02558 1.30724 1.79696 −0.17134
−5.67838 0.98820 0.88566 −1.37675 −1.07058 0.01072 −6.39290 0.83754
0.66610 −0.22403 0.06930 −0.06148 −6.28575 0.54318 0.42291 0.79741
1.47270 1.25679 −4.48705 0.39712 0.16209 1.75606 1.86385 0.60097
−5.14715 0.42466 0.18139 1.78845 2.34475 0.11327 −4.40330 0.46796
0.24368 0.74068 1.09174 0.02284 −6.31785 0.15406 −0.16245 −0.59176
−0.34872 −0.17134 −6.86633 0.68486 0.54012 1.01301 1.42213 −0.25806
−5.52822 0.65513 0.42492 1.48012 1.60758 −0.08355 −6.03939 0.66237
0.40397 0.82321 1.45551 1.53934 −3.82324 0.54139 0.35607 −0.71670
−0.41567 0.12024 −5.89123 1.00000 0.75499 1.45454 1.90758 0.23325
−4.43441 0.51142 0.30186 1.33122 1.61439 −0.28483 −5.26408 0.23772
−0.05779 −1.52288 −1.17393 −0.29243 −7.01382 0.91457 0.85585 −0.49485
−0.01773 0.34044 −5.97005 0.67003 0.50903 −0.95468 −0.34199 1.34335

−5.20176 0.68711 0.49421 1.13862 1.70436 0.07482 −5.39556 0.33330
0.13590 1.71893 2.16715 0.08849 −4.45647 0.38102 0.09040 −0.11464
0.18639 −0.19111 −6.55912 1.26512 1.15087 1.56811 2.02576 0.67916
−4.14273 0.54873 0.27742 1.01569 1.40824 0.23401 −5.22548 0.41706
0.18906 0.45939 0.70381 −0.47756 −6.69897 0.37824 0.19270 1.47082
2.09129 0.00260 −5.28225 0.27669 0.04116 1.54969 1.93166 0.15685
−5.19098 0.21640 0.03123 0.72526 1.19179 0.00561 −6.22526 0.34511
−0.01806 −1.85387 −1.46852 0.12581 −6.14933 0.87287 0.69100 −0.89279
−0.59176 1.22003 −5.82295 0.35185 0.11387 0.48742 1.18093 1.22848
−4.67244 0.22206 −0.01699 0.94606 1.57630 0.88812 −5.17121 0.63947
0.45084 0.59857 1.18820 0.62552 −4.97102 0.62766 0.39883 1.08264 1.49282
−0.25649 −5.96908 0.11971 −0.14259 −1.14267 −0.81816 0.76208 −5.27774
1.06926 0.92764 −1.72125 −1.17393 0.28892 −5.93843 0.85469 0.61476
−1.40894 −1.00000 −0.41341 −5.92758 0.90845 0.76967 1.50166 1.66891
0.70372 −5.23915 0.65992 0.49239 0.54654 1.10285 −0.11070 −5.93889
−0.02500 −0.34851 0.39724 0.81902 −0.30627 −6.21722 −0.04948 −0.60159
0.08493 0.63225 0.16286 −5.68636 0.88930 0.67322 0.33766 0.68699 1.02350
−5.11227 0.34386 0.18436 −0.59176 −0.29073 0.43008 −6.25898 0.76441
0.66636 1.13659 1.55900 0.30016 −5.21155 0.37187 0.17529 0.99087 1.15836
−0.47625 −5.86918 0.30122 0.08854 −0.93930 −0.60206 0.56455 −5.99388
0.73670 0.58780 0.28330 0.61909 0.11594 −6.19978 0.84759 0.55207 1.30033
1.53372 −0.39362 −4.96889 0.29105 0.16659 0.93651 1.35272 0.27554
−5.17659 0.38845 0.17097 −1.23657 −0.92445 0.15534 −6.05688 1.04299
0.80378 −0.56864 −0.38405 0.43933 −6.37233 0.66375 0.56664 1.75994
2.25489 −0.10073 −5.52608 0.33821 0.08123 1.48652 1.74868 −0.07676
−5.49949 0.37071 0.09549 1.14264 1.42943 −0.05948 −5.30706 0.58899
0.41982 −0.07988 0.16791 0.54295 −6.04934 0.50859 0.37948 0.93328
1.31672 0.02202 −5.50432 0.43072 0.19469 1.22614 1.77513 0.84942
−4.68089 0.42848 0.28129 0.77931 1.31538 0.10755 −5.52223 0.23532
0.01286 1.52218 2.00894 0.01242 −5.11873 −0.01026 −0.33908 0.74068
1.30449 0.65811 −5.13371 0.51550 0.31544 1.11793 1.43864 −0.52433
−5.97384 0.15226 0.02590 1.26557 1.89111 0.62315 −4.37799 0.72472
0.49034 −1.52288 −1.25964 0.27254 −5.79218 0.93583 0.80634 1.31269
1.73149 0.79071 −3.60014 0.71816 0.47888 0.45939 1.03918 −0.08938
−6.09501 0.09973 −0.16159 1.19535 1.86041 −0.01954 −5.91934 −0.40822
−0.73771 0.69827 1.04673 −0.03152 −5.64847 0.51461 0.16794 0.32469
0.50515 −0.38722 −6.66136 0.25826 0.03924 0.76042 1.79607 0.11727
−5.36251 0.10213 −0.08961 1.35763 1.66170 0.13640 −5.34228 0.55229
0.35468 1.03918 1.57036 −0.42022 −6.26400 0.01029 −0.13387 −1.20066
−0.73755 −0.21681 −6.24153 1.07238 0.81451 −0.34872 0.25334 −0.04866
−6.05393 1.04634 0.95104 0.73046 1.09621 −0.48678 −5.85608 0.69682
0.31866 0.25334 0.73560 −0.08302 −6.14026 0.33653 0.09029 0.29754
0.81050 −0.03105 −6.13419 0.43654 0.20923 1.60345 1.91101 0.50623
−4.47496 0.45255 0.26673 1.31538 1.78527 −0.04624 −5.58443 0.31429
0.02938 −1.04096 −0.51856 0.13513 −6.84472 0.17858 −0.03738 1.32862

1.76665 0.04727 −5.44033 0.29217 0.02697 0.64503 0.85926 0.07188
−5.76120 0.46316 0.29561 −0.59176 0.10721 0.49651 −6.27564 0.55798
0.34188 1.01030 1.40497 −0.18442 −5.64108 0.08211 −0.23784 1.21442
1.50079 −0.25571 −5.32597 0.91452 0.67295 1.34776 1.87437 −0.38616
−5.45783 0.40920 0.09230 0.77931 1.23915 0.81218 −4.99883 0.15747
0.02063 −1.22185 −0.88273 0.34577 −6.18456 0.87540 0.66845 −0.59176
−0.49485 −0.24109 −7.06616 0.53022 0.45623 1.70873 1.88391 0.11494
−5.37782 0.76000 0.61525 0.76042 1.24075 0.78376 −5.22541 0.23213
0.00168 −0.04769 0.37438 −0.16368 −5.87154 0.77029 0.50655 −0.89279
−0.04769 −0.11520 −7.12989 0.10456 −0.10323 0.17173 0.41731 0.37218
−5.37647 0.70629 0.47384 1.28619 2.04322 0.54617 −4.72677 0.47423
0.32594 −0.59176 −0.41567 0.03941 −6.30587 1.10089 0.97552 −0.65170
−0.40012 0.38039 −6.23493 0.87613 0.70642 2.03304 2.19888 0.55255
−4.78875 0.37337 0.13086 −0.49485 −0.07988 −0.39254 −6.64168 0.83032
0.21942 0.33766 0.82321 0.40364 −5.70329 0.35325 0.06533 0.96152 1.44666
−0.22768 −6.20025 −0.18945 −0.44191 −0.04191 0.15776 0.58081 −5.83372
0.64896 0.41028 0.41896 0.69827 −0.54061 −6.07593 1.06805 0.72360
1.43149 1.68585 −0.21681 −5.78270 0.78423 0.46389 1.34651 1.95417
0.72329 −4.34960 0.59008 0.30998 0.23754 0.50515 −0.47237 −5.95546
0.43180 0.26627 −0.11464 0.82321 1.06254 −4.94623 0.56471 0.33826
2.44035 2.55227 −0.15802 −4.97597 0.28094 0.02365 −0.04769 0.41896
1.21147 −5.13472 0.32295 0.11998 1.56358 2.05830 −0.02549 −5.13001
0.18233 −0.12155 −0.89279 −0.49485 −0.23210 −7.11982 0.37048 0.13867
−0.71670 −0.29073 0.32531 −6.53880 0.51720 0.36217 −1.25964 −0.79317
0.43249 −5.95187 0.88396 0.71339 0.76042 1.28330 1.48222 −4.22959
0.34445 0.18979 1.34776 1.44066 −0.07109 −5.85496 0.28917 0.04187
−0.89279 −0.71670 0.72908 −5.51563 0.96848 0.77893 −1.06048 −0.10018
0.01410 −6.79309 0.44913 0.32975 −1.10237 −0.72816 0.21272 −6.51465
0.54575 0.39186 −0.62525 −0.35458 0.15746 −5.81296 0.65031 0.48298
0.44963 0.68124 0.40398 −5.84001 0.15824 −0.07344 0.53857 0.95847
−0.52724 −6.39794 0.27599 0.13974 0.78390 1.27305 1.06453 −4.60119
0.50387 0.25559 1.23754 1.50688 −0.31876 −5.91417 0.61458 0.35726
0.98227 1.30724 −0.09637 −5.65330 0.59327 0.37065 −0.29073 −0.01773
−0.22330 −6.84076 0.51915 0.35133 0.83960 1.30311 0.95453 −4.24003
0.39035 0.15449 0.20412 1.50775 0.24601 −6.27100 0.12396 −0.16799
1.17540 1.69323 0.16613 −4.94900 0.72931 0.56462 1.39501 1.78981 0.23249
−5.17224 0.25409 −0.06888 2.20706 2.43914 −0.54516 −5.47996 0.43387
0.15734 −0.98297 −0.71444 0.21299 −5.92000 0.90331 0.82472 1.80269
2.27659 −0.20831 −4.43176 0.49819 0.27980 0.96755 2.23188 1.00173
−4.65922 0.53855 0.27630 0.20412 1.26102 0.54357 −5.60836 0.62629
0.44314 0.25334 0.85540 0.53020 −5.67728 0.32103 0.01940 1.44166 1.62572
−0.21467 −5.89106 0.27477 0.02431 2.01435 2.21204 −0.25649 −5.68006
0.61548 0.45551 0.97350 1.37555 0.07408 −5.48382 0.17285 −0.08002
0.86308 1.44267 0.06781 −5.42228 0.19461 −0.06209 0.01030 0.56961
0.51904 −6.30945 0.36014 0.19501 −0.55909 −0.11748 0.22660 −6.68212

0.51584 0.42403 1.09398 1.30587 −0.36856 −5.98657 0.71956 0.55224
0.61236 0.97644 0.09517 −5.74800 0.27605 0.12282 0.33766 0.78390 0.00130
−6.18509 0.00102 −0.35286 −0.46725 −0.01144 −0.09909 −6.74698 0.19474
−0.25723 2.14424 2.27864 0.40381 −4.99166 0.39974 0.13058 1.35518
1.75800 −0.06702 −5.50533 0.12105 −0.28856 1.07569 1.57703 0.08063
−5.73608 0.27757 0.02903 2.00097 2.50323 0.81644 −4.44045 0.62213
0.48390 −0.58004 −0.16685 0.15746 −6.48879 0.50742 0.28157 0.83556
1.18458 0.27989 −5.33451 0.73324 0.56795 0.89960 1.32732 −0.08145
−4.84336 0.68278 0.41285 0.48742 0.86308 −0.26600 −6.11008 0.16450
−0.16859 0.33766 0.65744 0.15290 −6.06318 0.47247 0.35647 −0.97062
−0.58838 0.09552 −6.18302 0.78198 0.54037 0.50515 0.89254 −0.29243
−6.10106 0.35518 0.12767 1.49282 1.69154 0.11959 −6.31345 0.81291
0.46893 1.18458 1.71413 0.55835 −4.02521 0.63920 0.44739 0.86308 1.24866
−0.46344 −5.41696 0.27288 0.15084 1.17909 1.32205 0.22583 −5.58369
0.32646 0.11097 −0.15243 0.14860 −0.24949 −6.79838 0.73478 0.59072
0.89254 1.30033 0.29403 −4.99114 0.22461 −0.04028 0.18639 0.53857
0.17696 −6.37603 0.16277 −0.10490 1.02102 2.08812 0.65369 −4.66186
0.39129 0.14641 0.53046 1.01837 0.96497 −4.95253 0.21136 −0.10590
0.33766 0.92012 −0.27246 −5.97449 0.00959 −0.30103 0.43965 1.38710
1.06055 −4.82927 0.62060 0.41220 0.06145 0.39724 0.65244 −5.74618
0.44333 0.28013 0.96454 1.52551 −0.31605 −5.56431 0.52842 0.30598
1.04673 1.35641 0.24130 −5.06344 0.52916 0.36139 1.35518 1.52135 0.16673
−5.32166 0.30507 0.05892 0.79741 1.08493 −0.10679 −5.91311 0.41099
0.27541 −0.71670 −0.71670 0.86368 −5.92632 0.82416 0.65525 1.24551
1.68008 −0.27327 −5.55382 0.34355 0.00032 1.73509 1.91237 −0.24033
−5.68897 0.06924 −0.17656 −0.52143 −0.33630 −0.20343 −6.42209 0.80450
0.59807 0.86308 1.42839 0.35430 −5.00353 0.46905 0.19956 1.02102 1.70873
0.52375 −4.48192 0.87282 0.66387 0.78845 1.19712 0.16465 −5.29337
0.37552 0.17175 0.22115 0.61236 −0.27491 −5.99340 −0.18432 −0.60666
1.57036 1.84954 −0.07935 −5.24818 0.65447 0.34278 −0.41567 0.03663
0.50215 −5.87903 0.91055 0.59344 1.93134 1.99903 0.60184 −4.94931
0.44305 0.28296 1.02366 1.41896 1.55125 −3.85855 0.59839 0.42474 1.12424
1.33510 0.29885 −5.55177 1.05448 0.77921 1.08493 1.46894 −0.23508
−5.28877 0.89615 0.73626 1.27453 1.68413 −0.44249 −5.81953 0.13959
−0.13028 −0.29073 −0.01773 1.03177 −4.96182 0.82902 0.61342 1.03663
1.44066 −0.41117 −5.86214 0.41983 0.15452 1.10065 1.51033 −0.01728
−5.25916 0.57599 0.43092 −1.50864 −1.13077 −0.29843 −6.54130 0.70292
0.57489 0.72526 1.23431 −0.27084 −5.63937 0.27700 0.04154 1.76522
2.03304 −0.17198 −5.06976 0.32767 0.06321 0.80182 1.31269 0.57368
−4.90035 0.47773 0.31470 1.40278 1.86612 0.51162 −4.73587 0.83363
0.66066 1.18093 1.58937 0.08243 −5.32203 0.53383 0.35599 1.73458 1.95694
−0.04287 −5.32523 0.19552 0.01407 −0.71670 −0.41567 0.15655 −6.08561
0.70235 0.57037 0.94290 1.68585 0.33021 −4.68219 0.97665 0.72498 0.98516
1.45939 −0.27327 −5.59322 0.32786 0.07596 1.72735 1.94732 0.89326
−4.20754 0.47630 0.27875 0.16791 1.10285 1.76553 −4.32799 0.48939

0.30777 1.80182 2.33252 0.00604 −5.34180 0.37723 0.22105 1.39836 1.82028
0.47407 −4.48387 0.67884 0.45087 1.00208 1.20758 1.43605 −3.93509
0.66757 0.52571 1.36847 1.48196 0.41979 −5.87335 0.51611 0.26114 1.34901
1.56358 0.14799 −5.21638 0.34297 0.06954 −0.11464 0.14860 0.25600
−5.40188 1.14838 0.94403 −0.49485 −0.29073 0.06595 −6.77009 0.66686
0.57714 −0.11464 0.31133 0.13450 −6.27482 0.35135 0.19710 0.18639
0.67541 −0.24565 −6.44570 1.66584 0.90649 1.45258 1.91845 −0.29843
−5.63097 0.50502 0.37427 0.58433 0.93003 0.44592 −5.27075 0.46736
0.11146 1.37905 1.83067 0.61109 −4.60537 0.35069 0.15684 1.35149 1.79829
−0.15181 −5.62104 −0.05899 −0.39089 1.20758 1.82696 −0.34872 −5.51509
0.08043 −0.19568 0.69827 1.12424 −0.42713 −5.98577 0.53585 0.34064
−0.71670 −0.11464 −0.12784 −6.39292 0.90395 0.54281 1.47920 1.73149
0.22246 −5.35527 0.25265 −0.00869 1.12633 1.56433 −0.39469 −5.28205
0.49326 0.22507 −1.15490 −0.37779 −0.20204 −6.85406 0.42154 0.14986
1.02629 1.55592 1.27885 −3.76891 0.73486 0.56983 −1.19382 −0.71670
0.29994 −5.95479 0.89589 0.71393 1.48833 1.72263 1.02082 −4.45124
0.47655 0.28150 0.14860 0.65744 −0.32148 −5.78656 0.24971 −0.09002
0.20412 0.59151 0.86046 −5.02770 0.53722 0.33676 −0.77728 −0.41117
0.26788 −6.19382 1.01962 0.67192 0.73046 1.37555 0.00604 −5.70536
0.51371 0.28052 0.81902 0.98516 −0.26520 −5.61004 0.05825 −0.10969
0.20412 0.53857 1.10527 −4.97277 0.41961 0.21715 0.65128 0.77466 1.28838
−4.62093 0.72032 0.53708 1.81350 2.48805 0.49108 −4.58737 0.57978
0.40085 1.03407 1.37321 −0.47886 −5.81801 0.29277 −0.11765 0.52218
1.16602 0.47741 −5.31876 0.48234 0.26529 −0.65365 0.01995 0.44685
−5.72407 1.16030 0.91332 1.13862 1.49192 −0.22841 −5.70336 0.41964
0.17980 0.89960 1.19888 0.35755 −4.92300 0.62882 0.47172 0.87064 1.56660
0.47770 −4.50475 0.72339 0.52634 1.50947 1.80357 −0.08407 −5.58107
0.63950 0.46610 −1.16115 −0.23807 0.07151 −6.52181 0.67956 0.59043
1.02629 1.71734 −0.10403 −5.52418 0.30312 0.01184 0.38435 0.46776
−0.02919 −6.21977 1.38458 0.80113 −0.49485 −0.23958 0.56644 −5.92398
0.68742 0.51454 1.19712 1.49460 −0.28735 −5.81197 0.31772 0.07869
1.03407 1.29614 0.60660 −4.95257 0.47256 0.28111 1.25334 1.72421 0.30211
−5.41360 0.83027 0.69426 2.11175 2.19053 −0.17718 −4.97400 0.60267
0.43651 −0.89279 0.06145 0.25066 −5.92846 0.56859 0.48358 0.55437
1.23915 0.71600 −5.03796 0.55949 0.32688 0.56205 2.44046 1.21397
−4.00349 0.80204 0.67806 0.26858 1.20930 0.47363 −5.63873 0.40101
0.19268 0.80182 1.42423 0.06707 −5.25204 0.47314 0.21872 −0.41567
0.44963 0.26316 −6.12263 0.65091 0.36314 0.61236 1.14662 0.41061
−5.26201 0.38931 0.14130 −0.59176 −0.15243 −0.05849 −6.58299 0.93379
0.55521 0.60552 0.85926 0.74749 −5.12349 −0.17418 −0.45237 1.40606
2.04070 0.22634 −4.57836 0.61694 0.47860 2.13537 2.43528 0.43600
−4.13656 0.38350 0.15349 1.31940 1.48379 0.22866 −5.60586 −0.05753
−0.39349 1.23106 1.77977 1.21344 −4.61013 0.42820 0.15925 −0.89963
−0.28233 0.14520 −6.61629 1.12008 0.77947 1.08034 1.33638 0.16167
−5.11160 0.42464 0.20866 0.22115 0.70381 0.07555 −6.25626 0.31078

0.05607 0.55437 0.79741 0.87105 −4.89073 0.74027 0.54210 0.14860 0.73560
0.38148 −5.23106 0.81136 0.69169 −0.71670 −0.29073 −0.12552 −6.99667
0.34633 0.17502 0.12840 0.80618 −0.21467 −5.45154 0.97752 0.67326
1.08721 1.61372 0.32531 −5.62169 0.38004 0.11098 −0.89279 −0.89279
0.09272 −6.43116 0.76483 0.61659 −0.04769 1.18639 0.57646 −5.72157
0.48633 0.26938 0.08493 0.25334 −0.16431 −5.93926 0.17137 0.01812
1.86423 2.24218 0.34104 −4.23717 0.69793 0.53800 0.83960 1.48923 1.25765
−4.04556 0.66041 0.50158 −1.03621 −0.22841 0.89708 −6.12854 0.58582
0.40782 1.25797 1.92212 0.11528 −4.94723 0.47929 0.26162 0.86688 1.34021
−0.02733 −5.77940 0.45026 0.22582 1.35641 1.73046 0.75136 −4.48466
0.50013 0.31907 0.61236 1.19535 0.06070 −5.26180 0.79682 0.61011 0.87806
1.18093 1.65720 −4.07526 0.52909 0.32323 0.43965 1.00484 0.32325
−5.50264 0.54695 0.31204 0.93651 1.58143 1.00668 −4.79929 0.51061
0.27653 1.30172 1.76427 −0.00833 −5.47010 0.21209 −0.04075 0.10721
0.40824 −0.07988 −6.07894 0.67145 0.60797 1.15448 1.48012 −0.31785
−5.27344 0.50917 0.28498 0.55437 1.06145 0.20790 −5.60801 0.47929
0.18768 −0.59176 0.01030 0.09307 −6.26528 0.82446 0.57165 0.89609
1.41577 0.19368 −4.90025 0.61977 0.42988 0.87437 1.27305 0.66238
−4.39335 0.70907 0.55855 1.17725 1.46324 0.08707 −5.34056 0.24575
0.08499 1.91542 2.31833 −0.24109 −4.68905 0.08561 −0.28826 0.42943
0.85540 −0.10679 −6.13283 0.87687 0.70260 −0.55129 −0.29414 0.04100
−6.69323 0.56213 0.31636 1.57259 2.08149 0.27669 −4.37201 0.48712
0.25598 0.93003 1.34147 0.32695 −5.19627 0.25072 0.00153 −0.78252
−0.39686 0.36511 −6.40882 0.81709 0.60772 1.07802 1.88789 0.84535
−4.39244 0.53012 0.27379 1.06625 1.34776 −0.10958 −5.36623 0.54205
0.35692 1.69154 1.90378 0.14426 −4.85571 0.14199 −0.16477 −0.37882
−0.14997 0.13322 −6.03872 0.95683 0.70390 −0.89279 −0.89279 0.29754
−6.17144 0.73068 0.59039 1.25950 1.98544 −0.13253 −5.59825 0.45874
0.18558 0.90309 1.35272 −0.37986 −6.05943 0.40514 0.08662 1.70764
1.87139 −0.07469 −5.34113 0.38053 0.10004 0.94606 1.31269 −0.27984
−5.74084 0.34693 0.21704 1.60206 2.13313 0.26031 −4.97593 0.42500
0.11438 0.86688 1.40278 −0.08040 −5.33734 0.57329 0.30444 1.62638
2.13087 0.18921 −4.56575 0.75473 0.63912 1.08493 1.60136 0.01703
−5.38437 0.36303 0.17244 0.51375 1.17909 0.13481 −5.65886 0.43500
0.15467 1.04922 1.36248 −0.14267 −5.99650 0.33224 0.06621 1.83271
1.92544 0.14520 −4.82597 0.63329 0.40705 0.80618 1.75214 0.77945
−4.91775 0.53717 0.30760 0.80618 1.15253 −0.49894 −5.62268 0.49070
0.22375 1.06145 1.29332 0.33766 −5.59312 0.46995 0.24793 0.28330 0.73560
0.21431 −5.87322 0.52836 0.31910 −0.59176 −0.34104 0.04571 −6.81556
0.66956 0.58734 0.74570 1.14860 −0.01592 −5.39556 0.49721 0.31240
1.76163 2.07604 −0.06854 −4.54165 0.05341 −0.31159 1.56129 1.84201
0.06670 −5.35793 0.59280 0.37251 1.50079 1.84520 −0.26922 −5.33096
0.47829 0.21618 1.25256 1.73483 −0.11520 −5.52331 0.40958 0.15935
0.59857 1.32469 0.00346 −5.47623 0.49165 0.40340 1.10938 1.31133 0.14082
−5.20107 0.78416 0.61498 1.16507 1.42995 −0.28150 −5.81759 0.42240

0.07612 0.95075 1.42160 −0.12552 −5.52398 0.33570 0.02987 0.97937
1.28619 0.22037 −5.16456 0.49584 0.27836 1.05903 1.46514 0.58297
−4.26696 0.48733 0.21808 −0.34872 −0.04769 0.76050 −6.10182 0.27927
0.08964 0.50515 0.92345 0.87737 −5.01950 0.54374 0.31788 1.43864 1.96710
−0.08355 −5.05904 0.32713 0.16643 1.24393 1.73227 0.19479 −5.08970
0.25899 0.00217 1.06386 1.35149 0.20575 −5.38775 0.34816 0.15062 1.13456
1.49859 −0.34199 −5.08063 0.50522 0.22679 1.72211 1.94919 0.39550
−5.25657 0.44755 0.25307 1.57777 1.91962 0.46642 −5.26352 0.27296
0.02729 0.60552 1.08493 −0.26520 −5.99371 0.24658 0.08914 −0.19382
−0.01773 −0.20135 −6.84448 0.49693 0.30882 −0.38934 0.00346 0.11561
−6.38998 0.75877 0.61752 −0.63451 −0.37263 0.30471 −6.45780 0.63023
0.46395 −0.23958 0.18639 −0.00261 −6.31363 0.62366 0.46833 0.80182
1.27453 0.61045 −5.89719 0.41819 0.16965 0.81050 1.25797 0.67117
−4.73353 0.44678 0.26429 1.14464 1.72026 −0.29757 −5.83556 0.29753
0.08358 1.55281 1.83149 0.04297 −5.20964 0.81564 0.54794 1.64941 1.91373
−0.01682 −5.22651 0.65023 0.45412 1.35025 1.70709 0.48615 −4.47940
0.73798 0.53996 1.11368 1.62308 0.58670 −4.65423 0.45312 0.28431 0.77931
1.14364 −0.45842 −5.38978 0.41124 0.24136 0.86688 1.31133 −0.22841
−5.87788 0.22288 −0.12150 1.78023 2.14633 0.04766 −5.16374 0.41667
0.15784 0.16791 0.50515 −0.17328 −6.03100 0.56637 0.24943 1.39947
1.80627 −0.04576 −5.70498 0.62580 0.40792 1.52051 1.88227 −0.25964
−5.51499 0.62786 0.52916 0.66351 1.06145 −0.28483 −6.06178 0.24930
0.02390 1.12633 2.16115 0.31218 −4.84934 0.57379 0.41456 1.31404 1.93571
0.14270 −4.76135 0.40649 0.23320 1.04673 1.44864 −0.15864 −5.48492
0.37446 0.05367 0.65128 1.13456 0.10003 −5.38796 0.43507 0.13954 2.00180
2.29367 0.14176 −5.01345 0.34686 0.12426 1.39051 1.52218 0.78362
−4.71023 0.67669 0.46436 0.61909 1.20151 −0.31426 −6.03133 0.64861
0.39621 1.40002 1.76641 −0.41567 −5.27034 0.43444 0.09429 0.75066
1.46084 0.32118 −4.84248 0.46841 0.29211 0.67541 0.91677 0.58320
−5.04479 0.58343 0.43034 1.51967 1.80813 −0.25806 −5.46353 −0.03439
−0.25818 1.01569 1.36127 −0.00305 −5.37976 0.19258 −0.08104 0.34400
0.69688 0.42830 −5.81043 0.25163 −0.13109 1.23915 1.62009 −0.41341
−5.76080 0.52893 0.35224 1.15836 1.50079 0.24403 −5.02632 0.15467
−0.12716 1.07569 1.55125 0.12057 −5.37789 0.22459 0.01827 1.31538
1.58070 −0.15802 −5.35566 0.71211 0.55676 1.59293 1.85072 0.09342
−4.98670 0.36466 0.13232 0.70381 1.19179 −0.09583 −5.38122 0.39316
0.18665 1.79829 1.95540 0.18441 −5.21254 0.29127 0.06367 1.75361 1.91945
0.50893 −4.66454 0.59369 0.36650 1.91169 2.28977 0.26951 −4.70785
0.29838 0.04397 1.55398 1.95122 0.07954 −5.24390 0.22012 −0.05779
0.78390 1.25022 −0.02919 −5.19132 0.76157 0.52938 0.94606 1.37438
0.01115 −5.79218 0.30956 −0.01753 0.79295 1.14063 0.40756 −4.99991
0.60531 0.44568 1.70981 2.01850 0.21985 −5.08650 0.44817 0.15596 0.89960
1.10938 0.93661 −4.83803 0.62634 0.46251 1.49237 1.74794 −0.13966
−5.77459 0.23128 −0.11456 0.96454 1.47550 0.19368 −5.36603 0.22522
−0.08271 0.97350 1.39277 0.50406 −5.02678 0.13915 −0.10389 1.67424

2.12933 0.37621 −4.06819 0.55149 0.34959 1.48923 1.68756 0.11860
−5.20051 0.62381 0.36104 1.55976 2.18203 −0.13312 −4.69124 0.28769
0.04714 0.25334 0.53046 0.34025 −5.87582 0.74263 0.60296 0.23754 0.76997
0.85169 −5.36472 0.49937 0.32938 1.08948 1.76831 0.13799 −5.05700
0.28884 0.03013 0.42943 1.41471 0.04336 −5.45232 0.30040 −0.00993
0.80182 1.20758 0.17955 −5.39019 0.52481 0.29925 1.09621 1.54615
−0.26600 −5.34294 0.37987 0.13649 1.01837 1.48561 0.29798 −5.20754
0.19506 −0.07532 1.06386 1.55281 0.03463 −5.50059 0.49136 0.28626
1.06625 1.42736 −0.22768 −4.95048 0.93289 0.75627 1.15253 1.52343
−0.23508 −5.21169 0.35353 0.14174 1.23754 1.65866 0.06108 −5.12016
0.28300 0.03453 0.67541 1.13659 0.09726 −5.38923 0.38029 0.05604 1.51375
1.76283 0.60821 −4.64056 0.49887 0.27184 1.35395 1.76090 0.46075
−4.62237 0.58307 0.34843 0.90998 1.33766 0.28556 −5.19682 0.50562
0.31269 0.95231 1.26708 0.29842 −5.12610 0.57015 0.39330 1.94227 2.09644
−0.04287 −5.26616 −0.07731 −0.46852 0.89609 1.54258 0.09447 −6.19736
0.52612 0.28557 0.51375 1.05415 −0.28904 −5.67001 0.49825 0.41127
1.67365 1.80813 −0.34582 −5.18827 0.50178 0.12829 0.90655 1.31269
−0.38091 −5.78662 0.45158 0.22798 0.93651 1.33766 −0.00043 −5.52014
0.55612 0.31156 1.42943 2.25789 −0.17070 −4.21456 0.56083 0.42454
1.05169 1.74144 −0.26281 −5.23018 0.44011 0.20960 1.12215 1.50035
−0.34679 −5.38927 0.19194 −0.07345 1.18276 1.72942 0.20222 −4.83221
0.84510 0.70821 1.47457 1.65560 0.54045 −4.65738 0.50758 0.27664 1.15057
1.48652 1.02189 −4.59312 0.53039 0.34024 2.02129 2.11973 −0.01547
−4.75962 −0.23374 −0.52617 0.82737 1.04673 0.18213 −5.42992 0.72809
0.51909 1.09621 1.52218 0.82360 −4.29090 0.63233 0.48286 0.70381 1.14563
−0.06753 −5.49174 0.44448 0.35287 0.03342 0.16524 0.18724 −5.88636
0.76136 0.65656 1.20758 1.62965 0.26293 −4.65868 0.56675 0.32862 0.97644
1.49058 −0.16431 −5.32938 0.51822 0.34242 1.37905 1.71252 0.36586
−4.31646 0.72095 0.50946 0.82737 1.17540 −0.10568 −5.22214 0.30837
−0.08600 0.75066 1.00621 0.01242 −5.51235 0.36838 0.14934 1.73865
2.20620 0.45500 −4.78545 0.33113 0.12605 0.87806 1.23431 0.11361
−5.74899 0.41118 0.17343 1.14264 1.66742 −0.10958 −5.42806 0.48892
0.27578 0.45939 0.71466 0.14706 −5.52856 0.64047 0.43840 1.68813 2.00525
−0.15490 −4.98640 0.29175 −0.03664 1.70873 1.94763 −0.24949 −4.93052
0.67332 0.44480 1.07100 1.51630 0.66408 −4.46763 0.70643 0.50711 1.07335
1.22614 −0.28483 −6.09794 0.60955 0.43346 1.34526 1.70627 −0.37986
−5.30246 0.05142 −0.31812 0.98802 1.51925 0.08565 −5.53062 0.12936
−0.11107 1.09621 1.47270 0.62242 −5.21098 0.50678 0.28667 1.03149
1.57556 0.45117 −4.25438 0.65666 0.42718 0.95231 1.42943 −0.04721
−5.36522 0.44052 0.17334 1.36609 1.69604 0.22660 −4.99887 0.80677
0.64323 1.04922 1.42736 0.24428 −5.23151 0.53058 0.31650 1.48287 1.87825
−0.03905 −5.10807 0.45707 0.24047 0.58433 1.04171 0.01870 −5.66312
0.35194 0.08351 1.30033 1.65066 −0.14026 −5.05159 0.82119 0.71568
0.90655 1.11793 0.36530 −5.32716 0.34895 0.23058 1.04673 1.41471 0.23019
−5.50197 0.41001 0.15557 0.99651 1.56015 0.38256 −4.85393 0.36828

0.10006 0.87437 1.31605 0.10653 −5.14997 0.49390 0.29551 1.42213 1.85326
−0.02965 −4.38405 0.38131 0.14312 0.59857 0.93972 0.02776 −5.67627
0.43912 0.16956 1.62835 1.92511 0.14333 −4.98661 0.32394 0.06070 1.47457
1.68671 −0.01011 −5.04661 0.66408 0.23880 0.69267 0.98227 0.34753
−5.29065 0.48470 0.22253 0.61909 1.05169 0.12516 −5.43214 0.31961
0.00720 0.67541 0.97644 −0.06702 −5.68039 0.23935 −0.03944 0.92345
1.57259 0.34596 −4.96505 0.48796 0.27367 1.17540 1.56886 −0.02000
−5.03096 0.58081 0.38500 1.13252 1.55125 −0.12090 −5.12680 0.22268
−0.12812 0.98516 1.27527 0.05994 −5.71280 0.40672 0.11402 0.97644
1.68441 0.16702 −5.19702 −0.19002 −0.49240 0.66950 1.12943 −0.21254
−5.73542 0.14442 −0.21445 0.92675 1.51967 0.30771 −5.33583 0.63604
0.46982 1.16602 1.50558 −0.12668 −5.22352 0.08355 −0.22022 0.81902
1.25720 −0.29499 −5.60794 0.56544 0.18987 1.53291 1.75752 −0.21254
−4.75863 0.69075 0.54684 0.98516 1.30380 −0.28819 −5.11114 0.14800
−0.07520 0.33766 0.82530 0.43185 −5.62599 0.44825 0.13788 1.22115
1.80661 −0.18509 −5.39534 0.48050 0.22310 1.07218 1.92129 0.30535
−4.98318 0.38992 0.16637 1.22115 1.65159 −0.00261 −5.21875 0.49955
0.15767 1.41949 2.00194 −0.15802 −4.80022 0.51220 0.27183 0.78390
1.31672 −0.11070 −6.00113 0.01277 −0.22786 0.82737 1.66502 0.30384
−5.32892 0.35406 0.03593 0.80618 1.32073 0.20058 −5.78566 0.24002
−0.08555 1.14464 1.73611 0.05690 −5.28433 0.34988 −0.06008 1.67069
2.30256 0.16077 −4.44454 0.65723 0.47956 1.35395 1.80357 0.48615
−4.83538 0.56872 0.34155 1.18093 1.68528 0.31450 −4.85608 0.55680
0.34064 1.26102 1.73509 −0.34582 −5.63358 0.43485 0.23259 0.79071
1.30518 0.10278 −5.55624 0.31134 0.20807 0.96152 1.66109 0.41061
−5.48879 0.37461 0.13664 1.08948 1.55746 0.44824 −4.52765 0.66526
0.45257 1.42160 1.63323 0.23019 −5.42800 0.14006 −0.21195 1.27305
1.62965 0.05767 −5.14249 0.32055 0.14164 0.89960 1.36006 −0.24795
−5.91017 0.70066 0.55079 1.05538 1.42684 −0.11295 −5.94218 0.48076
0.34877 1.31739 1.71332 −0.24185 −5.12956 0.55908 0.26755 1.29684
1.78686 −0.13847 −5.23657 0.46592 0.12980 0.68124 1.11581 0.03463
−5.46889 0.51561 0.22727 0.38596 0.77466 0.32797 −5.79442 0.26208
−0.02495 1.00208 1.39836 −0.03198 −5.53639 0.25128 0.02702 1.11687
1.59822 −0.08197 −4.93401 0.28783 0.01264 0.77232 1.58469 −0.15802
−5.87572 0.43744 0.25169 −0.05552 0.35870 0.10653 −6.21161 0.59333
0.38597 0.65973 0.98802 −0.12378 −5.52946 0.89357 0.46983 1.27156
1.82861 0.18611 −4.75412 0.81832 0.56934 1.63805 1.96076 0.18441
−5.16775 0.38130 0.17990 1.11474 1.57296 0.11826 −5.71783 0.41983
0.03948 0.97644 1.60206 0.55121 −4.73356 0.49951 0.16082 0.89254 1.23188
−0.24949 −5.33089 0.53383 0.36516 0.61236 1.20238 0.12090 −5.55898
0.30667 0.03954 0.56205 1.42160 0.11793 −5.87719 0.36204 0.10406 1.04297
1.24155 −0.07624 −5.69680 0.45134 −0.11854 0.73560 1.17447 0.03463
−5.47873 0.39638 0.28286 0.96755 1.47457 0.68502 −4.91055 0.53582
0.32656 1.69380 1.94873 −0.09745 −4.97515 0.40295 0.05464 0.89960
1.42839 −0.13847 −6.05384 0.31300 0.05877 0.56205 1.13252 0.47813

−5.55878 0.49790 0.21703 0.96152 1.36668 −0.17457 −5.63077 0.05410
−0.43932 0.22115 0.78390 0.63959 −5.55658 0.48033 0.26725 0.70927
1.22778 0.23019 −5.52915 0.52288 0.17776 0.62572 1.12005 0.43088
−5.38563 0.51892 0.25700 0.77931 1.42318 0.63114 −4.99478 0.44397
0.19733 0.58433 1.07452 0.45939 −5.11907 0.57103 0.37935 0.72526 1.28834
−0.05453 −5.59556 0.32414 0.01454 0.54654 1.10175 −0.25181 −6.61780
0.33622 0.08674 1.38710 1.92610 0.34459 −5.27860 0.38836 0.07662
−0.38091 −0.26440 0.44886 −5.98093 0.84479 0.71904 1.48923 2.12142
0.15381 −5.32258 0.40707 0.21303 1.73534 2.03110 −0.41567 −4.93935
−0.54501 −1.11478 0.47828 1.05415 −0.17198 −6.34256 0.34121 −0.02307
1.09286 1.48787 0.04336 −5.76165 0.48386 0.29284 0.96755 1.47828
−0.08884 −5.70890 0.34992 0.17143 0.45939 0.75557 −0.08197 −5.95762
0.66348 0.46748 0.28330 0.83149 0.36568 −5.86025 0.07629 −0.35646
0.20656 0.62003 −0.22548 −7.03488 0.30718 0.08428 1.08493 1.70353
0.36773 −4.88074 0.50578 0.33171 0.43965 0.93651 0.00087 −5.89688
0.40477 0.15141 0.90998 1.64440 0.04336 −5.23890 0.35817 0.13093 0.23754
0.92345 0.44747 −5.96605 0.05666 −0.29586 0.68124 1.48968 0.17840
−5.61973 0.24667 −0.02691 1.02366 1.09621 0.32428 −5.71366 0.43144
0.14838 0.71466 1.08721 0.02284 −5.88048 0.17764 −0.03796 0.22115
0.59151 0.22167 −5.73938 0.31605 −0.19707 1.10503 1.55204 0.56773
−4.73967 0.42281 0.23601 1.12005 1.64723 −0.33068 −5.46693 0.27404
−0.07431 1.09174 1.63869 −0.37366 −5.83806 0.25316 −0.03103 1.25022
1.82238 −0.13253 −5.54645 0.55434 0.18395 0.71466 1.38482 0.64246
−4.92919 0.49259 0.28308 0.84361 1.31672 0.36586 −5.21396 0.30864
0.03365 0.72787 1.14364 0.00732 −5.90115 0.52822 0.30192 0.87064 1.26255
0.44295 −5.56257 0.22645 −0.15737 0.87064 1.41577 0.60606 −5.05670
0.42770 0.12850 1.54891 1.90896 0.03383 −4.89333 0.45288 0.30129 1.32732
1.73585 −0.49080 −6.07942 0.39246 0.17583 1.50688 2.06242 0.05538
−5.00485 0.35106 0.02145 1.31940 2.04322 0.30038 −4.66637 0.54368
0.29579 0.53046 1.01970 0.04727 −5.72086 0.22253 −0.25252 0.85150
0.99791 −0.06298 −5.55225 0.69954 0.67139 0.88173 1.43508 −0.30627
−5.14803 0.45871 0.21273 0.18639 0.45939 0.77034 −5.50128 0.54883
0.40764 1.21442 1.62933 −0.05948 −5.34046 0.13108 −0.13378 0.81050
1.30172 0.35641 −5.33106 0.11926 −0.14932 1.02102 1.40606 0.17984
−5.24887 0.40362 0.03938 0.44963 0.78390 0.29048 −5.55705 0.62627
0.39029 0.51375 0.99930 0.46150 −5.38802 0.42393 0.15023 0.13988 0.47188
0.55847 −5.54837 0.59850 0.33993 0.84954 1.30242 0.11860 −5.57988
0.46884 0.21935 1.04171 1.49593 −0.30627 −5.37427 0.47288 0.25995
1.37905 1.49947 −0.03716 −5.41795 −0.09588 −0.42317 0.66351 1.05415
0.10517 −5.32075 0.79490 0.59078 1.07569 1.63548 0.13354 −5.43711
0.47169 0.28598 1.44864 1.88063 −0.14691 −5.22856 0.52006 0.35629
1.55514 1.84757 −0.07160 −4.95082 0.52309 0.25518 0.71466 1.43355
−0.10182 −5.83262 0.15989 −0.21112 1.62900 1.99497 0.14333 −4.98376
0.23117 −0.12203 0.90655 1.41256 0.14737 −5.44965 0.36956 0.12794
1.12215 1.36609 0.11760 −5.46597 0.41564 0.14049 0.93328 1.35763 0.53517

−5.45300 0.25748 −0.03639 1.30929 1.71225 −0.33913 −5.12761 0.43924
0.08286 0.60552 1.12424 −0.11238 −6.24314 0.28532 −0.04183 1.21611
1.88063 0.45133 −5.21617 0.39245 0.14762 0.82737 1.54456 −0.03621
−5.49801 0.54147 0.17432 0.70927 1.38938 0.63377 −5.07109 0.47569
0.24705 1.22778 1.55476 0.00860 −5.51230 0.33749 −0.02199 0.31133
0.59151 0.45849 −5.90686 0.45289 0.25834 −0.11464 0.14860 0.03663
−6.30697 0.79454 0.69992 1.37203 1.81669 −0.11862 −4.90682 0.48429
0.19227 1.03149 1.46036 0.41497 −5.08019 0.69126 0.46491 0.13862 0.47276
0.14644 −5.62501 0.89558 0.69575 −0.01773 0.17725 0.32181 −6.15565
0.73603 0.63646 1.05660 1.74144 0.29248 −5.39664 0.38200 0.04698 0.76522
1.26557 −0.04144 −5.56719 0.14804 −0.31859 1.12215 1.34021 0.40654
−5.33245 0.32012 0.01884 1.12005 1.81007 0.42797 −5.19491 0.51050
0.26727 0.99087 2.17167 0.00260 −5.18138 0.63344 0.44758 1.42684 1.85092
−0.09259 −5.19111 0.30480 0.03672 0.51375 1.41949 0.32715 −5.47314
0.42176 0.13597 1.56320 2.07300 −0.16877 −5.49363 0.15805 −0.24993
0.56538 0.88536 0.40909 −6.10375 0.06107 −0.23888 1.38596 1.69660
0.00689 −5.36653 0.36054 0.14573 1.05660 1.44217 −0.24489 −5.86485
0.39198 0.27555 0.42943 1.16029 0.24748 −5.72241 0.45986 0.22651 0.83960
1.71520 0.37162 −5.13018 0.46973 0.18960 0.56961 1.25256 0.56407
−5.16222 0.37377 0.09977 0.87806 1.53046 0.22220 −5.35369 0.33389
−0.00773 0.85926 1.37203 −0.20066 −5.41183 0.63517 0.50620 1.62835
1.88518 0.02938 −5.30257 0.11437 −0.36120 0.76997 1.42213 0.21906
−5.43368 0.32825 0.09573 0.38596 1.09844 0.34811 −5.53340 0.60260
0.40777 1.26406 1.89925 0.09899 −4.88446 0.59579 0.38228 0.63225 1.22365
−0.26122 −5.97482 0.87988 0.64164 1.30861 1.60032 −0.05306 −4.87877
0.61008 0.31548 1.04922 1.39501 −0.20343 −5.81741 0.13637 −0.22507
0.12840 0.90655 0.38881 −6.03138 0.51941 0.30183 1.45843 2.07849 0.07041
−5.15527 0.45690 0.13604 0.68699 1.10721 −0.19791 −6.02050 0.47704
0.22176 0.93972 1.49371 0.37310 −5.01968 0.52526 0.23528 1.31538 1.71574
0.05115 −4.92053 0.63859 0.38198 1.05169 1.96964 0.37658 −4.48062
0.46187 0.23548 0.62572 0.96755 0.11361 −5.65679 0.62130 0.40066 1.48470
1.86423 −0.47108 −4.92416 0.81156 0.59172 1.10721 1.50341 0.27416
−4.70115 0.71486 0.47865 0.73560 1.34526 −0.02182 −5.35421 0.66477
0.36560 0.99930 1.34776 −0.29499 −5.66961 0.34907 −0.04202 1.19712
1.69380 −0.11014 −4.66158 0.77779 0.49876 1.04673 1.49549 −0.08619
−5.68647 0.58529 0.24733 0.69267 1.12005 0.20656 −5.37064 0.43382
0.13405 0.51375 1.21611 −0.02136 −5.82871 0.31267 −0.00774 1.09398
1.61236 −0.02965 −5.59007 0.46069 0.13762 0.81902 1.52135 0.14176
−5.31849 0.30855 0.01639 1.42108 2.18184 0.05994 −4.70360 0.52490
0.19011 1.32601 1.70271 −0.05404 −5.53077 0.41011 0.04992 0.50515
0.99087 0.24748 −5.74211 0.49125 0.26048 1.40824 1.60136 −0.12552
−5.63367 0.11098 −0.18684 1.18820 1.72421 0.35717 −4.71661 0.27753
−0.02027 1.74719 1.89855 −0.37263 −5.49010 0.02262 −0.42359 0.35025
1.12005 0.66219 −5.59074 0.55539 0.31177 0.55437 1.12424 −0.05799
−5.48020 0.44944 0.21292 1.30997 1.68528 0.24650 −5.19997 0.28213

−0.05441 1.00484 1.33122 0.29579 −5.67129 0.26797 −0.06701 0.41896
0.78390 0.59028 −5.41488 0.44580 0.25332 0.56961 0.97053 0.19728
−5.72446 0.38517 0.12078 1.06145 1.41896 0.06781 −5.25142 0.46394
0.18887 1.28330 1.78161 −0.31605 −5.84762 0.12288 −0.18749 1.38824
1.69380 0.24428 −5.03517 0.51353 0.24687 0.63869 1.07569 −0.15243
−5.95967 −0.05812 −0.45880 1.15448 1.72942 0.02572 −4.96879 0.57093
0.24776 0.70381 1.17909 −0.17134 −5.83732 0.69401 0.47941 1.35518
1.84559 −0.43297 −5.08948 0.75549 0.48102 0.63225 1.00208 0.05423
−5.90215 0.67936 0.48340 0.08493 0.28330 0.29601 −6.35853 0.30103
0.03208 1.03407 1.51630 0.08099 −5.35057 0.18439 −0.15856 0.69827
1.07335 0.13481 −5.46306 0.49175 0.12146 1.20585 1.53453 −0.11805
−5.53432 0.48977 0.22971 0.70927 1.01030 0.02735 −6.25342 −0.25326
−0.66626 1.21611 1.69827 −0.16622 −5.42435 0.24232 −0.10818 0.80182
1.15057 −0.17914 −6.11008 0.89634 0.77944 −0.29073 −0.01773 0.04415
−6.81147 0.47596 0.31962 0.71466 1.20930 −0.10624 −6.40358 0.06495
−0.21876 0.75066 0.89960 0.11193 −5.31211 0.87959 0.60532 0.60552
0.91339 −0.04191 −5.32459 0.25033 −0.04705 1.38596 1.63677 0.34183
−5.20405 0.22438 −0.09643 0.56961 1.02102 −0.03905 −7.12581 0.33914
0.18100 0.06145 0.38596 0.48487 −5.96102 0.18593 −0.00789 0.26858
0.43965 −0.19997 −6.83412 0.59443 0.57083 1.30449 1.52964 −0.13077
−5.37847 0.20094 −0.14055 0.68124 1.04171 −0.25727 −6.21789 0.40363
0.01622 0.78390 1.36369 0.07004 −5.40144 0.59613 0.31299 1.01030 1.79962
0.43233 −4.89187 0.41414 0.17767 1.17354 1.48287 −0.27819 −6.27100
0.39794 0.29771 1.35885 1.95262 −0.30103 −6.04605 0.56978 0.35389
0.43965 0.94290 0.19893 −5.75649 0.33482 0.09288 1.45454 1.54891 0.01536
−5.45390 0.47127 0.26448 1.48470 1.85111 −0.19654 −5.16412 0.24870
0.00219 1.53453 2.00948 0.06258 −4.59107 0.65808 0.36173 0.25334 0.47828
0.78972 −5.35232 0.60376 0.31220 0.51375 1.10938 −0.09909 −5.97294
0.38159 0.17090 0.78390 1.36127 −0.28400 −5.99371 0.59607 0.26550
1.19535 1.68699 −0.08831 −5.29022 0.34616 0.11711 1.20238 1.51204
−0.02091 −5.14667 0.05544 −0.39123 1.34274 1.85964 0.07372 −4.90675
0.42888 0.15425 0.56961 1.12005 0.40620 −5.62351 0.15743 −0.23868
0.98516 1.53127 0.51733 −4.87906 0.50409 0.25994 0.94606 1.35518
−0.17393 −5.63551 0.33833 −0.08960 0.86688 1.57110 0.30125 −5.19915
0.26242 −0.01360 0.49638 0.88897 −0.04096 −6.07930 0.27176 −0.02798
1.30311 1.51883 0.05729 −5.56406 0.22544 −0.12548 1.41041 1.76331
0.11561 −5.36524 0.24247 −0.10032 0.80618 1.35395 0.62107 −5.33677
0.40678 0.11243 0.39724 0.68699 0.49150 −5.84634 0.18806 −0.15717
0.55437 1.30861 0.09552 −5.69708 0.23733 −0.07128 0.76997 1.16221
−0.41117 −6.04048 0.34088 −0.01733 1.12424 1.56585 −0.44733 −6.06951
0.59524 0.49783 1.48742 1.81007 0.02325 −4.92125 0.56436 0.44210 0.65128
1.18276 0.11193 −5.17361 0.87057 0.64773 1.53372 1.83108 0.15927
−5.05340 0.34447 0.01814 1.57110 2.13374 −0.21896 −4.98126 0.51414
0.24395 1.25334 1.55746 −0.11351 −5.33904 0.42320 0.14859 1.09844
1.46514 −0.24872 −5.51120 0.29468 −0.02182 0.77931 1.46705 −0.19111

−5.58867 0.29449 −0.08401 1.15253 1.49282 −0.12090 −5.75412 0.60921
0.33440 0.88536 1.31940 −0.03763 −5.64699 0.43244 0.22094 0.85150
1.25643 −0.20971 −5.74875 0.79401 0.50081 1.48287 1.72158 −0.18575
−5.54837 0.54842 0.35513 0.93972 1.14860 0.27715 −5.83476 0.13242
−0.24428 0.75557 1.56811 −0.11407 −6.25571 0.84807 0.68805 0.97937
1.47920 −0.19997 −5.50546 0.35462 0.09826 0.56961 0.85150 −0.19314
−6.09756 0.98268 0.63231 0.28330 0.81050 −0.08672 −6.86547 0.30340
0.22119 1.05660 1.68182 0.26788 −4.58767 0.51827 0.17946 1.33252 1.62835
−0.24260 −6.49264 0.54380 0.45958 0.33766 0.81902 0.04179 −5.97029
0.73606 0.53173 1.75557 2.00291 −0.04576 −5.05670 0.21287 −0.10135
1.25178 1.37085 −0.22330 −5.79849 −0.18168 −0.59132 0.23754 0.52218
−0.42597 −6.36151 0.80811 0.61396 1.62308 2.03561 −0.18509 −5.25383
0.29896 −0.08355 1.15448 1.67658 0.36717 −5.16711 0.41693 0.20726
1.29048 1.61574 0.26150 −5.35281 0.38047 0.09425 0.63869 1.19535 0.62346
−5.24512 0.47171 0.25259 0.90998 1.20585 −0.10513 −5.49625 0.22820
−0.06892 0.97644 1.49460 0.36642 −4.90160 0.52504 0.32148 1.08721
1.42527 −0.01100 −5.42969 0.36573 0.02348 1.05415 1.51799 0.07004
−5.03003 0.56050 0.27289 0.90309 1.48742 0.14208 −5.38605 0.29386
0.02895 1.58794 1.94258 −0.09151 −4.97671 0.22288 −0.16875 0.99651
1.35763 0.00775 −5.70865 0.38435 0.10940 0.29754 0.46894 0.40312
−5.65614 0.75208 0.51838 0.41896 0.84757 0.37014 −5.93104 0.16022
−0.27288 0.86308 1.59364 0.04883 −5.40044 0.36152 0.07036 0.81902
1.24393 0.04883 −5.44624 0.38951 0.00571 1.36489 2.05733 0.19173
−4.93550 0.39842 0.14449 1.00208 1.40932 −0.07988 −5.47160 0.39273
0.06467 0.78390 1.24234 0.10789 −5.40960 0.30765 0.02378 0.32469 0.84361
0.42797 −5.96807 0.33858 −0.01621 1.01030 1.59857 0.02036 −5.01233
0.61111 0.35375 0.74068 1.22282 0.36866 −5.20239 0.26176 0.01468 0.53046
1.00208 0.00860 −5.76105 0.25750 −0.07008 1.16979 1.54018 −0.38091
−5.99400 0.67296 0.32384 1.16791 1.55592 −0.04431 −5.21063 0.45632
0.25028 0.35025 0.85540 −0.11182 −6.00502 0.32162 0.15809 0.38596
0.48742 0.22376 −5.40856 0.82097 0.60400 1.38938 1.94637 −0.27984
−5.48598 0.68421 0.57256 1.28040 1.80182 −0.13608 −4.89729 0.56331
0.26507 1.30449 1.62704 −0.02319 −5.25720 0.30960 0.06625 0.63225
1.04171 0.14953 −5.83050 0.26100 0.00731 1.28762 1.60552 −0.10513
−4.98263 0.84036 0.68596 1.40824 1.93619 −0.19179 −5.23538 0.37265
0.06427 0.56205 1.10503 0.09202 −5.79132 0.51377 0.28332 −0.04769
0.14860 0.53326 −5.58168 0.70964 0.48349 0.51375 0.96755 0.64385
−5.45930 0.38648 0.14960 1.10503 1.55359 −0.33536 −5.88971 0.76656
0.71236 −0.11464 0.08493 0.20763 −6.15113 0.79149 0.62388 1.28185
1.65066 0.23905 −4.98615 0.59951 0.37116 0.99087 1.54258 0.03383
−5.39718 0.31319 0.00314 0.59151 1.08493 0.29048 −5.63506 0.74265
0.63512 0.62572 0.87437 0.41497 −5.86385 0.35698 0.13654 1.04673 1.53046
0.30406 −4.65701 0.42615 0.12718 1.71680 2.34021 −0.13608 −5.18376
0.28619 −0.01967 1.12005 1.49549 0.03822 −5.49053 0.36338 0.11438
0.45939 1.28475 0.54716 −5.35992 0.37108 0.13098 0.29754 0.85150 0.03862

−6.02287 0.26995 −0.03657 1.02366 1.15836 −0.00966 −6.40694 0.52051
0.36373 0.82737 1.11581 0.25983 −5.39445 0.35171 0.15263 0.56205 1.02890
0.03222 −5.97102 0.45839 0.27559 0.81478 1.17725 0.16524 −5.61782
0.27549 −0.12519 1.31133 2.03484 0.31387 −4.68742 0.48566 0.26524
0.49638 0.64503 0.44295 −5.83830 0.51347 0.32559 1.51290 1.94574
−0.25964 −5.15286 0.51642 0.24444 1.23593 2.01810 0.00604 −5.52395
0.12794 −0.22231 0.97937 1.52218 −0.33348 −5.69672 0.43737 0.09098
0.96755 1.61372 0.04218 −5.35203 0.69622 0.43727 1.19357 1.73713 0.08063
−5.12773 0.53097 0.27211 −0.17718 0.03302 0.52543 −5.86816 0.71055
0.52189 1.14464 1.35149 −0.03386 −5.49080 0.21171 −0.04665 1.45649
1.71946 0.13830 −5.34285 0.04923 −0.32120 0.69827 0.88897 0.01953
−6.11126 −0.02996 −0.37239 0.66351 1.31269 −0.03810 −5.67940 0.40666
0.12456 0.95231 1.38367 0.44044 −5.43168 0.37318 0.13479 0.84757 1.66532
0.11561 −5.65807 0.45948 0.25733 1.52717 1.78618 −0.12321 −5.12854
0.19971 −0.16021 1.35272 1.73201 0.08493 −4.95393 0.49794 0.28335
1.22115 1.99819 0.40960 −5.15945 0.40246 0.15708 0.97053 1.72578
−0.03953 −5.28452 0.48741 0.16402 0.61909 1.27747 0.25840 −5.45107
0.44943 0.03796 1.22282 1.78161 0.14551 −5.22570 0.53937 0.23130 1.72735
2.03278 −0.24413 −5.09114 0.27854 0.01181 1.31672 1.80488 0.08991
−5.14606 0.02107 −0.33083 0.96152 1.72158 −0.09366 −5.28923 0.54991
0.20941 0.83556 1.56205 −0.12843 −5.59640 0.56959 0.31972 1.50166
2.01569 0.15927 −4.88884 0.46577 0.26627 1.15057 1.69492 −0.16494
−5.84363 −0.06473 −0.35380 0.53857 1.08493 0.65552 −5.33320 0.53784
0.32689 1.93036 2.32706 −0.22185 −5.38585 0.02455 −0.55024 1.41041
1.82571 0.07151 −5.29260 0.36732 0.10545 0.87064 1.31404 −0.28400
−5.80385 0.75224 0.51601 1.43762 1.80094 −0.03810 −5.50038 0.15942
−0.12272 1.33510 1.81860 0.10823 −5.01332 0.30068 −0.01073 0.94606
1.60136 −0.14327 −5.70025 0.18859 −0.08889 0.71466 1.27305 −0.09474
−5.79452 0.08376 −0.30004 1.28185 2.02550 0.21112 −4.67919 0.55772
0.36492 0.89609 1.31940 −0.11520 −5.81172 0.25548 0.06053 1.13046
1.44267 −0.24872 −6.16128 0.47164 0.06542 0.75066 1.34274 0.06967
−5.84670 0.44486 0.14304 1.72211 2.34160 −0.34679 −5.48652 0.20071
−0.12383 0.32469 0.77466 0.29776 −5.77775 0.20293 −0.06068 1.24551
1.51119 0.30211 −5.13781 0.33211 −0.01770 0.97053 2.04673 0.32572
−5.48096 0.39077 0.07902 1.25022 1.78161 −0.33255 −5.37387 0.06597
−0.33489 1.41896 1.98111 −0.11691 −5.15552 0.25387 0.03027 1.26406
1.59435 −0.36051 −5.97159 0.26541 −0.02933 0.50515 0.98227 −0.11919
−6.09637 0.41595 0.27395 0.53046 1.14264 0.46923 −5.31976 0.42751
0.20019 0.38596 0.94290 0.43680 −5.42102 0.30678 0.01906 0.19535 0.88536
−0.03953 −6.23830 0.84669 0.64016 1.18276 1.45454 −0.00261 −5.36917
0.50862 0.09596 1.19888 1.71198 −0.03668 −5.53463 0.35726 −0.01404
0.99370 1.49638 −0.45346 −5.49677 0.43357 0.16226 0.83149 1.19888
0.14489 −5.72874 0.58929 0.49238 1.27894 1.82321 −0.25727 −5.90191
0.02379 −0.35245 0.92012 1.27747 −0.43415 −5.98126 0.07591 −0.10551
0.82737 1.39947 0.31931 −5.16954 0.23573 −0.09641 1.41790 1.71680

−0.06248 −5.20866 0.52737 0.21732 0.88173 1.36966 0.43807 −5.33781
0.28131 −0.04149 0.46075 0.63022 0.05154 −6.56970 0.22265 0.12665
0.87806 1.69548 −0.34775 −5.76861 0.65067 0.38408 0.92345 1.26557
0.08778 −5.40012 0.54366 0.33061 1.11368 1.42632 −0.35458 −6.00659
0.21588 −0.05045 1.04673 1.68870 0.13577 −4.94275 0.58009 0.27786
0.94919 1.62043 −0.17070 −5.90617 0.28625 −0.10577 1.09398 1.42527
0.41963 −5.46686 0.43094 0.13894 0.96755 2.09866 0.24304 −5.53542
0.19511 −0.17376 0.97350 1.49991 0.19145 −5.23374 0.52456 0.28562
1.55204 1.83840 −0.07572 −5.18910 0.21711 −0.01914 1.15253 1.71466
0.41280 −5.09082 0.34229 0.05172 0.80618 1.36609 −0.20691 −6.63782
0.07887 −0.29253 1.49013 1.77373 −0.17914 −5.35576 0.32825 0.15288
1.49192 1.88100 −0.07520 −4.77588 0.74815 0.51156 0.69267 1.24075
0.25237 −5.59912 0.42279 0.13648 1.03407 1.86650 0.33405 −4.64533
0.34190 0.08516 0.53046 1.19712 −0.02136 −5.90767 0.43173 0.09070
1.36609 1.87026 −0.03905 −5.07217 0.05702 −0.28798 0.93003 1.37321
0.30920 −5.27032 0.29173 0.04697 1.32732 1.82943 −0.25885 −5.17993
0.31660 0.14582 0.76042 1.28330 −0.39254 −6.06877 0.24104 −0.16381
0.45939 0.93651 0.37254 −5.44904 0.56169 0.36130 −0.00087 0.40002
0.19838 −6.01854 0.73609 0.56049 1.05903 1.33510 0.15320 −5.46180
0.21989 −0.11049 0.20412 0.69827 0.72624 −5.59827 0.23944 −0.01357
1.70655 2.06458 −0.22040 −5.87933 0.27208 0.04777 0.94919 1.47270
0.32160 −5.34056 0.31157 0.00120 1.15643 1.63096 0.20978 −5.49377
0.54407 0.28245 0.65744 1.42002 0.48572 −5.04148 0.29401 0.00685 1.24709
1.85926 −0.04866 −4.84579 0.22425 0.06396 1.10938 1.64124 0.42959
−4.96086 0.46745 0.25603 0.99370 2.00976 0.13162 −5.18456 0.46235
0.16067 1.00758 1.56660 −0.14935 −5.36288 0.46173 0.01679 1.74169
2.35062 −0.28988 −5.83224 0.39879 0.12466 0.87806 1.57630 0.03383
−5.56671 0.25514 0.03305 0.91677 1.67306 −0.24795 −5.65538 0.09070
−0.27497 0.50515 1.38482 0.25864 −5.58895 0.44642 0.14410 −0.01773
0.26858 0.43616 −5.95742 0.79139 0.70537 1.27600 1.81817 −0.20135
−5.40545 0.85326 0.49611 0.93328 1.22943 0.19285 −5.84603 0.06178
−0.36574 1.16602 1.81093 −0.34969 −5.87429 0.61382 0.43949 1.50341
2.46950 0.27531 −4.57004 0.44649 0.19941 1.27156 1.53209 −0.01323
−6.00577 0.24977 −0.18753 0.16791 0.53857 0.50893 −5.60263 0.78937
0.64516 1.85501 2.36525 −0.21681 −4.61762 0.04254 −0.39039 0.94606
1.34526 0.52802 −5.14856 0.13360 −0.16637 0.62706 1.01837 0.42619
−5.01002 0.57334 0.37239 0.89025 1.48524 0.32777 −5.23875 0.47376
0.19860 0.85150 1.22282 −0.00218 −5.31659 0.50990 0.27258 1.08721
1.51204 0.17493 −5.15261 0.54745 0.23993 0.46894 1.15643 0.06819
−5.79390 0.49186 0.28179 0.94919 1.24709 0.00087 −5.50369 0.13556
−0.20947 1.16602 1.64503 −0.09259 −5.84863 0.39621 0.11283 1.42002
1.86003 0.20276 −4.81417 0.43764 0.12085 0.92675 1.32732 −0.23136
−5.80110 0.47640 0.26405 1.52881 1.89573 0.04297 −5.63786 0.34368
0.10549 1.16411 1.43046 −0.32239 −4.96353 0.97636 0.68979 0.56205
1.46132 0.30792 −5.80277 0.22892 −0.14231 0.64503 1.51883 0.48458

−5.84257 0.45858 0.26247 1.42318 1.91474 0.26126 −4.80981 0.20585
−0.10173 0.16791 0.55437 0.43838 −5.32659 0.90698 0.72325 0.66115
1.26557 0.46613 −5.00917 0.61525 0.34535 0.99651 1.59080 −0.26201
−5.95900 0.32416 −0.02430 0.82321 1.05415 0.14457 −5.73539 0.09009
−0.21196 0.58433 1.01301 0.16584 −5.16501 0.86164 0.72624 0.58433
1.16411 0.62449 −5.02018 0.48439 0.18979 0.57703 1.17540 −0.16622
−6.15983 0.30549 0.03443 1.18820 1.56660 0.06707 −5.23092 0.12086
−0.24934 −1.61979 −0.66555 −0.42713 −5.77461 1.36462 0.92495 −0.90309
−0.65170 0.97011 −5.34133 0.78209 0.61297 −0.71670 −0.11464 −0.15989
−6.72283 0.66945 0.54111 −1.05061 −0.60555 0.04805 −6.43045 1.07954
0.81695 0.01030 1.28330 2.02145 −4.17660 0.77636 0.53859 1.96394 2.18874
1.21932 −3.77004 0.63194 0.51369 −1.39794 −0.75203 −0.11407 −6.69607
0.90826 0.77120 −0.64016 −0.40782 −0.05899 −6.35497 0.69175 0.55615
−0.88941 −0.42022 0.10072 −6.50612 0.69706 0.51663 0.95847 1.23106
−0.23657 −6.16046 −0.30672 −0.62167 −0.23958 0.06145 −0.06702
−5.90173 0.84198 0.73597 −0.69250 −0.22040 0.14364 −6.50059 0.94642
0.76717 1.71305 1.79161 0.66191 −5.03198 0.45209 0.22062 −0.49349
−0.18310 0.47929 −5.33497 1.15300 0.91619 −0.91009 −0.02182 0.15594
−6.45051 0.58449 0.45392 1.48104 1.95478 0.90875 −4.54607 0.47502
0.28893 −1.01773 −0.64975 0.97685 −5.54726 0.79435 0.63395 2.39187
2.66463 0.28126 −4.41930 0.46956 0.32087 −0.59176 −0.34872 −0.09963
−6.84780 0.98924 0.73803 −1.19382 −1.19382 −0.55596 −6.79001 0.83976
0.73229 −0.95078 −0.56543 0.17754 −6.24073 0.99751 0.73835 −0.60555
−0.21610 0.90037 −5.74724 0.58462 0.39578 −1.49485 −1.05552 −0.21681
−6.31349 0.71050 0.64140 −0.01773 0.39724 −0.39147 −6.37947 0.68499
0.20949 0.86688 1.34526 1.60633 −3.41848 0.68396 0.53455 −1.05552
−0.09474 0.69073 −5.74347 0.82502 0.62479 −1.61979 −1.28400 −0.33348
−7.03044 0.69769 0.54394 −0.69897 −0.24565 0.06819 −6.69732 0.48648
0.33743 0.25888 0.61648 −0.32057 −5.92775 0.52320 0.32963 −0.57840
−0.30452 0.46776 −5.59076 0.95097 0.74777 −0.07988 0.28330 −0.43652
−5.77393 −0.27356 −0.63921 −0.89279 −0.71670 −0.26360 −6.93133
0.24474 0.05659 0.25334 0.71046 0.62449 −5.29964 0.07722 −0.27015
−1.19382 −0.89279 −0.17005 −7.20525 0.54034 0.52066 0.26623 0.39270
−0.36151 −5.80831 0.20139 −0.05134 −0.07988 0.44342 1.55617 −4.73414
0.45127 0.26800 −0.35655 −0.11464 −0.12205 −6.11569 1.01954 0.98523
−1.01773 −0.68194 −0.22841 −6.50307 0.98771 0.66116 −0.19928 0.13322
−0.04721 −5.95417 0.30013 0.15200 −1.19382 −1.19382 −0.23062 −6.80000
0.20282 0.01934 −0.89279 −0.47366 0.16107 −6.22871 1.04178 0.72200
−0.69897 −0.29073 0.15806 −6.18698 0.90647 0.78471 1.39724 1.46514
−0.28483 −5.27500 0.99772 0.66247 −0.95078 −0.44491 0.33062 −6.49689
0.65410 0.41448 −0.34872 0.14860 −0.10791 −5.86297 0.05110 −0.09467
−0.95078 −0.69897 1.25892 −5.47283 0.63090 0.45550 −0.50585 −0.16749
0.13767 −6.01291 0.84168 0.60873 −0.71670 −0.59176 −0.16115 −6.78600
0.53703 −0.12735 1.41471 1.87101 0.95027 −3.69476 0.68310 0.52203
−0.89279 −0.89279 −0.42022 −6.61527 0.40700 0.13907 0.47828 0.68124

1.40776 −4.70597 0.35694 0.08212 −0.89279 −0.59176 0.05956 −6.06329
0.79296 0.28282 −0.89279 −0.59176 0.65562 −5.74434 0.68782 0.56774
−0.86646 −0.59176 −0.05404 −6.97078 0.81024 0.68600 −1.19382 −0.75449
−0.10791 −6.08041 1.28515 0.88053 0.03663 1.08679 1.50907 −4.55925
0.67203 0.43731 −1.05552 −0.52871 0.77041 −5.54316 0.76990 0.52654
−1.92082 −1.55284 −0.29757 −6.75726 0.73464 0.54360 −0.31876 0.15927
0.28870 −6.09243 0.35773 0.17796 0.06145 0.36248 −0.37469 −5.53748
0.12546 −0.15031 −0.89279 −0.59176 0.13194 −6.75088 0.47280 0.30524
−0.04769 0.33766 −0.20691 −5.78752 0.27645 0.04215 −0.71670 −0.59176
1.09409 −4.79458 1.13318 0.85957 0.40603 0.73046 −0.04287 −5.58902
−0.36850 −0.51929 −0.89279 −0.69897 0.01912 −6.50714 1.01986 0.80738
−0.78516 −0.72584 −0.13312 −7.02914 0.88773 0.49712 −0.71670 −0.41567
0.94949 −4.84795 1.15266 0.92390 0.15259 0.30728 0.40976 −5.04735
0.91451 0.66843 0.20003 0.70458 0.68690 −5.14255 0.40583 0.13755 0.37438
1.25178 0.09377 −5.64917 0.22113 −0.12277 0.25334 0.66950 −0.06298
−5.46517 0.08220 −0.21345 −0.31876 −0.05948 0.10483 −6.13141 0.80866
0.64090 0.36248 0.84361 0.60423 −4.81460 0.80065 0.62643 −0.95078
−0.63451 0.28240 −6.09263 1.21781 0.94976 0.42943 0.94606 0.91777
−4.68763 0.56143 0.34308 0.95540 1.69154 1.37563 −3.63603 0.69890
0.58395 −0.10679 0.50705 0.84807 −5.39094 0.40644 0.17126 −0.49485
−0.34872 −0.15490 −6.35821 0.74518 0.59905 0.04297 0.38202 −0.04528
−5.90278 0.63229 0.49230 −1.19382 −1.19382 −0.11464 −6.69724 0.72192
0.53148 −0.61979 −0.23582 −0.06956 −6.44535 0.81115 0.59704 −0.79588
−0.16749 0.80284 −5.52443 0.85600 0.70128 −1.39794 −1.19382 −0.01100
−6.88995 0.63862 0.46019 −0.09745 0.20385 0.20439 −5.75132 0.42422
0.19825 −1.35655 −1.07572 −0.19518 −7.37217 0.66056 0.63955 −0.84164
−0.34104 0.88857 −6.10930 0.25213 0.02232 0.55047 1.25873 −0.33161
−5.48680 0.58620 0.36327 −0.89279 −0.49485 0.04805 −6.00879 1.33717
1.04511 −0.59176 −0.34872 0.33526 −6.21360 0.82140 0.70261 0.60552
1.30792 0.76140 −4.59235 0.52299 0.32711 −0.71670 −0.49485 0.83181
−5.42800 0.84423 0.68951 −0.04769 0.53046 0.12969 −6.02715 −0.15782
−0.51020 −0.73518 −0.16749 0.56879 −5.60940 0.90597 0.70568 −0.86646
−0.50585 −0.11126 −6.73152 0.31005 0.08867 −0.58336 −0.19586 0.12840
−6.64466 0.51698 0.38351 −1.14267 −0.61979 0.00130 −6.24365 0.77784
0.60190 −0.59176 −0.23958 0.26458 −6.28108 0.71816 0.47747 −1.05552
−0.44370 0.58320 −6.30601 0.60013 0.39751 0.19089 0.60552 −0.19997
−6.16915 0.33296 0.16270 −0.59176 0.66950 0.78923 −5.44900 0.70439
0.49218 −0.98297 −0.15243 0.31723 −6.27826 0.72194 0.60499 −1.37675
−0.92082 0.02490 −6.04357 1.09867 0.93701 −1.74473 −1.31876 −0.13253
−6.89644 0.68186 0.56157 −0.55284 −0.33348 0.13033 −6.09956 0.91351
0.81949 −1.05552 −0.75449 0.09552 −6.48969 0.73405 0.70335 −1.55284
−1.25181 −0.17522 −7.01002 0.56291 0.43041 −0.95078 −0.64975 0.21245
−5.77827 1.10679 0.76709 0.29754 0.71466 −0.27165 −5.91172 −0.01602
−0.23395 −0.49485 −0.36452 0.07188 −5.97417 1.07918 0.88139 −0.33348
−0.01055 0.49346 −5.38724 0.80172 0.66361 −0.86646 −0.63451 0.07335

−6.56891 0.88737 0.72413 −0.51713 −0.10127 −0.04431 −6.04973 0.92453
0.69268 0.78845 1.11687 −0.32057 −5.72563 −0.28063 −0.67523 0.20925
0.61773 0.41313 −5.20572 0.50040 0.23841 0.53857 0.82321 −0.44009
−5.63746 0.57856 0.41165 0.12156 0.29292 −0.06956 −5.60838 0.45113
0.32241 −0.89279 0.25358 1.14070 −4.91087 0.78044 0.56245 0.10721
0.69267 0.26623 −5.25211 0.55329 0.26836 0.45939 0.87437 −0.14267
−5.17556 0.61810 0.31914 −1.05552 −0.68194 0.80140 −5.53106 0.72815
0.54956 −0.89279 −0.19382 0.81164 −5.86066 0.75714 0.53970 −1.39794
−1.01773 −0.08249 −6.39405 1.09080 0.91608 −0.84164 −0.21467 0.14051
−6.57059 0.92514 0.82326 −0.11464 0.14860 −0.19723 −5.82348 0.63942
0.57243 0.12840 0.53046 −0.41005 −6.82609 0.55471 0.27153 −1.46852
−1.10791 −0.01412 −6.80699 0.69394 0.56247 0.06145 0.70381 0.01284
−6.40782 0.64964 0.51486 0.02366 0.37822 0.22917 −5.31574 0.99089
0.69807 −1.01773 −0.56543 0.38561 −6.09039 0.69810 0.49038 −0.49485
−0.29073 −0.16877 −6.52169 0.64658 0.25436 −1.69897 −1.28400 −0.28567
−7.41274 0.56647 0.47232 −0.21183 0.09830 −0.13430 −6.16501 0.92942
0.64876 −0.35655 0.00689 0.27921 −6.26376 0.62095 0.36037 −1.28400
−0.93554 −0.14327 −6.42099 0.84734 0.80465 −0.71670 −0.41567 0.36380
−6.54745 0.66527 0.53516 −0.75449 −0.46344 0.40500 −6.24672 0.89977
0.58913 −0.54061 −0.21610 0.21643 −6.38998 0.78032 0.56491 −1.25181
−0.93554 1.00574 −6.18970 0.48457 0.29973 −1.92082 −1.16749 1.34670
−5.38700 0.80297 0.63707 −0.77469 −0.46344 −0.19179 −6.85344 1.64679
1.28454 −1.01773 −0.61979 0.06521 −6.91147 0.71132 0.57183 0.52634
1.29684 −0.22475 −5.52824 0.44974 0.16989 −0.08566 0.16465 −0.15739
−6.33536 0.95191 0.68915 −0.57512 −0.19246 −0.14813 −6.42499 1.03233
0.82974 −0.04769 0.14860 0.01995 −6.15150 0.49986 0.05171 −0.68194
−0.38091 0.32056 −6.15642 0.90842 0.76562 −0.86012 −0.57512 0.01494
−6.99559 0.87180 0.66762 −0.29073 0.20412 0.08743 −6.08542 −0.11231
−0.43061 −0.20551 0.15806 0.12450 −5.79461 0.98798 0.73580 0.18639
1.32272 0.65629 −5.69144 0.38837 0.11441 −0.21610 0.26102 0.53161
−5.62534 0.88440 0.69878 0.08493 0.54654 −0.14086 −6.07989 0.35728
−0.43398 −1.09691 −0.64207 0.02202 −6.90236 0.30636 0.08086 −0.59176
−0.15243 0.59879 −6.11896 0.60655 0.44342 −0.52871 −0.11014 0.72616
−5.98134 0.40654 0.21638 −1.44370 −1.11919 0.60217 −5.82948 1.03706
0.89872 −1.11919 −0.75449 0.32777 −5.53167 0.95707 0.66410 −0.71670
−0.41567 0.28937 −6.38764 0.65123 0.33202 −0.77469 −0.27084 0.12123
−6.95691 0.60820 0.50301 −0.49485 0.11428 0.61899 −5.38007 0.99588
0.82216 −1.09691 −0.25181 1.04198 −5.29620 0.73869 0.54334 −1.09691
−0.49485 0.39146 −6.01927 0.70470 0.54093 −0.59176 −0.34872 −0.05849
−6.08999 1.22515 1.06274 −0.75449 −0.27165 0.29732 −6.21477 0.76489
0.70471 −1.19382 −0.79588 −0.20482 −7.42763 0.49822 0.27110 −0.95078
−0.47366 0.32675 −5.77940 0.92523 0.62185 −1.35655 −1.11919 −0.01233
−6.61941 0.83837 0.76714 0.38596 0.73560 0.37676 −5.46276 0.06149
−0.42062 −0.20482 0.50974 0.35813 −6.12067 0.24441 −0.13144 0.01030
0.62572 0.75220 −5.35962 0.67152 0.38591 −0.52433 −0.02733 0.46434

−5.81096 0.81933 0.65660 −0.89279 −0.89279 −0.17070 −6.82658 0.46343
0.24661 −0.75449 0.23955 0.46285 −5.77728 0.66950 0.38306 −1.55284
−1.22185 −0.02365 −6.66033 0.86045 0.68935 −0.12378 0.17114 −0.16052
−6.52240 0.67282 0.43756 −0.41567 −0.11464 −0.21896 −6.90679 0.61684
0.42465 −0.89279 −0.71670 0.62138 −5.89630 0.90953 0.70615 −1.79588
−1.44370 −0.26360 −7.05562 0.65290 0.48721 −1.25181 −1.05552 −0.02965
−7.21767 0.20681 −0.00632 −1.31876 −0.87943 −0.24489 −7.27417 0.33096
0.30378 −0.71670 −0.33348 0.50161 −5.87162 1.03915 0.93156 −0.23958
0.23477 0.26811 −6.02305 0.02648 −0.30319 −0.59176 −0.23958 0.36736
−6.43711 0.56106 0.36437 −1.19382 −0.89279 0.52362 −6.40550 0.70924
0.56732 −0.21610 0.06744 0.63094 −5.35851 0.98419 0.76324 −0.89279
−0.89279 −0.04001 −6.84179 0.51704 0.36401 −0.71670 −0.34872 0.66210
−5.75427 0.79968 0.59624 −1.19382 −0.71670 −0.28735 −6.52770 1.40358
1.01147 −0.89279 −0.41567 0.09272 −6.81217 0.84714 0.70897 −1.01773
−0.33348 0.54283 −6.23582 0.73779 0.47291 −0.25181 0.07335 −0.04144
−6.42848 0.62188 0.47040 −0.75449 −0.51713 0.41531 −5.58403 0.88208
0.74047 −0.59176 −0.41567 0.18921 −6.13407 0.80351 0.71114 −0.71670
−0.30452 0.86451 −5.52672 0.88830 0.76307 −0.34872 −0.07988 0.84572
−5.03685 0.90689 0.70532 −0.71670 0.65744 0.83174 −5.57167 0.62804
0.40083 −0.29073 −0.04769 0.00087 −5.97201 1.27355 1.19880 −0.29073
−0.00656 −0.07727 −6.64933 0.54983 0.43923 0.75557 1.21187 −0.13430
−5.59157 −0.09513 −0.57792 0.38596 1.48287 0.68070 −4.66392 0.70626
0.50249 −0.08938 0.24502 0.39164 −4.92595 1.13745 0.85295 0.11793
0.40824 0.12189 −6.37862 0.01741 −0.35281 0.25334 0.58433 −0.40230
−6.29294 0.53806 0.26044 −1.01773 −0.71670 −0.14388 −7.16616 0.83772
0.77715 −0.56384 −0.11407 −0.14752 −6.80156 0.79420 0.71637 −0.36452
−0.20482 0.12646 −6.08842 0.81758 0.64403 −1.25181 −0.86646 0.44201
−6.41499 0.94332 0.68429 −1.19382 −1.19382 0.00647 −5.35770 1.46447
1.01353 0.59506 1.23024 −0.44249 −5.93230 0.37640 0.20848 −0.44733
−0.08619 0.65312 −5.17122 1.02242 0.83255 −0.64975 −0.06803 0.29732
−5.89238 0.81134 0.63265 −0.59176 0.00647 0.44342 −6.20087 0.78114
0.62097 −0.34872 −0.11464 0.68708 −5.28441 1.13388 0.89047 −0.59176
−0.34872 −0.06500 −6.50086 0.89125 0.69779 0.14860 0.57703 −0.21824
−5.61441 0.84811 0.65935 −0.29073 −0.04769 0.22531 −5.68348 1.18865
0.76773 −1.49485 −0.98297 0.86876 −6.34785 0.26658 0.04149 −1.01773
−0.79588 0.17696 −6.43445 0.98496 0.83481 −1.25181 −0.34872 0.44809
−5.75948 0.96097 0.73959 0.41896 0.89254 −0.45593 −5.81141 0.50312
0.20163 −0.95078 −0.71670 −0.12205 −6.72670 0.84782 0.76793 −0.52724
−0.22330 0.45682 −5.75132 0.82161 0.63874 0.23754 0.86308 −0.23136
−5.95705 −0.24683 −0.72623 −0.29073 0.03663 0.63296 −5.55474 0.83487
0.63677 −1.18709 −0.77728 0.26741 −6.60954 0.76357 0.60266 −1.79588
−1.39794 −0.15552 −7.12720 0.51112 0.21857 0.83960 2.06649 0.26670
−4.57636 0.61737 0.37703 −0.53462 −0.08407 −0.17914 −6.54661 0.32744
0.16761 −0.71670 0.03663 0.52905 −5.65317 0.96352 0.83372 −0.26520
0.04139 −0.04915 −6.03423 1.17313 0.87818 −0.11464 0.39724 −0.22475

−6.71790 0.38866 0.14920 0.76042 0.94290 −0.52433 −6.03795 −0.05213
−0.49282 −0.34872 0.10721 0.27091 −5.98801 0.86430 0.71271 −0.07988
0.88897 0.27416 −6.17211 0.70516 0.27919 −1.31876 −0.17070 −0.03152
−7.00056 0.51130 0.33597 −0.45717 0.05918 0.07041 −6.23026 0.63956
0.48640 −0.60033 −0.13847 −0.00524 −6.27925 0.74295 0.31275 0.39724
0.79295 0.29358 −5.69949 −0.11387 −0.43300 −0.41567 0.63225 0.30492
−6.14038 0.55877 0.45269 −0.90658 −0.68403 0.04883 −6.76319 0.32188
0.05576 0.22115 1.57923 0.79302 −5.56076 0.31837 0.05263 −0.43063
−0.14813 0.08135 −6.08948 0.67062 0.54778 −0.89279 −0.19382 0.09726
−6.26223 0.47358 0.21609 −1.16115 −0.73049 0.85588 −5.43632 0.93896
0.68530 0.26858 0.60552 −0.27246 −5.74870 1.09540 0.70049 −0.11464
0.92345 −0.18977 −6.39105 0.70695 0.53330 1.13046 1.48652 −0.28904
−5.39835 −0.39873 −1.03620 −0.76447 −0.33724 0.41913 −6.37997 0.70682
0.53243 −0.34872 0.12840 0.08314 −6.56209 0.67542 0.44096 0.25334
0.93328 0.28847 −4.98809 0.19139 −0.18052 −1.19382 −0.71670 0.06032
−5.03560 0.09977 −0.29132 −0.59176 −0.04769 0.25479 −5.99499 0.92115
0.49623 −0.71670 0.59151 0.07846 −5.76881 0.37526 −0.05332 −0.71670
0.33766 0.04650 −6.35605 0.90987 0.60270 −0.71670 −0.19382 0.19312
−5.87859 0.90465 0.72669 −0.49485 0.10721 0.31597 −6.38964 0.74266
0.49754 −0.11464 0.68699 −0.12610 −6.79315 −0.16389 −0.80522 −1.20066
−0.70115 0.39985 −6.10713 0.73655 0.55744 −0.71670 0.29754 0.16584
−6.24885 0.64314 0.50010 −0.29073 0.48742 −0.29499 −6.07335 1.12131
0.65188 −0.29073 0.89609 0.18583 −5.73974 1.04414 0.67329 0.46894
0.99930 −0.47108 −6.12362 0.01270 −0.38244 −1.04096 −0.69465 0.35965
−6.67046 0.53817 0.29444 0.91677 1.80050 −0.37059 −6.34312 −0.27354
−0.71024 −0.19382 −0.01773 0.10346 −5.88924 0.87886 0.65392 −0.23958
0.29754 0.21696 −5.76726 0.91801 0.78098 0.42943 0.82321 −0.44491
−5.37363 0.55051 0.41332 −0.07988 0.38596 −0.09909 −6.62428 0.26077
−0.03484 −0.01773 0.28330 0.43648 −5.49990 0.89369 0.69865 −1.23657
−0.21325 −0.16368 −7.36653 1.02478 0.49760 −0.59176 −0.34872 0.00173
−6.74921 0.93145 0.84426 −1.20761 −0.04528 −0.01909 −6.64308 0.57849
0.38117 −0.41567 −0.11464 0.15351 −6.70163 0.68556 0.52390 −0.26680
0.12057 −0.00745 −6.67826 0.74417 0.31971 −0.59176 −0.04769 0.32490
−6.64064 0.71020 0.62801 −0.34872 −0.07988 0.17667 −5.90858 0.99582
0.69877 −0.29073 0.29863 0.76193 −4.67615 0.42522 0.21288 0.01030
0.53046 0.58546 −5.43177 0.44856 0.14671 −0.04769 0.47828 −0.00656
−6.65695 1.18282 0.60186 0.03663 0.94919 0.43521 −5.84109 0.50239
0.25845 −0.15243 0.46894 0.17260 −6.12999 0.64411 0.43698 −0.71670
0.79295 0.88207 −5.60350 0.66346 0.43974 −0.71670 −0.59176 −0.07366
−7.01868 0.45057 0.30741 −0.59176 0.18639 −0.08197 −6.18491 0.75696
0.17711 −0.07988 0.52218 0.59671 −5.02168 0.46755 0.20789 −0.54363
−0.26761 −0.24489 −7.14661 0.56810 0.42390 −0.89279 −0.59176 0.09552
−6.27524 0.56781 0.46196 −1.37675 −1.12494 0.53212 −6.37330 0.68912
0.57810 0.06145 0.35025 −0.29930 −6.11369 0.49003 0.18244 −0.19382
0.26858 0.89669 −5.65987 0.10680 −0.26315 0.08493 0.63869 0.10857

−6.47211 0.22726 −0.08249 −0.15243 0.22115 0.83467 −5.00463 0.78608
0.45230 0.29754 0.83149 −0.19860 −5.95870 −0.25057 −0.52979 −1.79588
−1.49485 −0.17783 −7.04998 0.47337 0.36527 −0.07988 0.01030 −0.05750
−6.23137 0.64163 0.34337 0.10721 0.83149 0.46300 −6.03386 0.04365
−0.23787 −1.53760 −1.01323 −0.35360 −6.87641 0.67029 0.46440 −1.08619
−0.53910 0.09026 −6.71981 0.63999 0.39104 −0.25337 0.44963 0.46090
−5.89425 0.50076 0.34821 −0.52871 −0.11182 0.24576 −6.22048 0.80156
0.59070 −0.71670 −0.15243 0.69461 −5.92830 0.74195 0.48449 −0.23958
0.41896 0.30856 −6.22607 0.46833 0.27900 −0.71670 −0.29073 0.27068
−6.75355 0.84354 0.73633 −0.93930 −0.69897 0.42781 −6.15292 0.51085
0.33757 −0.07988 0.37438 −0.14448 −6.23552 0.66331 0.24196 −1.31876
−0.93554 1.25122 −5.38618 0.77113 0.54328 −0.71670 −0.59176 0.56597
−6.04340 0.78370 0.63831 −1.56864 −0.95861 −0.23807 −6.93405 0.31396
0.11951 −1.04576 −0.71670 0.18156 −6.37616 0.79470 0.35018 −0.90658
−0.54821 0.19562 −6.53398 0.71892 0.46633 −0.89279 −0.71670 0.20330
−6.08243 1.25559 0.84532 −0.23958 0.03663 −0.13966 −6.68315 0.89122
0.73507 −0.59517 −0.34486 0.45347 −5.99745 0.77868 0.62187 0.28330
0.65744 −0.34679 −5.46048 0.80266 0.55477 −0.01773 0.38596 0.07700
−5.94811 0.61297 0.41972 0.18639 0.81478 0.02407 −6.07551 0.36808
0.03684 −0.59176 −0.11464 0.35083 −6.20012 0.84860 0.72327 −1.30980
−1.03621 0.99822 −5.30565 1.00937 0.74394 −0.83565 −0.68613 0.46523
−6.27035 0.40772 0.22908 −0.07988 1.00758 0.59106 −5.28862 0.94338
0.66243 0.75557 1.30311 −0.37469 −5.68160 −0.22709 −0.78854 −0.71670
−0.01773 0.53970 −6.17940 0.76181 0.46911 −1.00877 −0.61979 0.60959
−5.79852 0.84619 0.65706 −1.19382 −0.71670 0.70088 −5.42962 0.80894
0.60908 −1.20066 −0.83565 0.14270 −6.66462 0.82487 0.50999 −1.19382
−0.89279 0.57646 −5.85038 0.97996 0.75734 −0.89279 −0.34872 0.32715
−6.69229 0.70429 0.46593 −0.66154 −0.39903 0.20493 −6.42251 0.83325
0.55688 −0.69897 −0.28150 0.14395 −5.89456 0.93283 0.62166 0.41896
1.30587 0.00389 −6.04756 −0.48174 −0.99363 −0.64016 0.07078 −0.07935
−6.16511 0.79976 0.60503 −0.43890 −0.01502 0.17493 −5.71863 0.96725
0.72514 −1.25964 0.31597 0.38489 −6.56443 0.49182 0.33188 −1.35655
−1.03621 0.13767 −6.73278 0.83858 0.68301 −0.52871 −0.11520 0.17056
−6.18339 0.85474 0.76963 −1.21467 0.20575 0.31471 −6.17334 0.95321
0.68103 −0.39147 0.79844 0.52479 −5.17819 0.99167 0.77014 −0.90309
−0.58170 0.42243 −6.59722 0.96692 0.74507 −1.11351 −0.81248 0.36493
−5.92347 1.32448 0.91829 −0.35853 0.05729 −0.08725 −6.09039 0.79613
0.67797 −0.44612 0.49122 0.89014 −5.38655 0.78722 0.56193 −1.04096
−0.43890 0.25285 −5.81135 0.98596 0.68823 −0.83268 −0.60380 0.86219
−5.03575 1.03065 0.81452 −1.36653 −1.00436 0.45240 −5.94873 0.77007
0.49867 −0.39903 −0.05061 0.07115 −6.61407 0.23183 −0.02395 −1.00000
−0.50724 0.64601 −6.16845 0.79970 0.59278 −0.40671 0.09587 0.63919
−5.13643 0.92708 0.70906 0.52218 0.98516 −0.47237 −6.14819 0.39740
0.06609 −0.92445 −0.39903 0.54083 −6.18582 0.90763 0.67412 −0.30016
0.59151 0.27554 −6.11532 0.68738 0.40014 −1.19382 −0.23958 0.29003

−6.23726 0.71724 0.37955 −1.67778 −0.40671 0.35295 −6.54356 0.52993
0.30268 −0.93181 −0.00568 0.46494 −6.64647 0.43191 0.27784 −1.58503
−0.32239 −0.14327 −7.10763 0.64238 0.54088 −0.76700 0.32634 0.22220
−6.83666 0.69854 0.60574 −0.41567 0.69267 0.49220 −5.14577 1.10538
0.73250 −0.04624 0.29181 0.30406 −5.42351 1.01908 0.80559 −0.45469
−0.07624 −0.25337 −6.85667 0.50337 0.26709 −1.19382 −0.41567 0.00000
−6.90018 0.64998 0.49227 −0.08092 0.29513 −0.36151 −6.10063 0.91319
0.77532 0.06145 0.87064 0.79810 −5.48745 0.46302 0.20275 −0.69037
0.00475 0.54543 −6.09157 0.57511 0.45068 −0.42597 −0.21896 −0.12610
−6.84817 0.72253 0.67058 0.55437 1.26858 −0.21183 −5.64606 0.46012
0.11840 −0.06905 0.37051 −0.10846 −5.86439 0.97692 0.87098 −1.85387
−0.37675 −0.32514 −6.94245 0.86905 0.73013 −0.76447 −0.10403 0.12710
−6.92948 0.49487 0.35200 −0.29757 0.80182 0.24329 −5.78944 0.77457
0.66754 −0.86646 −0.55909 0.21880 −6.58516 0.89420 0.74390 −0.71670
−0.41567 0.43933 −6.18071 0.68303 0.59230 −0.46344 −0.02457 0.74609
−6.00113 0.56463 0.42740 −1.14267 −0.92812 0.31952 −6.15079 0.55187
0.38597 −0.33536 0.00689 0.09342 −6.39631 0.71550 0.52468 −0.03433
0.51375 −0.09044 −6.26114 0.20412 0.02584 −0.95078 −0.75945 0.22479
−6.98338 0.44793 0.19873 −0.45842 0.51680 0.53148 −6.03673 0.52637
0.29448 0.06521 1.20529 0.84261 −5.77075 0.21941 −0.08402 −0.85387
0.28285 0.81311 −6.20356 0.32697 0.07625 −1.03621 −0.78516 0.59857
−5.64030 1.05027 0.79921 −0.25414 0.03862 0.54839 −5.24244 0.89323
0.66991 −1.38722 −0.90658 −0.07263 −6.89269 0.59243 0.21213 0.08493
0.63869 0.36078 −5.79024 0.12857 −0.20356 −0.04769 0.42943 −0.15181
−6.32474 0.41316 −0.03223 −0.68403 −0.24872 0.56348 −6.26440 0.27550
−0.09185 −0.29930 0.35545 −0.17718 −6.58416 0.58499 0.37704 0.22115
1.22282 −0.04287 −6.37182 0.25123 −0.28804 −0.89279 −0.29073 0.52427
−5.44969 0.89398 0.74262 −0.57187 −0.37469 0.16286 −6.85543 0.68660
0.40505 0.47828 1.12005 −0.23508 −5.54625 0.79462 0.56417 −0.80410
−0.41229 0.11294 −6.84351 0.41156 0.13611 −0.03198 0.42619 −0.27246
−5.80482 0.91447 0.76601 −0.98297 −0.61979 0.25624 −6.66675 0.64173
0.51491 0.16791 0.92012 0.75397 −5.05335 0.68036 0.51429 −1.11351
−0.38405 0.47538 −6.00895 0.77243 0.60892 −0.97062 −0.21610 0.14953
−6.16183 0.94153 0.64006 −0.47756 −0.27819 −0.15366 −6.96473 0.93119
0.87632 −1.19382 −0.89279 0.22037 −6.57937 0.79685 0.51134 −0.15243
0.18639 0.20898 −5.25209 0.85742 0.59204 −0.67366 −0.11464 0.07151
−6.48247 1.07938 0.77689 −1.60206 −1.10237 −0.17653 −6.97910 0.71793
0.65263 −0.36957 −0.04964 0.31513 −5.94161 0.90892 0.69090 −0.55440
−0.45223 −0.07314 −6.53211 0.61085 0.47390 −1.12494 −0.66154 0.10789
−6.39431 1.33368 0.92654 −0.11126 0.35946 0.08600 −5.79280 0.67059
0.49845 −0.19382 1.00484 0.43088 −5.79070 0.74512 0.50156 −1.29243
−0.91009 0.21590 −6.19078 0.97558 0.55269 −0.25259 0.00087 −0.22841
−5.83107 0.78663 0.53777 −0.71670 −0.41567 0.72558 −5.97220 0.92181
0.79355 −0.04769 0.14860 0.22840 −5.58548 0.86703 0.68354 −1.19382
−1.19382 −0.22915 −5.53548 −1.22128 −1.63928 1.81986 2.17651 1.63776

−3.17923 0.58369 0.13642 1.12005 1.62835 0.97896 −3.95043 0.81364
0.66751 1.43201 1.62143 −0.03905 −6.43418 0.38202 0.15262 1.31404
1.88026 0.77247 −4.18722 0.69897 0.50784 1.20585 1.78345 0.94022
−3.99191 0.72981 0.53220 0.20412 0.57703 0.93867 −4.90654 0.50536
0.34150 1.28762 1.36006 0.71600 −5.21382 0.33579 0.14346 2.03355 2.13700
−0.14813 −5.25688 0.36484 0.30712 1.14860 1.63096 0.77195 −4.07289
0.66020 0.46273 1.04922 1.22614 1.24309 −4.30601 0.53205 0.31828 1.45600
1.56698 0.57415 −4.80532 0.35790 0.14789 1.24709 1.65128 0.50799
−4.18889 0.72762 0.50971 1.18639 1.68642 1.15900 −3.97029 0.79638
0.57288 1.55359 1.94669 0.78003 −4.22141 0.51861 0.33858 0.22115 0.56050
0.87668 −5.04920 0.45317 0.25740 2.00069 2.25996 0.82885 −4.15434
0.63454 0.42003 1.12633 1.54258 1.51006 −3.10724 0.72042 0.40764 0.91677
1.82592 1.27119 −4.08767 0.54361 0.34293 1.51883 1.63933 0.36455
−5.31524 0.30920 0.05170 −0.49485 0.06145 1.15433 −5.13065 0.49849
0.25193 1.12633 1.44267 0.72181 −4.78265 0.74936 0.59076 0.69267 1.00758
0.55859 −4.41364 0.89929 0.79360 0.60552 1.65621 1.04513 −4.58586
0.71292 0.52665 1.15057 1.50775 1.09377 −3.96887 0.70599 0.56503 0.35025
0.71466 1.22084 −4.95323 0.44398 0.24067 2.29536 2.47712 −0.25337
−5.01095 0.19252 −0.15563 0.78845 1.60136 1.15640 −4.00279 0.70157
0.45243 1.31672 2.06193 1.23111 −3.99161 0.53083 0.31685 1.11368 1.65744
1.09608 −4.28929 0.39051 0.10454 0.42943 0.93328 0.42959 −5.28222
−0.12321 −0.71544 1.53453 1.88681 1.10680 −3.58195 0.65193 0.43124
1.77279 1.92511 −0.13253 −5.96887 1.25697 0.78601 1.30449 1.54258
1.07430 −4.77536 0.25357 −0.07996 0.63869 1.14264 0.79043 −4.76377
0.85431 0.61722 0.88536 1.52385 0.82556 −4.66993 0.48306 0.19776 0.53857
1.63996 1.02955 −4.75684 0.33203 0.04614 1.89004 2.05672 −0.09583
−5.07982 0.24773 −0.05692 1.86308 2.03548 0.51640 −4.79105 0.28433
−0.02039 2.23545 2.31444 0.32366 −5.85918 0.51624 0.34889 1.51290
1.79206 0.82924 −4.32868 0.72360 0.44619 1.96424 2.18412 0.13098
−4.83550 0.52562 0.28934 0.74068 1.60655 0.82988 −4.49268 0.52837
0.24617 1.59222 2.01247 0.65089 −4.96823 0.30821 −0.07348 1.21780
1.66592 0.56937 −4.08735 0.41289 0.12623 1.34147 1.76760 1.10473
−3.97094 0.57429 0.36088 1.42108 1.73509 0.96199 −4.49839 0.41027
0.14557 1.95138 2.17836 0.22115 −4.98447 0.37745 −0.01130 1.98458
2.52013 0.33965 −4.13888 0.40535 0.06358 2.13476 2.82855 0.93867
−4.43063 0.52653 0.34475 0.56205 0.96142 1.29212 −4.78281 0.66302
0.46349 0.57703 1.61032 1.34118 −4.24665 0.58778 0.34357 1.73201 1.77977
0.93676 −5.01180 0.42041 0.16401 0.93972 1.51925 0.82269 −4.30733
0.62458 0.42073 0.65128 1.14464 1.48022 −4.19647 0.46759 0.14599 1.77138
1.91912 −0.08672 −5.38899 0.87035 0.62601 0.81478 1.28330 0.72222
−5.06193 0.59877 0.33220 1.55125 1.77138 1.33381 −3.38856 0.74938
0.54889 2.68303 2.77051 0.07445 −4.62233 0.47114 0.18254 1.22943 1.77138
0.82027 −4.35384 0.64417 0.44618 1.70105 1.95509 0.39811 −4.74376
0.60879 0.42820 2.05268 2.15244 0.51904 −4.37376 0.60452 0.31263 1.44467
1.75654 0.75587 −4.79713 0.34357 0.05541 0.81478 1.02366 1.11066

−4.59288 0.50825 0.32558 1.39501 1.83819 0.48487 −4.48261 0.57845
0.36784 2.21187 2.31051 0.65157 −4.49405 0.51343 0.22873 1.51967 1.84658
−0.02780 −5.46983 0.70367 0.40849 0.97644 1.86079 1.40293 −4.31016
0.48718 0.26091 2.07194 2.12871 0.03782 −5.36835 0.47897 0.23074 0.61909
1.04423 0.74367 −4.78336 0.45469 0.13600 1.33122 1.81775 0.71609
−4.36967 0.46745 0.20765 1.59997 1.96030 0.48785 −5.33268 0.22392
−0.01118 1.12005 1.49815 0.64768 −4.16414 0.78329 0.60927 1.54812
1.70927 0.04376 −5.62690 0.87626 0.65925 0.42943 0.92675 1.14392
−4.84351 0.32010 0.03798 0.56961 1.61775 1.13650 −4.85208 0.45575
0.21271 2.02576 2.21408 −0.01189 −4.48633 0.48266 0.26217 0.92675
1.45551 0.59868 −4.62849 0.60664 0.35105 0.93003 2.23738 1.03019
−3.92698 0.78416 0.52463 0.59857 1.47270 0.91339 −4.55942 0.58100
0.34783 1.35272 1.76950 0.73640 −4.08087 0.67941 0.45084 0.62572 1.13659
0.65040 −4.79358 0.70321 0.41808 0.31133 0.95540 0.94310 −5.38648
0.13272 −0.19573 1.49638 1.71600 0.51441 −4.86621 0.36874 0.02632
0.92345 1.56205 1.74994 −3.65687 0.36447 −0.01113 1.77838 1.85150
−0.01502 −5.60067 0.44100 0.18912 0.37438 1.11687 1.20112 −4.43854
0.50463 0.29845 1.11368 1.87325 0.85339 −4.45210 0.53948 0.30629 2.39215
2.41412 0.17173 −5.02168 0.71359 0.45278 0.83960 0.97937 0.60184
−5.67293 −0.10233 −0.60105 1.02890 1.35763 1.12320 −4.10646 0.57144
0.39202 1.03407 1.37555 0.61669 −5.15286 0.42895 0.23107 0.39724 1.20585
0.91158 −4.97469 0.50139 0.25330 2.29894 2.42611 0.29403 −4.82385
0.53849 0.27298 2.19286 2.30145 0.30920 −4.83959 0.57645 0.24677 0.92012
2.23593 0.61836 −4.77788 0.61979 0.39747 1.17909 1.90655 0.57841
−4.58228 0.58459 0.34055 1.40932 1.86213 0.38310 −5.17153 0.55775
0.29994 0.12840 1.23269 1.37793 −3.95200 0.86780 0.64336 2.41960 2.47270
0.47114 −4.93059 0.44048 0.22232 1.58288 2.13781 0.44028 −4.90514
0.38011 0.11300 2.10361 2.40360 0.06221 −4.73065 0.43203 0.18297 1.17540
1.93812 0.97058 −4.03217 0.52006 0.25190 1.35763 2.14761 0.97479
−4.29269 0.56031 0.28586 1.27600 1.50515 −0.05502 −5.94006 0.72592
0.40870 1.27305 1.95586 0.53870 −4.28668 0.52729 0.32778 1.54258 1.86003
0.52956 −4.69607 0.47469 0.23659 0.42455 0.69504 0.73997 −5.48630
0.16757 −0.16807 1.47270 1.90309 0.15836 −5.01498 0.79155 0.57011
1.92840 2.27364 0.77452 −4.81687 0.30660 −0.02920 −0.10902 0.68931
1.33112 −5.13283 0.38211 0.17728 0.49638 1.10503 1.34033 −4.34814
0.62951 0.35976 0.25334 0.71466 1.12470 −4.39193 0.88997 0.69891 1.04922
1.30311 0.53542 −5.99598 0.71592 0.49341 0.92675 1.10721 0.93852
−4.94078 0.39737 0.08074 1.91441 2.36152 0.20790 −4.63481 0.26751
0.00504 1.67306 2.13190 0.62221 −4.07686 0.59683 0.37011 1.07802 1.88318
0.60466 −5.18883 0.50620 0.31221 0.67541 1.28905 0.65225 −4.83265
0.54718 0.32960 0.93003 1.19535 0.43393 −4.79140 0.78516 0.56527 0.33766
0.72526 2.25920 −3.58501 0.60161 0.39029 0.53857 0.97937 1.37816
−4.44201 0.52842 0.30295 0.59857 1.20063 0.92768 −4.77474 0.46631
0.20595 0.35025 1.20585 0.75105 −4.91318 0.52438 0.32506 1.75606 2.03841
−0.08991 −5.38321 0.43204 0.14447 1.40606 1.85345 0.66266 −4.14297

0.59697 0.41290 −0.04769 0.41896 1.05154 −5.44217 −0.08623 −0.43797
0.45939 1.00208 1.50875 −4.90925 0.39389 0.09176 −0.23958 0.35025
1.57490 −4.58937 0.65538 0.42865 1.41256 1.61708 0.45133 −4.99529
0.55438 0.39063 1.02629 1.60895 0.84739 −4.73016 0.63389 0.40136 0.50515
1.10503 0.70044 −5.28467 0.17586 −0.11817 1.41790 1.77838 0.50691
−4.66206 0.40415 0.05673 1.06386 1.49815 1.16388 −4.29982 0.44317
0.20662 1.35272 1.43046 1.84468 −4.72693 0.18347 −0.28558 0.89609
1.42632 1.15256 −3.68281 0.82596 0.59546 0.58433 1.21101 1.27616
−4.10508 0.64345 0.44627 1.67306 2.08812 0.36455 −4.12819 0.78239
0.55647 0.79295 1.22614 1.54565 −3.92768 0.67082 0.48416 1.44864 1.79696
0.15987 −5.61858 0.70382 0.56710 1.30587 1.79026 1.23676 −3.77529
0.66877 0.39943 0.88173 1.54733 1.79326 −4.03863 0.53374 0.31679 0.35025
0.60552 1.29671 −4.66815 0.67210 0.45136 1.53046 1.89111 0.59780
−5.22592 0.31246 −0.04092 1.19179 1.57482 0.58024 −4.52143 0.65548
0.43625 1.70709 1.77885 −0.09637 −5.47384 0.44527 0.21697 1.75214
2.08971 0.28691 −4.88556 0.01256 −0.28514 1.08034 1.82070 1.16967
−4.36896 0.42921 0.15542 −0.89279 −0.71670 1.66714 −4.36671 0.81338
0.56948 1.89679 2.25643 0.27068 −5.18125 0.27666 −0.06382 1.65866
2.20308 0.20439 −5.26321 0.73651 0.52845 1.12840 1.57556 0.86705
−4.32957 0.45801 0.20039 1.99002 2.17372 0.61836 −4.82845 0.25240
−0.06464 0.31133 0.81478 1.19607 −4.23813 0.69112 0.41995 1.04423
1.06863 0.00346 −6.83502 0.79727 0.54976 1.45160 1.91744 0.26387
−4.24978 0.63836 0.35925 1.64629 2.05879 0.42160 −4.52273 0.43536
0.16172 1.41041 1.60827 −0.32790 −6.74673 0.34416 0.10901 1.50688
1.81690 0.27346 −4.55238 0.57466 0.27840 1.40932 1.68297 0.42765
−4.66671 0.69452 0.47533 0.85926 1.16221 −0.33630 −6.19105 1.33702
0.86171 0.42943 0.81478 0.54667 −5.66930 −0.01207 −0.31703 0.37438
0.81902 0.90574 −5.17360 0.04483 −0.28821 0.92345 1.12633 0.42226
−5.42871 0.25302 −0.12807 1.61032 1.95355 0.72973 −4.26154 0.32828
0.04412 2.58375 2.78976 0.26269 −4.01936 0.58862 0.39709 1.81902 2.29048
1.24095 −3.57425 0.64983 0.39165 1.25178 1.62769 0.45530 −4.86009
0.34914 0.09744 1.61439 2.11730 −0.18776 −5.72640 1.00926 0.77561
2.53624 2.57747 0.42472 −5.82264 0.32895 0.06101 1.45258 1.68413 0.75266
−4.56035 0.59917 0.35485 0.76997 1.13456 0.21085 −4.37820 0.98109
0.81291 1.22282 1.56585 1.02584 −4.81096 0.44592 0.25292 1.46132 1.65375
1.05119 −3.98447 0.60492 0.36541 0.99370 1.36248 1.50043 −3.70198
0.65602 0.39621 1.24866 1.49460 0.99154 −4.25119 0.55001 0.28180 2.37379
2.43046 0.20167 −5.28300 0.37445 0.08484 1.52964 2.10634 0.38039
−4.09243 0.67642 0.51092 0.56205 1.63805 0.82698 −4.95849 0.35515
0.03809 2.45170 2.59037 −0.09691 −4.76861 0.36861 0.03534 1.45356
2.37940 0.85570 −4.07099 0.58121 0.29772 0.76522 1.37203 1.22528
−4.52666 0.51275 0.18772 −0.19382 0.06145 0.90009 −5.67733 0.22292
−0.18300 1.76283 2.02129 0.23249 −4.55859 0.64101 0.45641 0.76042
1.26858 1.12313 −4.08318 0.58474 0.33900 0.79741 1.44567 0.73656
−5.10768 0.31322 −0.02920 1.05169 1.60136 0.56891 −4.60462 0.85026

0.72035 1.66170 2.00401 −0.18910 −5.13880 0.96931 0.73484 0.88173
1.37672 0.67422 −5.02521 0.51249 0.21976 1.34021 1.59857 0.10789
−6.41074 0.67333 0.58812 0.97053 1.54575 0.69364 −4.97827 0.29350
0.01416 2.48615 2.61453 0.41263 −4.20558 0.54407 0.23787 1.71090 2.32509
0.18583 −5.20052 0.20412 −0.09869 −0.49485 1.41256 1.68214 −4.28951
0.90333 0.68966 1.30311 1.56585 0.74390 −5.17613 0.26907 −0.03779
1.03663 1.29754 1.35297 −4.44237 0.55342 0.38372 1.10721 1.45258 1.35468
−4.28375 0.53912 0.28694 0.56961 1.34651 1.19945 −4.78000 0.54836
0.38329 1.10065 1.67892 0.51720 −5.69914 0.39422 0.13291 1.96424 2.19623
−0.02780 −5.89022 0.62549 0.36501 1.62308 1.89325 −0.05355 −5.48825
0.80375 0.43733 1.17725 1.51883 1.69223 −3.31804 0.62481 0.41335
−0.19179 −0.01368 0.69992 −5.57468 0.99025 0.84129 0.67302 1.11581
−0.16368 −6.18954 1.70223 0.86096 1.94511 1.99819 1.13694 −3.82347
0.66893 0.45059 0.39724 1.44166 0.80291 −5.20979 0.53091 0.21615 1.08264
1.65252 −0.02503 −6.57561 0.83293 0.49401 1.14464 1.65560 0.41078
−4.98915 0.60854 0.36159 0.35025 0.48742 1.05053 −4.95660 0.78637
0.63298 1.80878 2.25704 0.61930 −3.73672 0.68359 0.50182 1.75165 2.02759
0.42210 −4.75928 0.25813 −0.01142 0.64503 1.01301 0.72337 −4.76667
0.58448 0.41076 −0.23958 0.22115 1.17780 −5.51089 0.44611 0.21309
1.74118 2.19339 0.35160 −4.64102 0.34082 0.12856 1.14860 1.68008 0.46938
−4.99516 0.52862 0.36948 1.29048 1.86801 0.51282 −4.65129 0.70024
0.40073 0.99087 2.24171 1.89433 −3.21382 0.54273 0.27192 1.74269 2.16335
0.10653 −4.26336 0.55026 0.35080 0.76042 1.17540 0.59583 −4.47823
0.73066 0.57431 1.42213 1.69604 0.43965 −4.97249 0.40526 0.14006 0.16791
0.49638 0.60778 −4.88409 1.07938 0.88069 1.42002 1.63741 0.14395
−5.10763 0.63261 0.32805 1.47457 1.80182 1.22084 −3.31253 0.77040
0.53073 1.41896 2.11432 0.74437 −4.23270 0.39637 0.12797 1.49103 1.88318
0.29645 −4.78291 0.60780 0.33373 0.85540 1.25178 0.53895 −4.95167
0.74072 0.54147 0.14860 1.13252 1.08693 −5.28861 0.40692 0.13278 0.99651
1.61372 0.74640 −4.25267 0.67284 0.48765 1.08948 1.95969 0.56808
−4.27630 0.30167 0.03882 1.39164 2.00236 0.60076 −4.40060 0.40983
0.16899 1.31538 1.75800 0.37254 −5.00432 0.58016 0.31766 1.31404 1.69436
1.36461 −3.51138 0.69020 0.47262 1.68528 1.91845 0.31133 −4.99106
0.36242 0.07483 1.22448 1.93876 0.66539 −4.47496 0.43019 0.11910 1.55746
2.51647 0.44809 −4.62198 0.39620 0.01513 0.65128 1.10503 0.48359
−5.16647 0.76358 0.61936 1.30861 1.63096 0.32858 −4.70518 0.51403
0.22578 0.88173 2.40475 1.06926 −4.38860 0.58271 0.36006 −0.11464
0.40824 1.06922 −5.48466 0.42358 0.18035 2.04847 2.11410 1.34561
−3.67605 0.43423 0.17896 1.72630 2.07779 0.14051 −5.81916 0.08215
−0.39667 1.50861 1.83271 1.00839 −4.68586 0.31213 0.04990 1.10938
1.59997 1.28212 −3.59174 0.75989 0.50533 0.97937 1.21272 0.58229
−5.12749 0.33281 0.12903 1.52551 1.84915 0.61140 −4.81220 0.00452
−0.40729 1.86876 2.09398 0.08849 −5.12453 0.67177 0.41084 1.93780
2.01623 0.01452 −5.44762 0.24778 −0.11127 1.37438 1.77885 −0.11919
−5.21243 0.92966 0.72949 1.65805 2.11878 0.13130 −4.28350 0.58079

0.37112 1.69041 1.96152 −0.03152 −6.50780 0.51503 0.48986 1.31133
1.63096 1.01275 −3.94378 0.64196 0.41217 2.00594 2.34337 −0.25259
−5.77826 −0.16779 −0.41010 2.10306 2.18874 0.67440 −4.03621 0.60662
0.42670 0.81902 1.28762 1.19371 −3.91024 0.80970 0.60114 0.74570 1.32073
0.56384 −5.14874 0.48910 0.16788 1.47920 1.84520 0.30190 −5.05779
0.00429 −0.31519 1.18093 1.89183 1.10537 −3.78952 0.85554 0.59578
2.07499 2.21544 0.30406 −5.09566 0.10478 −0.16295 1.54337 1.88861
0.66304 −4.58205 0.48500 0.20285 1.82112 2.11941 0.20656 −4.26472
0.48675 0.19599 1.66412 2.09599 −0.16813 −5.94892 0.58723 0.24841
1.34901 1.76997 0.43807 −4.49362 0.78642 0.50412 1.67010 1.81986 0.17289
−5.41364 0.49085 0.19180 1.15057 1.70215 0.45378 −5.39653 0.42240
0.08652 1.42839 1.91033 0.52166 −4.83890 0.73118 0.58118 0.85540 1.89785
0.59605 −5.34727 0.25029 −0.02498 1.92246 2.04996 0.36493 −4.54592
0.46795 0.22882 1.05660 1.72158 0.99176 −4.37748 0.58146 0.33743 1.68240
1.78390 0.15685 −5.09588 0.60623 0.41488 0.79741 1.18093 1.91432
−3.60010 0.45024 0.20885 1.20585 1.71413 1.04450 −3.93911 0.67957
0.44922 1.33510 1.82861 0.46746 −4.63924 0.34971 0.05575 1.57185 1.59576
0.18327 −6.06987 0.65497 0.32650 −0.29414 0.01662 0.98105 −5.69465
0.39921 0.20922 0.65128 1.17725 1.66064 −4.41016 0.50515 0.30675 1.59576
2.03432 0.01787 −5.77191 0.79290 0.59985 1.68355 1.97761 0.49290
−4.45136 0.51100 0.28712 0.98802 1.41790 0.74020 −4.52535 0.69793
0.55230 1.69548 1.87585 0.10380 −4.80044 0.38404 0.12216 0.88536 1.35518
0.95415 −3.96875 0.65217 0.40901 1.01837 1.70927 0.58872 −5.45420
0.21740 −0.01618 1.04423 1.35272 0.86016 −5.02118 0.28768 −0.01029
2.39175 2.51494 0.17926 −4.80746 0.33975 0.08343 1.24234 1.37789 0.70278
−5.03339 0.39794 0.16185 2.03714 2.12529 −0.14509 −5.32606 0.66533
0.52973 1.83394 2.02785 0.60065 −4.79218 0.63134 0.38632 2.06000 2.17092
0.05881 −4.97253 −0.05747 −0.36878 1.05903 1.31806 0.47334 −4.97457
0.39972 0.14722 1.76235 2.21577 0.75197 −4.05066 0.63904 0.36029 1.33893
2.13415 0.51878 −4.41274 0.31959 −0.05366 0.44963 1.19000 0.91297
−5.18217 0.29179 −0.02840 1.76139 2.06073 0.21906 −4.17822 0.74025
0.50702 1.34274 1.69883 0.50610 −4.86072 0.49463 0.30438

About the Authors

Tanuka Chattopadhyay is a Professor of Applied Mathematics at the University of Calcutta, India. Her areas of research are Astrophysics and Statistics. Professor Tanuka Chattopadhyay has published more than 25 research papers in national and international peer-reviewed journals. She has also published two books on Computer Programming and Astrophysics. Professor Chattopadhyay had visiting appointments at the Pennsylvania State University, USA and Institut de Planétologie et d'Astrophysique de Grenoble, Joseph Fourier University, France. Since 2002 she has been the Visiting Associate of Inter-University Centre for Astronomy and Astrophysics (IUCAA), Pune, India. She is a member of International Astrostatistics Association. She has received Major Research Projects Awards from University Grants Commission (UGC) and Department of Science and Technology (DST), India.
Asis Kumar Chattopadhyay is a Professor of Statistics at the University of Calcutta, India. He also worked as an Associate Professor at the Indian Statistical Institute, Kolkata during the period 2007–2008. Much of the material in the book was developed for class lectures at postgraduate level in Applied Multivariate Analysis and Statistical Inference. Professor Chattopadhyay has got visiting appointments at the University of Southern Queensland, Australia and Institut de Planétologie et d'Astrophysique de Grenoble, Joseph Fourier University, France. He has also been selected as Visiting Associate of Inter-University Centre for Astronomy and Astrophysics (IUCAA), Pune, India since the year 2005. Now he is the coordinator of IUCAA Resource Centre at Calcutta University. Professor Chattopadhyay has published more than 40 research articles in national and international peer-reviewed journals and two books. Professor Chattopadhyay is a member of International Astrostatistics Association, Calcutta Statistical Association, the Indian Association for Productivity, Quality and Reliability, Operational Research Society of India and the Indian Association for the study of Population.

Index

A.K. Chattopadhyay, T. Chattopadhyay, *Statistical Methods for Astronomical Data Analysis*, Springer Series in Astrostatistics 3,
DOI 10.1007/978-1-4939-1507-1

MIX
Papier aus verantwortungsvollen Quellen
Paper from responsible sources
FSC® C105338

If you have any concerns about our products,
you can contact us on
ProductSafety@springernature.com

In case Publisher is established outside the EU,
the EU authorized representative is:
Springer Nature Customer Service Center GmbH
Europaplatz 3, 69115 Heidelberg, Germany

Printed by Libri Plureos GmbH
in Hamburg, Germany